The Chemistry Classroom

The Chemistry Classroom

Formulas for Successful Teaching

J. Dudley Herron
Morehead State University

I. Dwaine Eubanks, Consulting Editor
Resources for Chemical Educators

AMERICAN CHEMICAL SOCIETY, WASHINGTON, DC   1996

Library of Congress Cataloging-in-Publication Data

Herron, J. Dudley (James Dudley), 1936–

The chemistry classroom: formulas for successful teaching / J. Dudley Herron.

p. cm.

Includes bibliographical references (p. –) and index.

ISBN 0–8412–3298–9 (cloth: alk. paper).— ISBN 0–8412–3299–7 (paper: alk. paper)

1. Chemistry—Study and teaching (Secondary)

I. Title.

QD40.H43 1996   95–44638
540'.71—dc20   CIP

About the Author

J. DUDLEY HERRON is a professor and the chair of the Department of Physical Sciences at Morehead State University in Morehead, Kentucky. He received his B.A in education from the University of Kentucky in 1958, his Master of Education from the University of North Carolina in 1960, and his Ph.D. from Florida State University in 1965. In between, he taught high school chemistry and general science for four years, one in the Army Dependent Schools in Kaiserslautern, Germany. He was at Purdue University from August 1965 through December 1993, where he held a joint appointment in chemistry and education, except for a two-year stint (1989–1991) as head of the Department of Curriculum and Instruction. In 1972–1973, Herron served as a training advisor at the Regional Education Centre for Science and Mathematics in Penang, Malaysia, and in 1982–1983 he studied curriculum research in Israel and Scotland.

Professor Herron has been active in curriculum development at many levels. He was on the author team that developed the Intermediate Science Curriculum Study, and individualized, laboratory-centered program for middle school science, and he coordinated the field trial of those materials in Indiana. He is the senior author of *Heath Chemistry,* a high school text (D.C. Heath, new ed. 1993) and the soul author of *Understanding Chemistry:* A Preparatory Course, a college preparatory course (Random House, 2nd ed. 1986) and has published more than 60 articles on chemistry education.

Dr. Herron served on the editorial board for the *Journal of Research in Science Teaching* for eight years, edited the High School Forum column in the *Journal of Chemical Education* for five years, and served as a reviewer for these and other journals. He has been active in the ACS Division of Chemical Education, Inc., serving on the Executive Committee, the Examinations Institute Board of Trustees, and various ad hoc committees.

His awards include Visiting Scientist of the Year (Western Connecticut Section of ACS, 1982); Catalyst of the Year (Chemical Manufacturers Association, 1983); Outstanding Science Educator (Association for the Education of Teachers of Science, 1985); and Lilly Endowment Faculty Open Fellowship (1982).

Contents

THINGS IN THE AFFECTIVE DOMAIN

APPENDIXES

INDEX

Preface

Many chemists want to be effective teachers, but they have little background in psychology or education on which to build successful practice. I wrote this book to help provide sound direction for these chemists.

THE BOOK'S INTENDED AUDIENCE

The primary target for this book is chemistry teachers at the secondary and tertiary levels, but others will also be interested. Everything I say should be equally appropriate for physicists, biologists, and other science educators. Virtually all of my examples come from chemistry because that is what I know best, but it is not difficult to see parallels in other sciences. Because the purpose of the book is to encourage readers to construct a new understanding of how *their* students learn, if the understanding that I have derived from teaching chemistry does not pertain to another field, it will soon be evident.

Although this book deals with teaching methods, it is not a textbook for a methods course. There are no chapters on lesson planning, test construction, classroom organization, discipline, how to run a laboratory, and other topics that beginning teachers must learn about. But once teachers have the skills needed to survive their first years, and once the feeling that students aren't understanding begins to gnaw at their souls, this book will help.

Few college chemists are exposed to methods courses, but those who work at teaching learn the basic skills through experience and the example of colleagues, eventually acquiring the same discomfort that engulfs high school teachers after their first few years of teaching. They begin to realize that students aren't learning as they should, and it is frustrating. They too can learn from this book. It will help them understand reasons that students aren't learning, and it will suggest things they can do to improve the situation.

Chemists at research universities are so insulated from students that they often retire without knowing how little of what they taught was actually learned. When they do see the light, they often assume that the problem lies solely in the heads of their students and that nothing can be done about it. This is a convenient lie, but it is a lie nonetheless.

It is becoming increasingly difficult to operate on the basis of such false assumptions. Books like Shelia Tobias's *They're Not Dumb, They're Different* and talks by Shelia and others on the subject have captured the attention of a few chemists on most university campuses. Those recently awakened souls who seek more specific information than Shelia gives, more evidence that she isn't blowing smoke, and ideas about what can be done to improve matters will find it here.

This book is not a review of research on teaching chemistry, but a great deal of research is reviewed. Chemistry educators who do research will find the book useful in two ways: First, a great deal of research is synthesized and summarized in an honest, down-to-earth manner. Even those who disagree with the synthesis should find that it helps sharpen their own understanding of what research is telling us. Second, much of the research cited in the book comes from literature that is not commonly read by science educators. The references in this book, along with the *Citation Index* for science and the social sciences, can provide a starting point for a comprehensive review of specific research topics.

ACKNOWLEDGMENTS

First acknowledgement must go to the Lilly Endowment, which provided the Faculty Open Fellowship that allowed me to spend the 1982–1983 academic year reviewing research in cognitive science and visiting overseas centers of research on science teaching. In addition, I wish to thank the Science Teaching Center at the Weitzmann Institute in Israel and the University of Glasgow in Scotland for hosting me during the sabbatical.

Conversations with Alex Johnstone and his students in Scotland; Ruth Ben-Zvi, Bat-Sheva Eylon, Avi Hofstein, Judit Silberstein, and their students in Israel; and conversations with my own students and colleagues at Purdue were invaluable in finding and interpreting research studies. To those named and unnamed, I extend hearty thanks.

Although work on the book began more than 10 years ago, the press of other work prevented me from finishing it until Purdue University granted me a second sabbatical in 1992–1993. This sabbatical provided the opportunity to review research in areas that I had not been able to cover in 1982–1983, to update the previous reviews, and to clarify the implications of that research for teaching chemistry. I wish to acknowledge the financial support of Purdue University during both sabbatical leaves. I could never write a book while trying to teach, so it is no exaggeration to say that I could not have written the book without Purdue's support.

Many people at Purdue and elsewhere read and commented on early drafts of some or all of the book during the years it was on and off the back burner. I will not try to name them here, knowing that the effort would only result in offending someone who contributed significantly during that time but whose name escapes me now. I do, however, name and extend special thanks to Loretta Jones and her students: Waheed Akbar, Julie Henderleiter, John Kennaly, Patananya Lekhavat, Tsui-jung Liao, William T. Oulvey, S.J., and Angela Powers, who used the book at the University of Northern Colorado during the spring of 1994. I found their comments kind but helpful, and I made considerable use of their suggestions during final revision.

After 38 years of marriage, my wife has grown accustomed to vacant stares and absent thoughts when I am writing, but her tolerance is no less appreciated now than before. When authors routinely acknowledge the sacrifice of spouses and children in their prefaces, they aren't spouting platitudes. The sacrifice is real, and it is accepted with knowing love.

J. DUDLEY HERRON
Department of Physical Sciences
123 Lappin Hall
Morehead State University
Morehead, KY 40351

The People

MY MOST MEMORABLE STUDENTS

This book is about teaching science, about the things that we teach and the students we expect to learn. Why are some of them unable to learn? What can we do to help them? To begin, meet some students I have taught. Perhaps you have taught students like them.

SAM: THE CONCRETE THINKER

Sam was my first memorable student. Sam was the student who captures your heart and challenges your skill. Sam was the perfect student: kind and considerate, hardworking, and anxious to learn. But Sam did not learn.

Sam could memorize chemical symbols and describe events seen in the laboratory, but atoms were a mystery and equations were meaningless. If I *set up* a problem involving the mole, Sam could get the answer, but Sam never understood what I was doing when I translated a chemical equation into a mathematical statement.

Sam was not lazy. Sam came in before school, during lunch, and after school. Sam baby-sat with my children and always arrived early so I could help with chemistry. Sam was going to be a pharmacist, you see, and *had* to know chemistry. Sam was determined. So was I. But Sam did not understand. Sam was my first failure.

LYNN: THE INTELLECTUAL

Not all of my students were like Sam. Lynn was just the opposite; learning was no apparent effort. The only challenge that Lynn presented was for me to ask a question Lynn could not answer. I never did. Anything that I taught, Lynn learned— and then some! Lynn was a student who would thrill any teacher.

When Lynn went to college, I knew there went a future Nobel laureate, but Lynn flunked out of school the first term! Lynn fell in love. Lynn had never dated in high school, and once someone stole the heart, the mind was stolen as well. Lynn learned more than enough chemistry that takes place in the laboratory, but too little was learned about chemistry that takes place in the heart. Had Lynn been more like Sandy and Sandy more like Lynn, both would have managed better.

SANDY: THE ROMANCER

Sandy was a handsome student from South America who attended my lectures at Purdue. I have not said, "took my course"; what Sandy did could scarcely be described as taking a course. Romance was taken (and given!); the course was left behind. Sandy could have been as successful in chemistry as in romance, but chemistry was not important and romance was.

CHRIS: THE LATE BLOOMER

Chris was different. As a junior in high school, Chris realized that school had purpose and decided to learn. Chris was determined to try chemistry and was able to convince a best friend to come along for moral support.

Chris grew up on a farm and had never dreamed of anything other than taking over the family farm some day. Neither Chris nor Chris's parents had viewed formal education as important, and it is fair to say that Chris had not emphasized schoolwork. Chris's grades in mathematics, biology, and English were mediocre at best and accurately reflected Chris's understanding. I was not optimistic when Chris signed up for my class, but every student deserves a chance.

Chris and I hit it off splendidly. Chris asked questions, intelligent questions. Chris asked to do experiments of personal interest—to check different motor oils for corrosive effects, to compare composition of various brands of gasoline, and to determine why hot water freezes faster than cold (It does, you know, if the conditions are right[1]). I was delighted, and I gladly substituted these experiments for ones I had scheduled. Chris shared what was learned with the rest of the class, and Chris learned a lot.

Chris's experience in chemistry was so good that Chris signed up for physics the following year. Chris even won the award for Outstanding Physics Student. The last I heard, Chris was successfully pursuing a degree in agricultural engineering while working full time to cover college expenses.

KIM: THE HELPLESS CHILD

Chris's determination stands in stark contrast to Kim's impotence. An only child and father's pet, Kim complained of "not knowing how" the moment something new was faced. "Show me how to do this problem", or "I don't understand that chapter", or "What does it mean?", or even, "I'm just too dumb to learn science!": These were the constant laments.

Kim had no hint of any of Sam's difficulties in understanding; Kim just did not try. With the determination of Sam or Chris, Kim would have done splendidly, but I never convinced Kim of that.

THE MANY FACES OF FAILURE

I have only listed a few of my students, but they remind us that teaching and learning are complex. When someone asks, "Why don't students understand chemistry?"

---

[1]Observations concerning this phenomenon apparently go back for centuries. Thomas Kuhn (1970) cites Francis Bacon as saying "Water slightly warm is more easily frozen than quite cold", and refers the reader to Chapter IV in Clagett (1941) for a partial account of the early history of the observation.

I hesitate to answer. Almost as many answers exist as do students. And how do we know when a student *truly* understands?

FRANCIS: A SUCCESS WITHOUT UNDERSTANDING

Francis is one of the most unusual students I have met. Francis completed an undergraduate degree at a small college and worked as a laboratory technician for a large chemical company before coming to Purdue for graduate work in chemistry. Although the undergraduate record was not outstanding, it was certainly acceptable. Nor did the work record suggest that Francis would have any serious problem in graduate school.

During the first semester at Purdue, Francis's physical chemistry instructor stopped me in the hall to express concern. Francis was not doing well in his class. During an office visit Francis had been unable to explain why salt is spread on ice in winter, how a candy thermometer lets us know that candy will be hard when poured out to cool, or why a pressure cooker must be used to cook beans at high altitudes. These questions were this professor's tests of understanding, and Francis failed them.

Shortly thereafter, Francis made an appointment with me to discuss some of the "unreasonable expectations" encountered in physical chemistry. We talked, and Francis decided to drop the course and take it again the following term. The second time Francis passed.

Francis's work was marginal in some courses and excellent in others. One professor wrote a letter commenting on Francis's outstanding performance; others questioned Francis's ability to complete a graduate degree.

Not until Francis's course work was finished and the final exam was graded did the problem came into focus. Francis complained that several questions on the Master's written exam *certainly* had not been discussed in class. For example, in one question Francis was asked to comment on the acid–base phenomena that are important in determining the pH of various salts in water. Among the salts were $Co^{2+}(ClO_4^-)_2 \cdot 6H_2O$ and $Co^{2+}(OAc^-)_2 \cdot 4H_2O$. Neither compound had been discussed in class.

Francis was correct, of course, because the purpose of the question was to test students' ability to apply general principles to new examples. I explained this to Francis and got a puzzled look. "How is that possible?" Francis asked.

The subsequent conversation revealed a number of interesting facts. "How do you study for physical chemistry?" I asked.

Francis had a very efficient system: "As I read material in the text, I make out 3 x 5 cards with pertinent facts and formulas on each card. I put them in a file box, and when I do homework, I systematically go through the file to find the pertinent information, and apply the relevant facts and formulas to the problem at hand. Before an exam, I review the most important facts and formulas in the file so I can remember them. Don't other students do something similar?"

Francis kept a large number of facts in memory—far more than I have retained—but Francis had few of those facts integrated into a sensible *story* about the physical world. In an attempt to explain phenomena that were new, Francis pieced together bits from memory that must have been sensible to Francis, but they were incomprehensible to me.

We explored what we believed about atoms and how we used that understanding to explain bonding. I asked Francis to elaborate on an earlier statement that "the sharing or transfer of electrons between atoms holds them together". I asked for an explanation as to why this would result in a bond, and Francis's response was as follows:

> [M]ost chemical bonds fall between a completely covalent bond and a completely ionic bond. Now I would contend that if an electron were totally transferred from one species to another (Ex: $Na + Cl \rightarrow Na^+ + Cl^-$ one would have ions, not a molecule. Complete transfer of electrons would be most likely to occur in solution or in the gas phase. Thus, for bonding to occur, there must be some sharing of electrons.

Francis's expression signaled satisfaction; mine, I am sure, signaled unaltered confusion. Nothing Francis had said told me *why* the sharing of electrons should hold atoms together.

In a later discussion of the difference in a chemical and physical process, Francis made the following remarks:

> Thermal expansion is a common example given for a physical change. Cement is still cement summer or winter. Thus, expansion would be considered a physical change. No outward change in the chemical entity is quickly observed. However, as the cement is heated by the summer sun, the energy of each molecule increases. As the molecules vibrate at a faster rate, the piece of cement expands slightly. This vibration could be considered a chemical change.

Almost all of Francis's sentences make sense, but when the sentences are strung together they fail to tell me the difference between chemical and physical change.

For over a year I worked with Francis by trying to elicit logical chains of inference.[2] Anytime I asked Francis to explain a phenomenon, Francis immediately searched for the solution in a book. When the explanation involved several principles found in separate references, Francis's explanation would contain verbatim quotes pieced together in a fashion similar to the previous answers; when the explanation could be found in one coherent reference, the answer was excellent, but it was not produced by Francis.

Only when I managed to ask questions for which Francis could find no answer in a book did Francis appear to make progress. First, I asked why the moon changes its apparent shape, and when Francis gave the expected response that the earth casts a shadow over part of the moon's surface, I asked Francis to provide data to prove that answer is correct.

Francis searched books but found no answer. Then Francis made observations and returned a month later with the correct explanation *and truly rational arguments to support it.*

Next I asked Francis if the moon's orbit around the earth is in the plane defined by the earth's equator, the plane defined by the earth's poles and a point on the equator, or some other plane. Again, Francis collected data and presented a rational explanation. Still, when I asked questions about chemistry, Francis dragged pieces from memory and placed them together in an illogical patchwork that Francis called *explanation.*

---

[2]A striking similarity exists between Francis's responses to logical problems and those given by adults in "primitive" societies. (*See,* for example, the discussion on pages 160–168 of Cole and Scribner, 1974.)

What happened to Francis during high school and undergraduate courses in chemistry, during work in industry, and during graduate courses at Purdue? In what sense did Francis understand?

SEEKING RATIONAL ANSWERS

Is Francis a Sam who did not give up? Do determined students who do not understand resort to the only avenue left, rote memory? And do they then come to believe that this is what education is about, as Francis apparently did? After getting average marks through rote memory for many years, are they incapable of discarding their well-developed habits and taking up a pattern of rational thought? I think so, and I think that many school practices encourage just such miseducation.

GENDER BIAS

When you read through the descriptions of students, you undoubtedly noticed that personal pronouns were not used, and you may have noticed that all of the student names—Sam, Lynn, Sandy, Chris, Kim, and Francis—are names that are often used for males and females alike. Which students did you assume were girls? Or boys? What shaped those assumptions? Were your assumptions guided by established facts, unfounded stereotypes, or some personal need to affirm that one sex does better in science than the other?

Research has shown that science teachers, male *and* female, *know* that boys are better at science than girls, and this knowledge often affects teaching in ways that reinforce this bias. Because teachers expect boys to understand, boys are treated so that they do; because girls are not expected to understand, they are treated so that they do not. Because ''boys are good with mechanical things'', they manipulate equipment in the lab while girls watch.

Research on ways that cultural, racial, and gender stereotypes affect our interactions with students and our expectations concerning their learning is extensive, and much of this research has been translated into suggestions about teaching that are well worth following. Undoubtedly, the absence of at least one chapter dealing with such issues is a major omission in this book. Still, I have little expertise in this area, and I must leave the matter to others. (*See* Kahle et al., 1993, and Scantlebury and Kahle, 1993, as well as companion articles in the same journal issues for recent discussions of these topics.)

In reviewing early drafts of this book, some readers expressed concern that by identifying the sex of students who are used to illustrate particular difficulties or gifts, some readers could incorrectly infer that the characteristic is somehow related to gender. Much as I detest the idea of political correctness and chafe at the awkward language that it frequently engenders, the point is well-taken. Gender does not matter when a student learns science; understanding, attitude, and potential are what we must pay attention to. We must recognize that every person is unique, and everyone—male or female, black or white, old or young—can learn chemistry.

Throughout this book genders of students have been obscured by using names that are commonly used for boys and girls and by avoiding personal pronouns wherever possible. Behaviors described in the illustrations are observed in males and females; whether the particular student in the illustration is one or the other is irrelevant.

AN INVITATION TO PARTICIPATE

DESCRIBE YOUR STUDENTS

I have described some of my students, and I have alluded to their problems with learning. As you read this book I hope that you will recall these students and their problems as we struggle to develop a model of learning capable of explaining their problems and suggesting a means of overcoming these problems.

My list of students is not exhaustive, nor is their list of problems. Perhaps my list is not even representative. Are your students like mine? What students do you remember, and what problems do they have? I invite you to think about these questions quite deliberately for a while, and I encourage you to write down your responses.

My purpose in this book is not to tell you how students learn or fail to learn. Rather, it is to share my current thinking about these questions and to invite you to think along with me. I can share my experience as I understand it, but if your understanding is to go beyond my own, you must add your experience to mine. You must ask the kinds of questions that I have asked: Given all of the things that stand in the way of successful education, how can we improve our enterprise? How can we identify our students' problems in time to be of help? How can we help once we understand their problems? A useful first step is to think about your students and describe your understanding of their successes and failures.

SUGGESTIONS FOR THOSE WHO DO NOT KNOW STUDENTS

Readers who have never taught will not be able to describe their students because they have never had any. They can, however, describe themselves and their friends. Describing your own responses to different courses, different instructors, and different learning environments can provide insights that go beyond those suggested by the vignettes in this chapter.

Some who have taught—particularly chemists in large universities—will have difficulty describing their students because they have had little opportunity to observe them. Their teaching has been limited to lectures for large groups where there is little opportunity for one-on-one interaction. Perhaps insight into student difficulties in such an environment can be obtained, but I do not know how. If you do not devote a portion of your time to individual student conferences, small group discussions, interactions with students during laboratory sessions, or other activities that provide opportunities to gain information about individual student learning, I encourage you to change your schedule.

2

A Framework for Discussion

In Chapter 1, I introduced several of my students. Many of them failed to understand chemistry, and they failed for different reasons. If we are to improve instruction in chemistry, we must understand the various reasons students fail, we must identify specific difficulties students have, and we must develop practical ways to assist students in overcoming these difficulties.

Saying that students "failed to understand" implies that we agree on what understanding is. Do we? Perhaps we need to find out.

WHAT CONSTITUTES UNDERSTANDING?

Just what does it mean to "understand" chemistry? Such a question has many answers. One answer is to look at differences in what we call *knowledge*. Knowledge can be described in many ways. Whether or not actual differences exist in the way categories of knowledge are held in our heads is immaterial; we think about thoughts in different ways. Three useful categories are described in the following sections: declarative or figurative knowledge, what Jean Piaget called operational knowledge, and what many cognitive scientists call procedural knowledge. Each category represents a different kind of understanding.

DECLARATIVE KNOWLEDGE

Does the student who answers these questions understand?

1. What is the symbol for sodium?
2. What is the formula for water?
3. When did Herbert C. Brown win the Nobel Prize in chemistry?
4. Which elements are gases at ordinary temperatures?

Declarative knowledge is knowledge that we can declare: Graphite is black. My name is Dudley Herron. Chicago is a city in the United States. Two plus two is four. Sodium is a soft, shiny metal that reacts with water. Water can act as an acid or base. Declarative knowledge is what we usually think of as factual information. It is the kind of knowledge required to answer questions like those given above.

Declarative knowledge represents an important kind of understanding. Some students do not know this part of chemistry, and they fail at complex tasks because they have never memorized such simple facts. Greenbowe's case study of "Sue" provides an excellent example (Greenbowe, 1984, pp 187–198). However, most students seem capable of such learning when they make the effort. Is acquiring such knowledge all that is involved in understanding?

OPERATIONAL KNOWLEDGE

What about the following questions? Are they the same as the others? Should students be able to answer them to prove that they understand?

> 5. Here is some water. What would it look like if you could magnify it so you could see its smallest piece? Does it have a smallest piece? Would it look different as a solid or as a gas?
> 6. When a direct electric current is passed through water containing sulfuric acid, the water gradually disappears, and two gases, one hydrogen and the other oxygen, appear in its place. If the hydrogen and oxygen gases are mixed and lit with a match, the mixture explodes, and there is water again. How can we explain what is happening? How can we describe the pieces that are present at each stage of the process described?
> 7. If you could see the things represented by $H_2O(g)$, $H_2O(l)$, $H_2O(s)$, H^+, $H_2(g)$, 2H, and H, what would they look like?

These questions differ from the first four. They require declarative knowledge, but they require something more. Part of the declarative knowledge involves a particular kind of mental image, and we expect students to see it as we do. Many of them do not. Sam, one of the students described in Chapter 1, did not. Neither do many others. This mental imaging is apparently difficult to do, or we have not learned how to make it easy; either way, these two statements say the same thing.

The ability to think in terms of abstractions such as atoms and molecules involves a particular kind of knowledge, called *operational knowledge* by Piaget. Operational knowledge is what we use to "operate" on other knowledge to transform it in some way. Most people would say that questions 5–7 require "reasoning", and that idea of reasoning captures the essence of what many people call *generalized intellectual skills*, of which operational knowledge is a part. Many cognitive scientists describe other generalized intellectual skills as *procedural knowledge*.

PROCEDURAL KNOWLEDGE

Suppose I hand a student a piece of metal and a rock and ask the following questions:

8. Which of these is iron and which is limestone?
9. Which of these is a metal and which is not?
10. Which of these is an element and which is a compound?

These questions require students to perform some kind of "procedure", either physical or mental, to put objects into categories. A person might, for example, strike the objects named in question 8 with a hammer. If the person has the declarative knowledge that iron is malleable and limestone is brittle, that procedure will produce an answer to the question.

Declarative, operational, and procedural knowledge are conveniently thought of as distinct and separate things, but they are not. Knowing what to do (procedural knowledge) is of little use without the factual information and logical operations required to make sense of the data that result when procedures are carried out. Similarly, a great store of factual knowledge and excellent reasoning are of limited value without procedures for getting new data, organizing it in new ways, and revealing new relationships.

Even though questions 8–10 all require procedural knowledge, question 10 is more difficult than question 9, and question 9 is more difficult than question 8. Students can easily be taught to answer the first two questions correctly; the third question is difficult to teach them. Why is this? If you think about these questions now, the detailed analysis in later chapters may make more sense.

We often say that students have difficulty with math. Do they? Do they have difficulty with all math?

11. $2x + 4 = 10$: Solve for x.
12. $2x + 2.5 \times 10^{-21} = 4.6 \times 10^{-22}$. Solve for x.
13. $A = V/2 + 2Bx$: Solve for x.
14. $PV = nRT$: Solve for V.
15. If the values of V, n, and R in the above equation remain constant, what will happen to the value of P when T is increased by 50%?
16 Calculate the pressure of 14.2 mol $H_2(g)$ in a 25.0-L cylinder at 40 °C.
17. What is the molecular mass of a gas if its density is 3.79 g/L at STP?
18. The chloride of an unknown metal is believed to have the formula MCl_3. A 2.395-g sample of the chloride is dissolved in water and treated with excess silver nitrate solution. The mass of the AgCl precipitate formed is found to be 5.168 g. What is the atomic mass of M, the unknown metal?

Most of my beginning students can answer the first of these questions with no difficulty; many graduate students have difficulty with the last one. At what point is the difficulty "a math problem", and at what point is it something else? What *kinds* of math problems do students find difficult? Which problems must they solve to show that they understand math?

We teach other skills that students do not learn. We expect them to use a balance, read a buret, light a burner, conduct an experiment, and write an intelligent report about what they did. Which of these things do we have in mind when we ask, "Why don't students understand chemistry?"

Qualitative differences exist among the various things that we try to teach and the tasks we use to gain evidence that students have learned. These differences interact with differences that exist among students (background, interest, and intellectual skills) so that a particular student learns some things easily and well but learns other things poorly, even with extra effort.

We can never make chemistry understandable without considering how learning takes place and the interaction between the nature of what is taught, how it is taught, how the student is equipped to learn it, and what we take as evidence that the student does understand.

VARIABLES THAT AFFECT LEARNING

Learning is complex, and the factors that influence what we learn are numerous. Several researchers (Bransford, 1979; Hertz-Lazarowitz and Shachar, 1992; Jenkins, 1979) describe four categories of variables and represent them in a diagram similar to Figure 2.1. Following their example, I show four clusters at the corners of a tetrahedron, but I add the teacher at the center because, as Hertz-Lazarowitz and Shachar (p 78) point out, "Teachers are perceived as instructional designers who can engineer classroom functioning in order to increase coordination among these four dimensions."

Chemists will, no doubt, detect in Figure 2.1 an allusion to carbon and its active sites. As in the case of carbon and its compounds, the sites are independent but interactive. Again quoting Hertz-Lazarowitz and Shachar (pp 77–78), "[l]ack of coordination between [variables at the four sites] during teaching with detract from the effectiveness of the classroom, while adequate coordination of these dimensions serves to improve classroom functioning". What happens at one site affects the others. We talk about each constituent in carbon compounds and each classroom dimension as though it behaved independently, but we ignore the rest at the risk of misjudging the molecule or classroom as a whole.

Throughout this book I will talk about characteristics of the learner, characteristics of learning materials, the nature of learning activities, and the nature of the criterion tasks used to assess learning. *Characteristics of learners* may include fleeting, individual considerations such as an empty stomach, the emotional aftermath of a family fight, or a search for love. But we normally think of more global issues like social maturity, intellectual aptitude, prior knowledge, or how people learn. *Characteristics of learning materials* include such factors as the quality of text and illustrations, the level of abstraction, and the number of new ideas presented in each lesson. The *nature of learning activities* refers to such things as whether students work individually or in groups and whether material is presented by lecture, direct experience, or through reading. The *nature of criterion tasks* refers to ways students are asked to

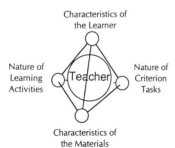

FIGURE 2.1. Variables that affect learning: materials and tasks. (Adapted with permission from Bransford, 1979, p 8.)

demonstrate understanding. It is important to keep the connections among these clusters of variables clearly in mind. Changing one set of variables certainly influences others.

Figure 2.1 is a visual model for the classroom milieu. The teacher is at the center to emphasize the central role that teachers play. Teachers are responsible for directing learning. To be successful, they must be aware of the surrounding variables and know how the variables interact to influence learning. They must also be able to manipulate those variables to mediate learning.

AN EXAMPLE FROM MATHEMATICS

The way that variables in Figure 2.1 interact can be illustrated by contrasting two approaches to mathematics. Mathematics Achievement through Problem Solving (MAPS) is a problem-centered mathematics program for grade 9 that was developed through the School Math and Science Center at Purdue University. Students work in groups to solve problems similar to the examples taken from the National Council of Teachers of Mathematics (NCTM):

> Twenty-eight children are going on a picnic. Four children can ride in each car. How many cars are needed? (NCTM, 1989, p 42)
>
> I have six coins worth 42 cents; what coins do you think I have? Is there more than one answer? (NCTM, 1989, p 23)

MAPS is predicated on certain assumptions about the characteristics of ninth-grade students and what they need to understand about mathematics. It is used successfully in many classrooms, but I had a disastrous experience with MAPS in one school. On none of the problems that I posed did students begin to make progress. They quickly began guessing. They combined numbers in the problem by addition, subtraction, multiplication, and division with no consideration of the problem's meaning or the appropriateness of the operations they performed. The students had no idea whether or not they were close to a correct answer, and they were frustrated because I would not tell them.

The lessons taught by the regular classroom teacher were very different from mine. On the day of my lesson, the regular teacher returned papers that students had completed the previous day. The worksheet, which was typical of the work these students were doing, looked something like this:

> Use the information given to solve the following 30 equations;
> $X = 3, Y = 5,$ and $Z = 2$:
>
> 1. $4 + X =$
> 2. $Y - 3 =$
> 3. $2X + Z =$
> 4. $2X - Z =$
> ...
> 30. $12 \times Z =$

Students were quite successful on this worksheet. Few scored below 70%, and many had perfect papers. Students felt good about their work, and so did the teacher.

Neither the classroom teacher nor I taught much mathematics in our lessons, but our failures occurred for different reasons. We made different assumptions about the variables in Figure 2.1, and those assumptions affected what we and the students did and what the students learned. They were the kinds of assumptions that all teachers make, implicitly if not explicitly, every day.

CHARACTERISTICS OF THE LEARNER

By focusing on the characteristics of the learner, we see the obvious differences between my assumptions and those of the regular teacher. I was doing this demonstration lesson because the classroom teacher *knew* that problem solving was beyond the capabilities of the students, and this teacher was concerned that chaos would erupt because students could not handle group work. That threat of chaos was a major reason that lessons were confined to individually completed worksheets like the one handed back. The students were kept busy at productive work that they understood, and the classroom was under control. The practice that the worksheet provided would reinforce the students' understandings of basic mathematical operations and strengthen their knowledge of number facts.

In contrast, I was concerned about neither the students' ability to solve problems nor their ability to work in groups. I had seen hundreds of other ninth graders—many in this school—working together on problems. No chaos existed in those classrooms, and students made reasonable progress on the problems. Furthermore, because learning occurs when students construct meaning in the context of problematic situations, I knew student interaction was important. By approaching problems in groups, differences in interpretation of a problem and varying strategies for solving it would naturally emerge, and the social interaction would provide opportunities for students to resolve misunderstandings about mathematical relationships and arrive at sensible solutions.

The assumptions that the regular classroom teacher and I made about these students' characteristics as well as their actual attitudes, competencies, and interests affected what the students were able to learn in the lessons we taught.

CHARACTERISTICS OF THE LEARNING MATERIALS

The tasks represented by the teacher's worksheet are very different from the word problems used in my lesson. On the worksheet, required computations are indicated; in the word problems they are not.

Notice how our assumptions about the students and our assumptions about what they will ultimately be expected to do (criterion tasks) interact with and shape the learning materials: I selected problems that mimic tasks that might be encountered in everyday life, and I consider the tasks on the worksheet to be abstract and removed from real problems. But the regular teacher prefers problems like those on the worksheet because students can do them, and she knows that similar items are part of the statewide testing program used to evaluate the school.

The differences in our learning materials were substantial, but, as we will see in subsequent chapters, differences far less significant than these can spell the difference between success and failure.

NATURE OF THE LEARNING ACTIVITIES

Not only were our materials different, so were the activities. Students completed their worksheets alone; I expected them to work in small groups. Groups could do whatever would help—role play, use coins, or just talk and figure. Activity on the worksheets was limited to what could be done alone with pencil and paper. Although talking with the teacher was permitted, talking with other students was not.

NATURE OF THE CRITERION TASKS

Differences in assumptions about the students in the class led to differences in what the classroom teacher and I looked for as evidence of learning. Though neither of us gave a test, we both had evidence of what the students were able to do mathematically. All of the students had done well on the worksheet, and the teacher was pleased that they understood. I was appalled! What did students know? They were told that a particular number could be substituted for a given letter. They apparently knew that when two numbers are connected by a "+", they are to find a sum, when joined by a "−", they are to find the difference, and so forth. However, they apparently made no connection between these operations and any real-world situation. How else could you explain their random connection of numbers when given the problems used in my lesson? Once again we must ask, "What do we mean by understanding?"

CRITERION TASKS AND UNDERSTANDING

Problems on the worksheet used by the regular mathematics teacher seemed trivial to me, and solving them did not assure me that students understood. But solving problems that are *not* trivial may not ensure understanding either.

The following question is typical of questions used in chemistry to test understanding of ideal gases:

> A certain mass of gas occupies 200 L at 95 °C and 782 mm Hg. What will be the temperature in Kelvin of the gas if the volume is changed to 176 L and the pressure is changed to 815 mm Hg? (Nurrenberg and Pickering, 1987, p 510)

What should one infer about learning when college chemistry students do well on questions like this one but do poorly on questions like the one in Figure 2.2?

The two questions about gases clearly test for different kinds of understanding: one quantitative and one qualitative. However, even questions that appear to test the same kind of understanding may not. In a study of reasoning, Wason (1966) presented this task (Figure 2.3):

> You are presented with four cards showing, respectively, "A", "D", "4", "7", and you know from previous experience that every card, of which these are a subset, has a letter on one side and a number on the other side. You are then given this rule about the four cards in front of you: "If a card has a vowel on one side, then it has an even number on the other side." Next you are told, "Your task is to say which of the cards you need to turn over in order to find out whether the rule is true or false." (Johnson-Laird & Wason, 1977, p 143)

Before reading on, answer the question yourself. Then answer the following question.

Imagine that you are a British postal worker engaged in sorting letters on a conveying belt (Figure 2.4). Your task is to determine whether the following rule has been violated: "If a letter is sealed, then it has a 5d stamp on it."

The following diagram represents a cross-sectional area of a steel tank filled with hydrogen gas at 20 °C and 3 atm pressure. (The dots represent the distribution of H_2 molecules.)

Which of the following diagrams illustrate the distribution of H_2 molecules in the steel tank if the temperature is lowered to –20 °C?

FIGURE 2.2.   Conceptual gas law question. (Reproduced with permission from Nurrenbern and Pickering, 1987, p 508.)

FIGURE 2.3.   Cards used in the four-card problem. (Reproduced with permission from Johnson-Laird and Wason, 1977.)

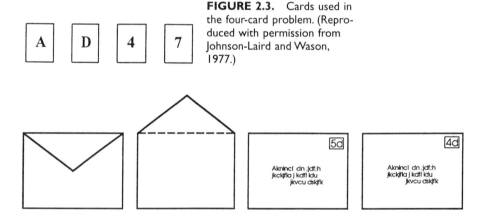

FIGURE 2.4.   Envelopes used in the four-card problem.

FIGURE 2.5.   Envelopes used in the four-card problem.

Select only those envelopes that definitely need to be turned over to find out if they violated the rule (Johnson-Laird and Wason, 1977, p 152).

The correct answer to the first question is A and 7, but most people respond A and 4 or only A. However, *22 out of 24 subjects presented with the second task responded correctly* that the first and last envelopes should be turned over to see if the rule was violated. When the same 24 subjects were presented with envelopes shown in Figure 2.5 and asked which envelopes should be turned over to test the rule, "If a letter has a D on the front, then it has a 5 on the back", only seven of the

24 subjects answered correctly! Logical isomorphs of this task have been used in numerous other studies, and the results leave little doubt that seemingly unimportant changes in the task can materially affect the results. Similarly, studies of test questions given in schools show that changes in wording and other alterations can affect response patterns (*See*, for example, Cassels, 1980).

SUMMARY

This chapter began with the question, "What constitutes understanding?" It then proceeded to describe declarative, operational, and procedural knowledge and the way these kinds of knowledge interact to produce understanding.

Understanding is a product of learning, whether formal or informal, and all learning is complex. It is influenced by tens or even hundreds of variables, some that characterize the learner, and some that characterize the material to be learned. Variations in learning activities and criterion tasks help determine the kind of learning that takes place and our perceptions of this learning. As seen by variations in the Wason four-card tasks, even minor changes in a task can have major effects on how well the task is completed, and the reasons for these effects are sometimes difficult to determine.

The kinds of knowledge and the variables that affect learning described in this chapter provide a crude framework for exploring how students learn chemistry. At times consideration of each variable in isolation will be convenient, but we must not forget that all variables are interconnected. We can no more answer questions concerning the effect of learning activities while ignoring the characteristics of learners or what we define as understanding than a chemist can predict what will occur at one bonding site on a carbon atom while ignoring the constituents at other sites.

Learning is complex, and we should not expect simplistic answers to questions about why students do not understand. But neither should we assume that learning is so complex that understanding is beyond reach. As the next chapter will make clear, most of us know a great deal about learning because of our personal experience.

3

▪

Eternal Verities

In Chapter 1, I described several of my students and talked about the problems that they had learning chemistry. I invited you to do the same. In Chapter 2, I discussed the concept of understanding and the variables that shape this concept. I invited you to think of other variables. The premise behind these activities is that we learn a great deal about teaching and learning through our experience with students. We make observations and we draw inferences from them. In the process we develop a rather diffuse mental model about teaching and learning that guides what we do. This model is not very explicit, and it contains a number of gaps and contradictions. If pressed to describe what we know about teaching and learning we may come up with a few platitudes, but the statements quickly degenerate into vague generalities that we have difficulty defending. Still, we believe them, and they do guide our action.

In 1978, I gave a speech at the "New Directions in the Chemistry Curriculum" conference held at McMaster University in Hamilton, Ontario. For that speech I summarized what I knew about teaching and learning in down-to-earth terms. It was a good exercise, and I recommend it to you. What follows is an updated version of that speech. Please pay attention. (There is an exam near the end.)

PRINCIPLES OF LEARNING

KNOWLEDGE IS CONSTRUCTED

No idea has more implications for teaching and learning than the realization that knowledge cannot be transmitted intact from one person to another. No matter how I try, I can never transfer an idea from my head to yours.

LEARNING IS DIRECTED TOWARD SURVIVAL GOALS

Human beings are subject to the evolutionary pressures of natural selection in much the same way that other living things are affected; changes that have survival value persist, and those that do not eventually fade away. In most organisms, the forces of natural selection operate almost exclusively through genetic transmission; in human beings, learning plays a major role.

Knowledge *is* power, as the history of the development of bronze and iron and gunpowder and atomic bombs aptly shows. Also, historical evidence suggests that the ability to commit large volumes of information to memory has diminished since printed books became common. But reading, an intellectual skill known only by aristocrats in an earlier age, is almost universal now. Learning is directed toward changes that enhance survival as we perceive it.

LEARNING OPERATES ON A PRINCIPLE OF MINIMUM EFFORT

Because energy is limited and learning is directed toward survival, we seek to learn with the least cognitive effort. When confronted with alternatives, each of which appears to have the same survival value, we select the alternative that requires the least effort. For example, if survival means passing a course and we are faced with the alternatives of comprehending ideas that are difficult or memorizing answers that we know will be tested, memorization makes sense because survival can be attained with less effort.[1]

This principle has important implications for teachers as well as students. Students must see value in what we ask them to do; teachers must see value in what students are allowed to do.

Although society as a whole may realize considerable value from citizens who have the technical sophistication required to deal with complex social and economic issues, students are concerned with themselves. Unless they can see how that same understanding will enhance *their* survival, they are unlikely to expend the effort required to develop the understanding. Similarly, teachers may perceive that independent, questioning students with a strong sense of purpose are good for society, but when independence and questioning are a threat to the teacher's survival in the classroom, these traits may be curtailed. All persons have a need to survive, and in a democratic society neither students nor teachers have the right to deprive others of survival goals as they seek their own.

PEOPLE CHOOSE PLEASANT EXPERIENCES
OVER UNPLEASANT ONES

Children who know they are in for a spanking or tongue lashing take more time getting home than those who expect a banana split. Students who find their studies fun get to them sooner than those who find them drudgery. (The same can be said for teachers.)

What makes things pleasant? For one thing, we enjoy things when we feel like we can say, "I did it!" Furthermore, we enjoy things that we are good at doing. For most people, a challenge is enjoyable—provided the challenge is not beyond one's capability. On the other hand, we avoid being forced to do exceptionally difficult and frustrating tasks.

We seem to find pleasure when things make sense and frustration when someone insists that something is so when it does not make sense. We may be forced to

---

[1]A reader of an early draft of this book took issue with this point: "If someone truly values what they are learning, they will expend the energy at the expense of something else." This statement is well taken, but it begs the question of why we value one thing over another. Is it not because of perceived survival value?

shrug our shoulders and accept something that we do not understand, but we do not feel pleasure when we do it.

PEOPLE FIND DIFFERENT THINGS INTERESTING

Several years ago, my family sat down to an appetizing meal of steak and all the trimmings. Keith, who was then in junior high, remarked as he put Worcestershire sauce on his meat, "I *love* Worcestershire sauce. I wish the oceans were filled with Worcestershire sauce."

Struck by the suggestion, I said, "Oh? How many bottles of Worcestershire sauce do you think it would take to fill the oceans?"

We were off and running. For the next 45 minutes we ate and figured. We estimated the percentage of Earth's surface covered by oceans. We pulled out the encyclopedia to find information that would allow us to estimate the average depth of the oceans. We calculated the surface of Earth in square miles, the volume of the oceans in cubic miles, converted cubic miles to cubic feet, and cubic feet to gallons. From gallons we got quarts and from quarts we got ounces, and finally, bottles of Worcestershire sauce. Just as we triumphantly announced our estimate, my wife rose angrily from the table and shouted, "*Who cares how many bottles of Worcestershire sauce would fill the oceans?*"

We were stunned. *Keith and I cared.* It had been a wonderful intellectual game, and we were proud that we had worked through it. To my wife it had been 45 minutes of boring trivia, and she was angered that we had wasted our time and hers.

Why did Keith and I find our dinner table calculation interesting whereas my wife found it boring? This question has no simple answer, but I think the answer is connected to the fact that our behavior is goal-directed, that we seek goals that have survival value, and that our perceptions are governed by what we already have stored in our heads.

Surely my wife was correct to suggest that knowing the amount of Worcestershire sauce that would fill the oceans has no value, but the ability to do such calculations does. Keith, who was just developing those skills, sensed their value and was anxious to develop them. As the consummate teacher, I was anxious to help. Joyce, who had the skills and practiced them in her work as a nurse, had far more need for polite conversation. All of our behavior was goal-directed, but we had different goals.

No simple formula guarantees interest. We cannot say that "students like to work with their hands rather than listen," or "students like to observe, but they do not like to analyze." *Students like what has value for them, and what has value depends on their immediate goals, how they connect new experiences to old, and the effort required compared to its perceived value.*

WE ARE MORE LIKELY TO LEARN THINGS THAT ARE TAUGHT

When the Physical Science Study Committee (PSSC) physics curriculum, the Chemical Education Materials Study (CHEM Study), the Chemical Bond Approach (CBA) chemistry programs, and other new curricula were introduced in the 1960s, everyone was interested in knowing how good they were. To find out, standardized tests such as the Cooperative Physics Test or the ACS High School Chemistry Exam were given to students taking the new and the old curricula (Hipshire, 1961; Heath

& Stickell, 1963; Rainey, 1964). Students in the traditional curricula did better. This outcome is not too surprising because the tests were designed for the traditional curricula.

When tests designed for the new curricula were given, students in the new curricula did better (Heath & Stickell, 1963; Ferris, 1959). The results seem to say that we are more likely to learn what we are taught (Herron et al., 1976) and that curriculum reforms designed to introduce something new usually leave out something old. We ought to be sure we are ready to give up the old.

A more important point needs to be made about the inability to learn what is not taught. We complain that students are poor at solving problems, but we devote little attention to teaching students to solve problems. The attention that we do give is of the wrong kind, as we will see in Chapter 8.

We seldom make deliberate efforts to teach laboratory skills that, much to our chagrin, students have never learned. By teaching, I do not mean telling students what to do or performing a quick demonstration in front of the class. I mean carefully prepared lessons, with clear statements of expectation, feedback to individual students so they can correct errors, and evaluation at the end of instruction to be sure that the lesson is learned. I am talking about doing what any good coach would do in teaching a psychomotor skill.

WE ARE MORE LIKELY TO LEARN WHAT OTHERS WANT US TO LEARN WHEN WE KNOW WHAT THEY WANT US TO LEARN

A great deal of research has addressed the effectiveness of giving students behavioral objectives in advance of instruction. Some of the research suggests that the lists have no effect; other studies show that there is considerable benefit[2] (Atkin, 1968; Baker, 1969; Consalvo, 1969; Cowan, 1972; Duchastel & Merrill, 1973; Gleit & Ellington, 1978; Haberman, 1978; Herron, 1971; Kibler et al., 1970; Koran et al., 1969; Lewis, 1965; MacDonald-Ross, 1973; Melton, 1978; Tyler, 1972; Varagunam, 1971; Wolke, 1973; Young, 1972).

Part of the problem is that the focus is on the wrong question. The question is not whether statements of objectives produce more learning; rather, it is whether students who know what is expected do better than those who do not. The answer is that they do, and rather than review the research to make my case, I will use analogy.

Suppose that I ask you to go downtown and learn all that you can about the banks that are there—a suggestion that seems a little like asking students to learn all that they can about acids and bases or equilibrium or just chemistry. I have no doubt that you would come back with a great deal of information, but if my intent is to rob a bank, I am not sure that you would come back with the most valuable information. I would feel more comfortable if I suggested that you observe the movement of guards, estimate the amount of cash on hand, and look for signs of burglar alarms. You would not necessarily learn any more, but you would learn more about what I consider to be important.

If you are an experienced thief, have robbed many banks, and are aware of the general goal of "our gang", I would not need to call these matters to your attention

---

[2]As suggested by the dates on the references, *behavioral* objectives went out of style in the late 1970s. Today "outcomes", "standards", and "proficiencies" are more stylish, and it is uncommon for teachers to write and distribute *behavioral* objectives, although more general statements of objectives are common. Regardless of what terms are used to describe statements of instructional intent, the point of this section is as valid today as it was when it was written in 1978.

explicitly. You might even be offended if I did. However, if you are new at the game—and our students usually are—you might welcome the suggestion and perform better because of them.

Statements of objectives represent only one way of providing information about instructional intent. Study questions, quizzes, and copies of old exams may serve as well or better. In any case, some means of communicating what students are expected to learn and why this selected information is important is a necessary part of effective instruction.

IF STUDENTS ARE TOLD TO LEARN TRIVIA, THEY WILL

A common criticism of behavioral objectives was that they encouraged students to focus on learning specific information suggested by the objective and not to attend to other important points that should be learned.[3] This criticism still has merit, and it should not be ignored. However, the criticism is one of inappropriate focus derived from objectives or test items that focus on trivia. To illustrate my point, take the following—free response, mind you—test:

1. How many times have I used the word "the" in this chapter? (An order of magnitude will suffice.)
2. What is my son's name?
3. What was the point of the Worcestershire sauce story?
4. Name four principles of instruction and give an example from your own experience to illustrate each.

I am confident that nobody can answer the first question, and I suspect that few can answer the second. The answers do not matter, however, because the questions are trivial. You have learned to construct knowledge in a meaningful way, and that construction requires you to ignore many details associated with the process of learning.

Although none of us admits to writing exam questions as trivial as these, they may *seem* as trivial to the beginning student. Beginning students often lack the frame of reference needed to separate the trivial from the germane. (I might mention that some of my colleagues intentionally ask trivial questions and argue that the truly good student will notice these things along with the major points. Perhaps, but I will bet that the "truly good students" reading this chapter did not notice the number of times I used the word "the.")

Research reviewed by Thomas Anderson (1980) indicates that questions interspersed in text are among the most effective aids to understanding. The general procedure is to ask one or two questions after two or three pages of text. These questions help students recall information of the *type* asked in the question, not just the specific information requested. For example, when the questions call for a numerical quantity or a proper name, other numerical quantities or proper names are recalled selectively. When questions call for a technical term, this sort of information is selectively recalled. The questions appear to tell students what is considered important, and thus, what they should remember.

Too much focus presents dangers. If I convinced you that you would be treated to a steak dinner if you could tell me the number of times I used the word "the" in

---

[3]Today this criticism is more commonly leveled at multiple-choice test questions and the standardized exams that are commonly used in statewide testing programs.

this chapter, you might be so intent on counting that you would not attend to more important matters. Such dangers are real. However, they are not dangers of giving objectives. Poorly conceived standards, outcomes, proficiencies, textbook questions, tests, or homework can be equally myopic. *How* students are told what is important appears to be less important than *what* they are told.

LEARNING IS MORE EFFICIENT WHEN YOU GET FREQUENT, UNAMBIGUOUS FEEDBACK CONCERNING PROGRESS

Learning is a constructive process that is goal-directed. Part of what we learn is how to interpret signals from the environment to determine whether we are getting closer to our goals. Feedback tells us how to modify behavior to get closer to goals.

The entire field of educational technology (programmed texts, audio-tutorial instruction, computer-assisted instruction, and Keller's personalized system of instruction (PSI)) is based on the principle that frequent and immediate feedback promotes learning among motivated students. Research on the effectiveness of these programs indicates that the most important factor in students' success is the quality and quantity of feedback provided (Kulik and Jaksa, 1977).

Notice that the *quality* of feedback is important. Students are told "you are getting an A" or "you failed the exam" but not why or how. Such feedback is inadequate.

> To be effective, feedback must a) be related to the learner's goal, b) provide specific information about the discrepancy between the goal state and the present state of the student, and c) indicate how behavior might be modified to be closer to the goal state.

Feedback may be provided externally (e.g., by the teacher) or internally (e.g., by self-analysis). Two of the most intractable problems in instruction are finding ways to provide such feedback on an individual basis in large group settings and teaching students ways to analyze their own progress so that external feedback is unnecessary. These problems are addressed later in this book.

VERBAL RESPONSE IS AN INADEQUATE MEASURE OF UNDERSTANDING

I have heard cooks who could describe the preparation of delectable dishes in exquisite detail only to serve a meal that was an offense to both palate and gut. Conversely some outstanding cooks are unable to explain how they do it. They are not being coy; they cannot put their knowledge into words. So much critical knowledge about cooking has been automatized that they are not conscious of important steps and fail to verbalize them. The master teacher has the same problem when teaching the beginner. A colleague of mine recently decided to program a computer to teach his students to balance equations. He was caught short when he realized that he did not know how he balanced equations. Until he was able to make that knowledge explicit, he could not program the computer.

Similarly, some students have no difficulty in distinguishing a chemical and physical change, but they cannot define either. Students who can give perfect definitions for boiling are unwilling to describe the bubbling of water under vacuum as boiling because the temperature is not 100 °C. Students who recite wondrous descriptions of atoms behave as though matter is continuous.

In spite of this phenomenon, responses to examination questions that require no more than verbalization of an idea are taken as evidence that students understand. Nothing could be farther from truth.

The next three principles are so closely related that they are better discussed together.

Two closely related ideas or facts are easily confused when they are taught at the same time.

When one of two closely related ideas or facts is thoroughly understood, the new one is learned quickly by contrasting it with the one we know.

Regardless of when two closely related ideas or facts are taught, learning is improved by focusing on important features.

When I was a child, I played rummy. Many related but slightly different rummy games exist, and if I taught you three or four different rummy games today, you would probably get them confused and make a mistake while playing one of the games next week. However, if you already know how to play one rummy game, I can easily teach you another by contrasting the new game with the one you already know.

We violate these principles in many places during science instruction. I offer three examples to illustrate, and I invite you to search for others. Molarity, molality, and normality are taught in rapid succession and often in the same class period. We talk about Arrhenius, Bronsted–Lowry, and Lewis acids in the same breath. The expression for rate constants is presented immediately after a discussion of equilibrium constants, and this order invites students to confuse the meaning of the two expressions that look alike.

One of the serious consequences of including too many ideas in introductory courses is the superficial treatment of closely related ideas presented in close proximity. We would do better to teach less, well. Rather than teaching every expression for concentration, we might focus on the general concept of concentration using one example (molarity, for example) and defer other expressions of concentration until they are required. Similarly, teaching one concept of acids and bases (Bronsted–Lowry, for example) and deferring other concepts until the one is thoroughly understood and practiced is more likely to result in understanding.

If two potentially confused ideas must be presented close together in time, we can help by focusing attention on the differences. When presenting the expression for the rate law, calling attention to the superficial similarity between this expression and the one for equilibrium constants and focusing attention on essential differences is helpful.

THINGS THAT MAKE SENSE ARE REMEMBERED FOR A LONG TIME, BUT NONSENSE IS NOT

When knowledge is meaningful, connections are made between ideas. The more connections, the easier information can be recalled when needed. Nonsense or rote material lacks these connections and is forgotten; we cannot retrieve it when we need it.

Notice that this principle has two aspects. Some information, such as nonsense syllables or a person's name, has little potential for being connected to other ideas that we hold in memory. However, potentially meaningful information may be learned by rote simply because the connections that *can* be made are not. How one idea relates to another and how an idea may be used in various contexts to solve important problems are important components of instruction if we expect students to retain and use information.

Simple repetition or drill has little effect on retention because few connections are made. However, review of ideas in a meaningful context *does* enhance memory because review enlarges the network of connections. This concept suggests that 100 homework problems on nomenclature are not nearly as effective as periodic practice in a new context—for example, giving students names of reactants and products rather than formulas when equations are to be balanced. Similarly, drill on the metric system in Chapter 1 is far less effective than practice in measuring and converting units in the context of experimental work. The same holds for teaching significant digits, applying gas laws in a lab designed to determine molecular weight, or balancing equations so it is possible to work a stoichiometric problem. All such practice increases the connections that make ideas meaningful.

LOGICAL VERSUS PSYCHOLOGICAL ORDER

As information is understood, it is normally placed in some logical order. However, David Ausubel (1963) made the point that *logical* order and *psychological* order do not necessarily coincide. Several people have argued that the science curriculum in secondary schools is inverted. As it is, biology precedes chemistry which precedes physics; logically, the order should be the other way around. Physics is needed to understand chemistry, and chemistry is needed to understand biology, but not the reverse. Ausubel argues convincingly that, psychologically, the existing order is correct.

Evidence related to child development, learning, and intellectual development all suggest that learning occurs from the near to the far, from what is close to personal experience to what is farther removed.

From birth we are confronted with observations of living things. We *are* living things, and we certainly know some things about ourselves. We do not have to be told that respiration is necessary for life. When we hear about the digestive system, circulatory system, or reproductive system, we already have numerous observations on which to build. One can scarcely live without knowing something about flowers and seeds and growing plants and plant cycles. Although one may argue that chemical reactions are involved in all of these changes, the nature of these changes is less obvious. Still farther removed from direct experience are the regularities in mechanical motion, heat transfer, and electromagnetic interactions that are the subject of physics. In fact, direct experience often interferes with one's ability to believe some generalizations we teach.

Tell a young child that light and heavy objects fall at the same rate. She knows better from experience. Explain that a moving object will continue to move in a straight line until some unbalanced force acts to change its speed or direction. Nonsense!

Pick any introductory chemistry text and look at it. Is it organized according to a logical order or a psychological order? Does it begin with phenomena that are closely related to the experience of students? Does it introduce abstract notions such as atoms, molecules, ideal gases, bonding, and kinetic theory only when the student senses a need for some way to explain what he has already observed? More likely it is developed logically and begins with some tools—the metric system, temperature scales (all of them), perhaps some chemical symbols and a few equations, some math skills, and significant digits—then proceeds with atoms in all of their glory, molecules, and bonding. This plodding takes a spell, but it certainly seems worthwhile because, once it is done, the chemical changes that we want students to see and to know are much easier to talk about.

The problem is that students cannot see where this information is leading, and it does not seem at all logical *to them*. "Why are you asking me to do all of these weird things?" is their unspoken question. We see the need for what we are asking them to learn, but beginning students do not. In fact, they cannot.

CLOSING REMARKS

In spite of recent achievements in cognitive science, much remains to be learned, and what we do know is difficult to put into practice. Still, we have every reason to believe that we can do better than we have done.

The suggestion has been made that the curriculum reforms of the 1960s resulted in nothing more than students memorizing meaningless abstractions instead of meaningless facts. This comment contains much truth, but we must use care in responding to it. As I pointed out elsewhere (Herron, 1975), "The temptation to return to a course based on the blind memorization of a catalog of descriptive chemical facts is as repugnant to me as the continuation of courses based on the blind memorization of inscrutable theory. The alternative, in my judgment, is to recognize why the theory is inscrutable." The problem is not that we have paid too much attention to theory and too little to descriptive facts. We have paid too little attention to how students learn. Unless we do that now, the educational reforms proposed in response to *A Nation at Risk* (1983) will have no more satisfactory results than those proposed years ago in response to Sputnik.

YOUR OWN LIST OF APHORISMS

As you read through my list of "eternal verities", several things may have happened. First, you may have disagreed with some of my aphorisms. Second, you may have thought of some eternal verities that escaped my list. Third, instructional implications other than those I described may have come to mind. Taking a few minutes to write down these reactions will prove helpful in the following chapters. If you take issue with one of my statements, say why. (As you read further in this book, you will find that I take issue with some of them myself!) What would you add to this list of eternal verities to describe your own understanding of teaching and learning? Write it down. Now think about the courses you teach and how you teach them. For those aphorisms that you truly believe, what are some instructional implications of these eternal truths? Make a list and keep it handy as you continue to read.[4]

---

[4]Loretta Jones, who used an early draft as a textbook in a graduate course in chemical education, found that making this list at the end of the course was particularly valuable for her students.

Some Useful Theory

4

Intelligence

Teachers frequently assume that students do poorly in science because they lack intelligence. Before we can judge the validity of this idea, we need to consider what we mean by *intelligence*.

We enter the world with some mental equipment that allows us to interact with our environment and begin the process of learning, and individuals differ in the equipment they inherit. Most cognitive scientists agree with this general statement. But dissension occurs concerning the amount of mental equipment that is innate and the amount that accrues through learning. Also, disagreement about the nature of what we inherit exists.

The prevailing view of intelligence is a capacity model: the greater one's intelligence, the greater one's capacity for learning. This view seems obvious on the surface, but learning can be limited in a number of ways, and how we respond to individuals with limited intelligence depends on our assumptions about the meaning of *capacity*.

MODELS FOR INTELLIGENCE

THE JUG MODEL

Capacity to learn is often viewed like the capacity of a jug. This "jug model" states that some people are born with very large jugs and can hold a great deal of knowledge. Others come equipped with smaller jugs and can learn relatively little. When the jug is full, nothing can go in without letting something else out.

The jug model views differences in aptitude as innate and unalterable. Unintelligent people are fundamentally incapable of mastering complex or sophisticated ideas, whereas intelligent people can. Thus, a major task of schools is to identify those students with high or low intelligence and to educate them accordingly. Curriculum tracks may be developed to tailor schooling to the intellectual capacities of students, and intelligence tests may be used to measure the capacity of each jug.

Many studies call into question educational practices that are based on the jug model, even though these studies may not sabotage the model itself. Rosenthal and Jacobson's (1968) celebrated book, *Pygmalion in the Classroom*, provides convincing evidence that teachers' assumptions about their students' limited intelligence can be self-fulfilling prophecies: Students who are assumed to be too dumb to learn do not learn.

Mary Budd Rowe's classic studies (1974a, 1974b, 1974c, and 1974d) on wait-time revealed that students who were identified by their teachers as *slow* were given less time to respond to questions than students who were identified as *bright*. When teachers were trained to increase the time that they waited for slow students to respond, the quality of the answers improved dramatically. The *slower* students gave longer and more thoughtful replies that were similar to the replies given by bright students.

Cole and Scribner (1974) described a number of cross-cultural studies in which unschooled adults in primitive cultures responded poorly to tasks commonly taken as indicators of intelligence. However, on tasks that were indigenous to the culture, these same adults often demonstrated very intelligent behavior. These and other studies point to serious dangers in assuming a limited capacity to think and learn.

THE FUNNEL MODEL

Other views of intelligence acknowledge innate differences, but the differences are seen in terms of learning rate rather than learning capacity. Think of this approach as the "funnel model".

In the funnel model of intelligence, no difference in total capacity is postulated. Rather, the difference is in the size of the funnel through which information must flow. Those people with high intellectual aptitude come equipped with very large funnels and are able to acquire information at a prodigious rate; those born with smaller funnels have lower intellectual aptitude and a restricted rate of learning, and they need a longer period of time to acquire a given amount of knowledge. The funnel model places no restriction on either the amount or kind of knowledge that a person may learn, but it does suggest that some people will learn faster than others.

Many readers will recognize the funnel model as Carroll's (1963) conception of intellectual aptitude that was used by Benjamin Bloom (1971) to formulate his model of mastery learning. The funnel model has acquired increasing acceptance in recent years, but it, too, has certain flaws.

Most teachers readily concede that some students catch on quickly whereas others take a long time to master an idea. But they also know that even the brightest child has difficulty visualizing atoms, infinity, a mathematical limit, or probability. Some children get lost in long chains of logical inference, and they are easily deceived into thinking that the amount of substance has changed when shape or position changes. Conventional efforts to teach such children where they are wrong lead to acquiescence but not belief. The problem appears to involve both learning *rate* and learning *capacity*.

If we assume that the rate of learning is more or less fixed, we see practical implications. Consider this problem:

Bloom (1971) claims that if we eliminate the extreme five percent of students who have been diagnosed as having organic learning deficiencies, the time required for the slowest student to master material is approximately six times the time required for the fastest student. Bloom suggests that if instruction becomes more effective, that ratio could probably be reduced to 3 to 1.

Now suppose that high school graduation requires mastery of the same material currently mastered by average students in 12 years of schooling. Assuming that students begin school at age six, what would be the age of the youngest and oldest students graduating from high school if the actual ratio of learning rates is the more optimistic 3 to 1?

I wish to make two points with this problem: First, society is not likely to accept students graduating from high school as early as age 12 or as late as age 24. If the funnel model is correct and the only difference in intellectual aptitude is the rate at which we learn, we are still unlikely to develop educational programs in which all students master the same amount of material.[1] The second point is related to the fact that fewer than half of my students (college seniors majoring in chemistry, physics, or geoscience) manage to solve this problem. Are they incapable because their jugs are too small or their funnels are too narrow? Can *you* solve the problem? If you have trouble, what causes it?

THE TOOL BOX MODEL

The "tool box model" of intellectual aptitude is most consistent with the ideas presented in this book and differs from both the jug and funnel models. This model suggests that we come into the world equipped with a few rudimentary intellectual tools that we use from birth to acquire and organize intelligence. We use these tools to interpret the world around us and construct new tools that allow new interpretations and new kinds of learning.

Intellectual tools vary from simple to complex. Newborn babies come equipped with intellectual tools that produce sucking in response to stimulation to their lips, prompt crying in response to discomfort, and random flailing of arms and legs. On the road to becoming adults, these rudimentary tools are used to fill the mental toolbox with increasingly complex tools: proportional and combinatorial reasoning, the ability to think in terms of possibilities as well as direct experience, and the ability to reorganize information to create new relationships. Thus, the toolbox model explains intelligence in terms of the number and kinds of intellectual tools that are available for use rather than the capacity to hold information or the rate at which information can be acquired.

MEASURING INTELLIGENCE

IQ TESTS

The jug, funnel, and tool box are metaphors for common ways of thinking about intelligence, but they are not descriptions of intelligence as measured by "IQ tests". A short history of intelligence testing may clarify the relationship.

---

[1]This statement assumes that the 3:1 ratio of learning rates cannot be overcome. Some educators believe that instruction can be improved to the point that essentially no differences in learning rates exist.

Intelligence testing originated in Paris at the turn of the century when the Minister of Public Instruction commissioned Alfred Binet and Theophile Simon to develop a test that would predict success in school.[2] The Binet–Simon test and all subsequent IQ tests have been shaped by the educational traditions of Western Europe and North America. As Jensen (1969) pointed out: "The content and methods of instruction represented in this tradition . . . are a rather narrow and select sample of all the various forms of human learning and of the ways of imparting knowledge and skills. The instructional methods . . . evolved within an upper-class segment of the European population, and thus were naturally shaped by the capacities, culture, and needs of those children whom the schools were primarily intended to serve." Jensen continues with a quote from O. D. Duncan: "Had the first IQ tests been devised in a hunting culture, 'general intelligence' might well have turned out to involve visual acuity and running speed, rather than vocabulary and symbol manipulation. As it was, the concept of intelligence arose in a society where high status accrued to occupations involving the latter" (cited in Jensen, 1969, p 14).

Several important implications here should not escape our attention. Using the tool box model previously outlined, we can say the following:

1. Intelligence tests measure only a few of the intellectual tools that are used to make sense of experience. If IQ tests measured a different constellation of intellectual tools, those people labeled *intelligent* and *stupid* would also change.
2. The abilities selected for measurement by intelligence tests are abilities that are important for success in schools dominated by a particular constellation of instructional practices. These abilities may not be equally important for success in schools using other instructional practices.
3. IQ tests do predict 30–40% of the variance in traditional school performance, and traditional school performance predicts success in the most prestigious occupations in Western society. Changing IQ tests so that they tap other intellectual skills would make little sense until school and societal practices change so that other intellectual skills predict success.

HERITABILITY OF IQ

Strong evidence exists that intelligence, as defined and measured by IQ tests, is partially—perhaps largely—controlled by genetics; it is inherited. Educators must understand that high heritability does not imply that environment can have little effect on the performance (e.g., success in school) that depends on intelligence. Jensen cites the interesting case of tuberculosis, which, at one time, had a very high heritability: Whether one died of tuberculosis depended largely on whether one inherited a constitution strong enough to resist the bacilli that cause the disease. However, changes in the environment—specifically, the development of diagnostic tests that allowed isolation of infected persons and the development of antibiotics that kill the tuberculosis bacillus—resulted in conditions such that heredity plays a negligible role in determining who dies from tuberculosis today (Jensen, 1969, p 45).

---

[2]Jean Piaget, whose ideas of intellectual development play a prominent role in this book, worked with Binet for a time, but he disagreed with Binet's approach and left to develop his own research program from a different perspective.

Environmental changes probably alter the intellectual effects of inheritance in much the same way that advances in medicine altered the physical effects of inheritance. To see how this change can occur, we need to take a closer look at differences in the kinds of tools that people inherit.

WHAT IS INHERITED? Innate intellectual tools probably differ from one person to the next, just as people differ in physical appearance, resistance to disease, and other characteristics. Minor differences in the kinds of information-processing mechanisms we are born with may affect the rate or efficiency with which sensory data are processed, just as minor differences in physical makeup may affect one's susceptibility to tuberculosis, breast cancer, and heart disease.

Cole and Scribner (1974) report a series of cross-cultural studies by Pollack that illustrate how genetic differences can affect the way individuals process information. Pollack's studies were designed to account for variations in susceptibility to the Müller–Lyer illusion shown in Figure 4.1. When asked to judge the relative length of the vertical and horizontal lines, most people perceive the horizontal line as shorter. However, people from some cultural groups are more susceptible to being fooled than are others.

Pollack's studies provide strong evidence that retinal pigmentation affects perception: the more dense retinal pigmentation, the less susceptible people are to the Müller–Lyer illusion. Pollack's studies suggest that retinal pigmentation affects other perceptions as well. A survey of color names in 126 societies produced a pattern of color naming paralleling the distribution of eye pigmentation (cited in Cole and Scribner, 1974, pp 77–78). These data suggest that differences in retinal pigmentation, a genetic trait, may cause people to *see* color differently. Little imagination is required to deduce that such differences could affect how individuals who differ in the amount of retinal pigmentation might respond to stimuli differently and develop certain competencies more quickly than others.

Even minor differences in the way stimuli are processed could result in different rates of learning and observable differences in behavior: the age at which a child learns to crawl, its response to various visual and aural stimuli, the advent of speech. We would see apparent differences in capacity due to differences in the rate that new intellectual tools are constructed.

Flavell, Miller, and Miller (1993) summarized recent research on innate abilities in these words:

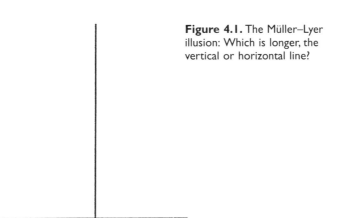

Figure 4.1. The Müller–Lyer illusion: Which is longer, the vertical or horizontal line?

[I]t is findings with regard to infant precocity that have been most responsible for the surge in nativistic theorizing in recent years. . . . [M]any . . . competencies are evident at such early ages that it is impossible to see how they could be instilled by the environment. . . . We seem to be biologically prepared to do very specific kinds of information processing and very specific kinds of learning, with no apparent links between one set of processing mechanisms (e.g., those for discriminating speech sounds) and another (e.g., those for extracting numerical information) (pp 336–337).

EFFECTS OF INNATE DIFFERENCES. As discussed previously, before the germ theory of disease and the development of antibiotics to combat germs, diseases such as tuberculosis were, in effect, *inherited*. Whether a person died of the disease was largely a matter of genetics. This is not to say that the disease itself was inherited, just as current theories concerning the heritability of breast cancer do not claim that the cancer itself is encoded in the DNA. Rather, the argument is made that some characteristic encoded in DNA affected the likelihood that a person would die of tuberculosis in the past or will die of breast cancer today.

The response to information about the heritability of disease is important for us to ponder: Rather than viewing persons with diabetes, sickle-cell anemia, hemophilia, and other genetically linked diseases as inferior beings who are relegated to menial jobs, we provide them with drugs and medical care that will enable them to function in virtually any occupation. Similarly, the response to news that some women may be born with "the cancer gene" is not to stop treating breast cancer because it is genetically linked and, consequently, nothing can be done about it. Rather the response has been to increase research aimed at identifying the gene, develop screening tests to identify women who carry the gene, and monitor them closely for early detection and treatment of cancer when and if it occurs.

This response to heritability of disease is in sharp contrast to the historical response toward evidence concerning the heritability of intelligence. The legacies of genocide, slavery, and racial prejudice hindered the study of how genetic factors lead to differences in the tools individuals use to learn. The fear—largely justified—that information derived from such research may be twisted by racists to justify discrimination makes funds difficult to obtain, creates danger for researchers, and makes biased data likely. Jensen (1969) argued that differences may exist in the *way*—not the amount—different racial groups learn. Jensen's argument caused such a furor that his suggestions have been largely ignored, even though they could conceivably lead to important improvements in education.[3]

In view of the narrow range of intellectual tools measured by IQ tests and the limited scope of teaching strategies that are used in schools, changes in the teaching and learning environment might reduce the heritability of *intelligence* (or at least school achievement, which is the variable of interest) in a manner analogous to the reduction in the heritability of tuberculosis. Even under existing school practices, IQ is often a poor predictor of achievement. All teachers know of "overachievers", whose learning is far better than predicted by IQ scores. Furthermore, teachers can

---

[3]Although I have not seen Jensen's work cited, recent research on learning styles may have been spurred by his review of "[emerging evidence] that there are stable ethnic differences in *patterns* of ability. . . . The . . . pattern of scores [is] distinctively different for Chinese, Jewish, Negro, and Puerto Rican children, regardless of their social class. Such differences in patterns of ability are bound to interact with school instruction. The important question is how many other abilities there are . . . for which there exist individual and group differences that interact with methods of instruction" (Jensen, 1969, p 109).

quickly identify many of the factors that account for their success: self-confidence, good work habits, favorable attitudes, a stable and supportive home environment, and strong interpersonal skills. Much can be done to improve school performance, regardless of inheritance. In particular, little credible evidence supports the belief, commonly held by college and secondary school science teachers, that only a select group of individuals who have inherited a high IQ can understand science and mathematics.

EFFORTS TO REDUCE HERITABILITY OF IQ. Just as antibiotics reduced the heritability of tuberculosis by making natural resistance unnecessary to fight the disease, changes in the way we teach may reduce the heritability of IQ. One of the few research programs aimed at reducing IQ deficits is that of Reuven Feuerstein in Israel.

While still a graduate student, Feuerstein began working with Youth Aliyah, the Israeli agency responsible for the integration of immigrating Jewish children into Israeli society. Many adolescents entering Israel came from primitive cultures and were severely *retarded* in terms of conventional measures of intelligence. They lacked the intellectual tools required for success in school, and Feuerstein's task was to help them construct those tools. Both the Learning Potential Assessment Device (LPAD), described in a later section, and Feuerstein's Instrumental Enrichment (FIE) program were developed in response to that challenge. The key ingredient in FIE is the Mediated Learning Experience (MLE):

> MLE is said to occur when an individual (typically a child) is shown or taught cognitive methods for interpreting information, for solving problems, or for learning something. For example, in interacting with a child an adult might illustrate the usefulness of categorizing a particular piece of information and then go on to demonstrate a technique for doing this categorizing. Feuerstein, like Piaget (1954), believes that children can learn from interacting with the environment; but, like Vygotsky (1962), he emphasizes the importance of the mediation of the child's learning by adults. . . . [O]ne of the anticipated results of providing children with MLEs is that they would become more aware of their cognitive processes and abilities; that is, they would exhibit an increase in their metacognitive activity (Savell, Twohig, and Rachford, 1986, pp 384–385).

"The Feuerstein Instrumental Enrichment program consists of more than 500 pages of paper-and-pencil exercises, divided into 15 instruments. Each instrument focuses on a specific cognitive deficiency" (Feuerstein, Rand, Hoffman, and Miller, 1980, p 125). These instruments deal with intellectual competencies ranging from identification of a geometric figure of a particular form and size within a field of dots, to orientation in space, to syllogistic reasoning.

Although FIE intentionally focuses on competencies that are not directly related to any school curriculum, the competencies are far more specific than those commonly addressed in conventional IQ tests. Relationships are easy to infer between the intellectual tools addressed in FIE exercises and competencies required for school subjects. The ability to discriminate size, shape, and orientation addressed in the exercises on identification of geometric figures, for example, would appear to be quite important in understanding projective geometry, some aspects of Euclidean geometry, and many areas of physical science.

Savell, Twohig, and Rachford reviewed the research on FIE in 1986. As with most school research, the studies that they reviewed are messy and leave many questions unanswered. Strictly speaking, Laughon's comment based (at least in part) on their review is correct:

> [R]esearch on Feuerstein's Instrumental Enrichment program . . . has failed to document a strong relationship between skills learned in the program and school achievement (Laughon, 1990, p 466).

Still, Laughon's comment is misleading. In general, the studies reviewed by Savell, Twohig, and Rachford are supportive of FIE when it is implemented as Feuerstein proposed. After describing several limitations of the studies, Savell, Twohig, and Rachford go on to say:

> Within the total set of studies, however, there is a subset that produced data that are striking and suggest that FIE may indeed be having an effect even though it is not clear . . . just what this effect means (p 401).

Among the generalizations and conclusions drawn by the authors are the following:

1. Several studies report statistically significant differences that favor FIE. Such studies have been done in Canada, Israel, the United States, and Venezuela; they have involved middle-class as well as lower-class social groups, and they have been done with normal as well as educationally disadvantaged and hearing-impaired groups.
2. The effects reported in these studies were usually on nonverbal measures of intelligence. The effect on such measures as self-esteem, impulsivity, classroom behavior, academic achievement, and course content were absent, inconsistent, or difficult to interpret.
3. Several commonalities run through those studies that show experimental and comparison group differences favoring FIE: The FIE instructors had at least one week of training. The students were generally exposed to FIE for at least 80 hours over a one- or two-year period. FIE was taught in conjunction with other subject matter that was important and interesting to the students (Savell, Twohig, and Rachford, 1986, pp 401–402).

FIE is unlikely to be applied directly in many science classes at the secondary or college levels. However, Feuerstein's work is important because it indicates that failure to learn is frequently due to lack of specific intellectual tools required to make sense out of particular kinds of experience. Instruction aimed at developing the missing tools can lead to substantial increases in meaningful learning, whereas the same amount of instruction focused on the information to be learned often leads to rote memorization without internalizing the information in a way that can be applied to new situations.[4]

OTHER TESTS OF INTELLIGENCE

In spite of the widespread use of IQ tests to predict success in schools, these tests have serious limitations. IQ is a measure of what has been learned (within the

---

[4]Although it is too early to tell, Brown and Campione's approach (introduced in the next section), which is closely tied to *meaningful* learning of conventional school tasks, may prove more effective than the more general approach developed by Feuerstein.

knowledge domains addressed by IQ tests) by a student compared with what has been learned by others of the same age. Because what we have already accomplished is a good indication of what we will accomplish in the future, this approach is a reasonable way to predict success in school. But under certain circumstances, this procedure does not work very well.

People from cultures that differ markedly from the one for which IQ tests were designed (and people from within that culture who are deprived of the stimuli in which intellectual tools develop) often learn less than individuals with similar genetic endowment who are in the cultural mainstream. In those situations, tests of intelligence would be more meaningful if they tapped learning *potential* rather than what was already learned.

Laughon (1990) reviewed three tests of learning potential: Budoff's Learning Potential Procedure, LPAD, and Brown and Campione's dynamic assessment. Budoff's test is designed primarily for educable mentally handicapped students, and I will ignore it. The other two, however, were used in school settings, and considering how these tests differ from conventional IQ tests is instructive.

THE LEARNING POTENTIAL ASSESSMENT DEVICE (LPAD). Whereas conventional IQ tests estimate the potential for learning by measuring what has already been learned, the LPAD measures the amount of learning that takes place during a lesson mediated by a skilled adult. Unlike ordinary intelligence tests, the LPAD allows the examiner to interact with the examinee using a test–train–test format. The examiner acts as a teacher, attempting to help the examinee improve on the test while the examiner observes the learning process. In assessing test performance, attention is focused on performance peaks rather than on errors (Laughon, 1990, p 462; Feuerstein, Rand, and Hoffman, 1979).

A study of Bedouin children by Rand and Kaniel suggests how the LPAD differs from more familiar IQ tests. Previous research showed that problem-solving and abstract reasoning are particularly difficult for Bedouin people. Conventional intelligence tests rely heavily on these intellectual tools, so Bedouin children appeared to be a good population on which to test the value of the LPAD. Rand and Kaniel used four LPAD subtests. After administering these subtests, they gave the children three hours of training followed by a posttest. Performance on the test improved substantially and approached the average score of Israeli children of comparable age on three of the four subtests (cited in Laughon, 1990, p 463).

Dynamic measures such as the LPAD are very new compared with conventional intelligence tests, and the research related to these measures is too limited to support conclusive statements about their value in school settings. However, two things are clear:

1. Factor analytic studies show that the intellectual tools tapped by dynamic assessment differ from those tapped by conventional tests.
2. The LPAD and Brown and Campione's dynamic assessment (described subsequently) provide information that teachers can use to remedy learning difficulties, something that conventional IQ tests do not do (Laughon, 1990, p 463).[5]

---

[5]Readers who wish to know more about the LPAD and other forms of interactive assessment should consult Haywood, 1992; Haywood and Tzuriel, 1992; and Lidz, 1987, 1991. These references contain extensive bibliographies of research in this area.

DYNAMIC ASSESSMENT. Brown and Campione's dynamic assessment is based on information-processing theory, research on individual differences, and research on transfer of learned information to new contexts. Dynamic assessment looks at the amount of instruction needed to learn and to transfer information (Laughon, 1990, pp 463–464). The goal of dynamic assessment is to identify important intellectual tools that the examinee uses poorly, if at all, and then to provide experiences that lead to the development of those tools. Early research has been encouraging.

Ferrara examined the ability of dynamic assessment to predict success on simple math word problems: "Results indicated that dynamic scores were more highly correlated with gain scores than were either ability or math knowledge scores. When static scores were extracted first, they accounted for 22% of the variance in gain scores. Dynamic scores accounted for an additional 33% of the variance, with transfer scores contributing most to prediction" (cited in Laughon, 1990, p 464). Laughon summarized the research on dynamic assessment with these comments:

> The predictive validity of Brown and Campione's approach has been demonstrated in part by the finding that learning and transfer measures were substantially related to improvement on the learning tasks. . . . Compared to static IQ measures, dynamic measures were more highly correlated with gain scores and contributed significantly in accounting for variance in gain scores.
>
> At present, the approach of Brown and Campione has provided only limited data regarding remediation. In a recent review, Campione and Brown (1987) presented preliminary evidence on a process termed reciprocal teaching, in which assessment is built into the teaching process and is conducted essentially by the teacher. Thus far, this technique has been utilized primarily in a laboratory setting, thus raising questions about its efficacy in the classroom (p 467; *See* Resnick, 1989, for a more recent discussion of reciprocal teaching).

Although far from conclusive, the results cited by Laughon suggest that dynamic assessment has the potential to tap intellectual abilities that are not measured by conventional IQ tests but are important in school learning. (For further information about dynamic assessment, see Lidz, 1987.)

SUMMARY

This chapter explored three aspects of intelligence that have important implications for teaching chemistry. First, three metaphors for intelligence were considered: the jug model, which explains differences in intelligence as differences in our capacity to be filled with knowledge; the funnel model, which assumes unlimited capacity but innate differences in learning rate; and the tool box model, which describes intelligence in terms of intellectual tools, some innate and some learned, that guide learning. The tool box model is the basis for arguments presented in this book.

Second, intelligence tests and what they measure were examined. Regardless of the metaphor used in thinking about intelligence, IQ is defined operationally by the tests used to measure it. Those tests evolved in a particular cultural context to predict success in a particular kind of school. Many intellectual competencies exist that are *not* tapped by standard measures of intelligence, and little is known about how those competencies might be used to benefit society.

Dynamic methods of assessing learning potential are being developed, and these testing methods clearly tap intellectual tools that are important in school learning and are different from the intellectual capacities measured by conventional IQ tests. The hope is that these assessment procedures will provide more specific information about weak or missing intellectual tools and that those weaknesses can be remedied, either through special programs such as FIE or in the context of normal instruction, as suggested by Brown and Campione. Whether that hope can be realized remains to be seen, but existing research is encouraging.

Finally, the nature of inheritance and its effect on what we think of as intelligence was discussed. The statement that intelligence is influenced by genetic makeup is generally accepted, but what that statement means for teaching and learning is hotly debated. The theory that guides discussion in this book holds that genetic differences may produce variations in visual acuity, aural discrimination, and perhaps other intellectual tools that serve as precursors to numeration. These differences in turn account for precociousness in music or language development, much as differences in retinal pigmentation influence susceptibility to the Müller–Lyer illusion.

From the teacher's perspective, how intelligence is influenced by inheritance is less important than how we respond to differences we observe. In particular, assuming that students who score low on conventional IQ tests can never learn chemistry is no more defensible than assuming that women who inherit genes that predispose them to breast cancer must die from the disease. Both assumptions result in an unnecessary loss of human potential that our society can ill afford.

5

How We Learn

ex · plain (ik splān'), *v.t.* **1.** to make plain or clear; render intelligible. **2.** to make known in detail: *to explain how to do something; to explain a process. . . .* **4.** to make clear the cause or reason of; account for: *I cannot explain his behavior* (Stein, 1975).

Explanations occur at various levels. Consider how a car operates. One explanation might begin with inserting the key in the ignition and continue with placing the car in gear, pressing the accelerator, and steering. Another explanation might focus on mixing fuel with air, injecting it into the cylinder of the engine, compressing the mixture before ignition, and continue with a description of how the power produced is transferred through the drive train to the wheels to propel the car. Still another explanation might give a detailed description of each system and subsystem in the automobile and point out how they are interconnected and tuned to propel the vehicle we call an automobile.

The kind of explanation needed depends on the purpose. A detailed explanation of the automobile's mechanical, electrical, and chemical systems is as useless to a beginning driver as basic instruction on "making it go" is to an engineer. Too much detail is confusing; too little is debilitating.

The discussion of learning in this book is based on *schemas.*[1] Schemas are one level removed from the overt observations of everyday life, but they imply nothing about the physiological processes that must ultimately explain how learning takes place. Other levels of explanation are available. *Connectionism*, the term applied to a theory that explains learning in terms of layers of neural networks, is currently in vogue. "Layers of neural networks" is a more apt description of brain structures than is schema. However, this book is meant for teachers rather than cognitive scientists, and schema is a powerful tool for discussing the behaviors we associate with learning. Schema theory will support intelligent discussion of how students learn without getting bogged down in microscopic models for neural networks.

---

[1]Although the dictionary says that "schemata" is the plural of schema, schemas and schemes are more often used in the literature on schema theory. I follow Bartlett's example and use "schemas" in this book.

SCHEMAS

WHAT IS A SCHEMA?

> schema (*pl.* **schemata**, also **schemas**) **1.** a generalized diagram, plan, or scheme. **2.** (in Kantian epistemology) a concept, similar to a universal but limited to phenomenal knowledge, by which an object of knowledge or an idea of pure reason may be apprehended (Stein, 1975, p 1177).

This definition is a good one for schema because it is used to describe learning. Anderson describes schemas as "large, complex units of knowledge that organize much of what we know about general categories of objects, classes of events, and types of people" (J. Anderson, 1980, p 128). Winograd (1976) elaborated:

> At the simplest level, a schema is a description of a complex object, situation, process, or structure. It is a collection of knowledge related to the concept, not a definition in the formal dictionary sense. A definition provides just the necessary information to distinguish a concept from others, while a schema contains a body of related knowledge to be used in reasoning (p 72).

The earliest use of schemas to describe learning appears to be by Bartlett in connection with his research on the recall of stories (Bartlett, 1932). Bartlett wondered how his British students would remember a story that did not fit well with their "cultural schemas". He had them read "The War of the Ghosts", a story from American Indian folklore. When students recalled the story later, they were prone to rationalize, reconcile, omit, or distort the story. The greatest distortions occurred in features of the story that would be unusual or foreign to the British readers (described in diSibio, 1982, p 150). Thus, Bartlett used schemas to describe *"existing knowledge, attitudes, or cultural orientations that influence the understanding of new information"*. This definition is very similar to the one I use here.

Schemas clearly refer to things that we have learned, so why do we not talk about knowledge or ideas? The reason is that knowledge and ideas usually have the connotation of declarative knowledge (*see* Chapter 2), such as, "The author's name is Dudley Herron," and "Matter is made up of atoms." We construct many schemas that are difficult to describe as information in the usual sense. We respond *emotionally* to events because of attitudes and values that we find difficult to describe as knowledge, and we automatically process information in particular ways without being able to name the idea that led to the interpretation. Schemas refer to such attitudes, values, and rules for processing information as well as to information.

KNOWLEDGE CONTAINED IN SCHEMAS

PHYSICAL KNOWLEDGE. The Swiss psychologist and epistemologist Jean Piaget made a distinction between two kinds of knowledge in schemas. Physical knowledge includes information, both implicit and explicit, about properties of objects and events. It includes things such as perception of size, color, texture, the heft of an object, and other information acquired through the senses as we interact with our environment. Piaget contrasts physical knowledge that is gained directly through sensory perception with logicomathematical knowledge, which is not.

LOGICOMATHEMATICAL KNOWLEDGE. Logicomathematical knowledge is abstracted knowledge about relationships that are independent of the objects themselves. To illustrate the difference between physical and logicomathematical knowledge, consider a group of five marbles on a table.

By interacting with the marbles you can learn (construct schemas) about color, size, feel, and how marbles roll when pushed. All of this is knowledge of the physical world.

By interacting with the marbles, you can also come to know that they can be grouped in various ways, and you can abstract from such grouping knowledge that 1 + 4 = 2 + 3 = 3 + 2 = 4 + 1. This knowledge has nothing to do with marbles per se; it is logicomathematical knowledge, and it represents some of the most important knowledge that we develop.

Logicomathematical knowledge does not record understanding of the environment; it indicates what kind of record is sensible. Logicomathematical schemas are the builders of intelligence. To what extent such knowledge is a part of a schema about marbles and to what extent it exists as a separate schema, disembedded from marbles or any other object and capable of generalization to all classes of objects, is a matter of ongoing research (Brown, Collins, and Duguid, 1989; Cognition and Technology Group at Vanderbilt, 1990).

SCHEMA CONSTRUCTION

CONSTRUCTIVISM I: KNOWLEDGE IS CONSTRUCTED

We generally think of knowledge as something that has an existence of its own. It is "out there," and our job is to get it inside our heads. Teachers, in this world view, have the job of transmitting knowledge to their students, and students have the job of retaining it. If we stop to think about this concept, we know that it is not so. As John Dewey said just after the turn of the century:

> [N]o thought, no idea, can possibly be conveyed as an idea from one person to another. When it is told, it is, to the one to whom it is told, another given fact, not an idea. The communication may stimulate the other person to realize the question for himself and to think out a like idea, or it may smother his intellectual interest and suppress his dawning effort at thought. But what he *directly* gets cannot be an idea (Dewey, 1916/1926, p 188).

A view of learning that comes closer to what must happen is *constructivism*. Constructivism asserts that "knowledge is not passively received but is actively built up by the cognizing [learner]" (Wheatley, 1991, p 10). Rather than knowledge existing outside the learner and teachers being the conduits of that knowledge to their students, constructivism holds that knowledge exists *only* in our heads where it is constructed by each of us in our own way. This point was made by Dewey, and it is a point that is generally accepted.

LEARNING AS ADAPTATION

We all know that we are born with our senses; we feel pain and respond to light. We do not, however, *see*. The sensors that allow us to receive information from the environment are present, but the intellectual capacity to make sense of the data is not. The newborn begins a long journey toward understanding equipped with only a few rudimentary schemas.

The newborn has a schema for sucking, for sight, and for grasping. Undoubtedly the newborn has others, but these three are among those postulated by Piaget.[2] With these schemas we begin to adapt to the world by constructing new schemas (i.e., we learn). Adaptation, according to Piaget, has two simultaneous and complementary aspects, *assimilation* and *accommodation*. Assimilation and accommodation are not separate events. Rather, the terms describe two aspects of adaptation.

Adaptation, Piaget argues, occurs in response to external and internal influences. Assimilation refers to adaptation in response to internal processes; accommodation refers to adaptation in response to external processes.

Clearly distinguishing assimilation from accommodation is difficult and unimportant. The important point is that in any learning event an interplay between external and internal influences exists. What we understand from the event is a result of this interplay.

Consider, for example, the interplay that takes place when I identify a playing card, which is the subject of research described in the next section. In viewing a card, I would notice many things: color, shape, number of figures, arrangement of figures, their size, orientation, texture, brightness, and the like. But if I gave equal weight to all of this information, I would not be able to interpret what I *see*. To make the image fit some idea in my head, I must attend selectively to what matters. Existing schemas tell me what matters and what does not. Fitting the image to a schema is what Piaget described as assimilation, and the selective attending required to make it fit is accommodation.

IDENTIFYING PLAYING CARDS: AN EXAMPLE OF ADAPTATION. Many studies suggest that interpretation of our environment involves complex interactions between the information coming through our senses and the schemas already in place. In one such study (Bruner and Postman, 1949), experimental subjects were asked to identify playing cards that were presented to them. Most of the cards were normal, but some, such as a *red* six of spades and a *black* four of hearts, were not. In each experimental run, cards were presented, one at a time, for a short, controlled period of time, and the subject was asked to name each card immediately after it was presented. In successive runs, the period of exposure was increased until every card was correctly identified on two successive runs.

Even when the time of presentation was short, most subjects identified the cards, whether normal or abnormal, with little hesitation. The normal cards were identified correctly; the abnormal ones often were not. The red six of spades, for example, was identified as the six of hearts about as often as it was identified as the six of spades. The anomalous image was made to fit one or another existing concept: a red six of hearts or a black six of spades.

Let us consider what is going on in terms of Piaget's theory of adaptation. The first point to consider is the necessity of some existing schema that includes the concepts six of hearts and six of spades as categories of playing cards with particular attributes. Among the attributes used by that schema to place cards into categories are color, number and arrangement of figures on the card, and shape of the figures. Let us call this our *card-naming schema*. Without such a schema, Bruner and Postman's experiment could not be done. Each card would appear like a mystery object

---

[2]Recent research postulates the existence of innate structures for language processing, perception of objects, perception of causality, recognition of the animate–inanimate distinction, and abstraction of number. However, the kind and number of innate schemas is still uncertain. (*See* Flavell et al., 1993, pp 336–339 for a summary.)

displayed in a natural history museum—something designed and understood by another person in another age but a total mystery to us. The prior existence of some kind of schema is essential for any interpretation of our environment. In other words, all learning is driven by schemas.

To name cards whose images are projected onto a screen, the information contained in the image must be assimilated—it must be fit to some schema we have previously constructed. But it is silly to think that it is only the internal schema that controls how we respond to the images projected onto the screen. The nature of the image and other external factors matter too. Suppose that the projected image is a photograph of Aunt Matilda rather than a playing card. The information in that photograph would not fit our card-naming schema at all. The image could not be assimilated to our card-naming schema because the information just is not right.

Even when an image is assimilated, some selective attending is normally required. The color, size, or shape of the image may differ from what our card-naming schema has learned to expect, but the schema can still accommodate the image if we ignore the unusual features. In accommodating the image of a normal card to make it fit a concept in our card-naming schema, we are required to ignore such things as the size of the card, its orientation, and the brightness of the image. In effect, our schema tells us that these things do not matter. But we are required to attend to such things as color, shape of the figures on the card, and the number. In effect, our schema tells us that these things do matter. What happens if some of these things fit but others do not? We can force a fit by paying attention to some things that matter while ignoring others, or we can modify our card-naming schema so that all of the relevant information fits.

When presented with a *red* six of spades, subjects readily identified it as a six of hearts or a six of spades. No problem. The projected image was accommodated by ignoring the shape of the spots (in the first instance) or ignoring color (in the second instance) so that it could be assimilated to the card-naming schema.

ADAPTATION AS LEARNING. The discussion thus far may explain how intelligent people so easily make perceptual errors, but it does not explain how we learn. Other aspects of the Bruner and Postman experiment do.

As Bruner and Postman increased the time that anomalous cards were exposed, the subjects became more hesitant. Kuhn (1970) described this phenomenon as follows: "Exposed, for example, to the red six of spades, some would say: That's the six of spades, but there's something wrong with it. The black has a red border. Further increase of exposure resulted in still more hesitation and confusion until finally, and sometimes quite suddenly, most subjects would produce the correct identification without hesitation. Moreover, after doing this with two or three of the anomalous cards, they would have little further difficulty with the others" (p 63).

Something happened so that these subjects were able to learn. They no longer made the same mistakes. Their card-naming schema had some new instructions: "Look out for a mismatch between color and shape!"

Piaget's theory of adaptation provides a useful model for learning as an alteration of schemas, but the model provides no information about the mechanism of change. Our description of adaptation does not explain, for example, the fact that a few subjects in the Bruner and Postman experiment never adjusted their schemas so that they correctly identified the anomalous cards, even when those cards were exposed for relatively long periods of time. According to one subject: "I cannot make the suit out, whatever it is. It did not even look like a card that time. I do not know what color it is now or whether it's a spade or a heart. I'm not even sure now what a spade looks like" (Kuhn, 1970, pp 63–64).

In recent years, research on misconceptions in science has been substantial. The research has received considerable attention because knowledgeable adults are invariably amazed that students hold such naive ideas, even after formal instruction in science. The number and variety of misconceptions are too numerous to name, but even beginning graduate students in chemistry believe that the bubbles seen in boiling water are hydrogen or oxygen, that rapidly boiling water is at a higher temperature than gently boiling water, and that mass changes when matter melts or boils (Bodner, 1991). Our natural reaction to such misunderstanding is, "How could graduate students be so confused?" The answer, apparently, is, "Because they are trying to make sense out of the world, and all they can do is use the schemas that worked in the past." The confusion that we observe as students struggle to explain new (and to them, very strange) data is similar to the confusion experienced by subjects in the Bruner and Postman experiment when the images they saw did not fit the card-naming schema they had previously built.

CONSTRUCTIVISM II: THE QUESTION OF REALITY[3]

"OUT-THERE-NESS." Constructivism is a worldview that has affected the philosophy and sociology of science as well as science education. In all of these domains, constructivists have had difficulty dealing with the question of "out-there-ness." The problem is basically this: If we all construct knowledge under the influence of our own schemas, how do we know what is *really* outside of our heads? How do we know, for example, that the bubbles in boiling water are not hydrogen and oxygen gas, as many students believe? Most scientists respond something like this:

> The criteria of validity of claims to scientific knowledge are not matters of national taste and culture. *Sooner or later, competing claims to validity are settled by the universalistic facts of nature, which are consonant with one and not with another theory* (Merton, 1942, p 554; cited in S. Cole, 1992, p 4; Cole's italics).

The positivist (realist) position stated by Merton has been the norm in science since the time of Bacon. However, recent work in the philosophy and sociology of science casts doubt on its truth, and when Merton's essay was reprinted in 1973, Merton altered the last sentence to read: "Sooner or later, competing claims to validity are settled by universalistic criteria" (cited in S. Cole, 1992, p 4). Many constructivists see things in even more relativistic terms:

> The function of cognition is adaptive and serves the organization of the experiential world, *not the discovery of ontological reality* (von Glasersfeld, 1987; cited in Wheatley, 1991). Thus *we do not find truth but construct viable explanations* of our experiences.
>
> From a constructivist perspective, *knowledge originates in the learner's activity* performed on objects. But *objects do not lie around ready made in the world* but

---

[3]The material in this section deals with an obtuse point that is important to radical constructivists. Graduate students in science education should read and understand the section because it will influence understanding of recent research in the field. Others, however, can skip this section without serious loss in understanding.

are mental constructs. . . . In contrast to a realist's perspective, a constructivist believes that *knowledge is not disembodied but is intimately related to the action and experience of a learner*—it is always contextual and never separated from the knower (Wheatley, 1991, p 10; my italics).

In contrast [to constructivism], information processing as well as behaviorism separate the individual from his or her environment. The image of ourselves . . . is one in which we live in an objective propositional reality that is external to and independent of us. *To the realist, the world is viewed as containing information or patterns existing prior to the organizing activity of the person* (Wheatley, 1991, p 12; my italics).

Comments such as those italicized in these quotations cause confusion among science educators. Constructivists seem to be saying that *ontological reality does not exist*! According to Cole (1992): "[Social constructivists] adopt a relativist philosophical position and deny the importance of nature as an objective external which influences the content of scientific knowledge. Nature does not determine science; instead, they say, the social behavior of the scientists in the laboratory determines how the laws of nature are defined" (pp 4–5). If scientific knowledge is not influenced by an "objective external", what is the meaning of *misconception* in the previous section, and what in the experiential world does cognition organize? Exactly what *is* out there?

Cole (1992) addressed this same question from the point of view of sociology, and I quote liberally from his work.

Data generated from the empirical world play an important role in the decisions of the scientific community about the acceptability of particular solutions. That the Watson and Crick model is likely to be "ultimately" seen as "wrong" . . . does not mean that *at the time it was proposed* that any other theory could have come to be accepted as "right". . . . Even though in some "ultimate" sense there may be no way to determine whether one paradigm is a better approximation to the "real" laws of nature than another, the exclusion of nature and the empirical world from our model of how scientific knowledge grows makes it difficult to understand why some knowledge enters the core and most does not. Thus it is on practical sociological grounds that I select my realist perspective (pp 24–25).

Nature poses some strict limits on what the content of a solution adopted by the scientific community can be. By leaving nature out, the social constructivists make it more difficult to understand the way in which the external world and social processes interact in the development of scientific knowledge (p 27).

The interaction between social processes and the external world determines the nature of scientific knowledge, and the interaction between an external world and our own schemas determines the nature of personal knowledge. *If we are to understand learning, the only viable position to take is that an external reality exists, even though the understanding of it may differ from one person to another and from one point in time to the next.*

SEPARATING REALITY FROM CONSTRUCTION. As I type this section, light is coming through a window and shining on a row of books: some red, some blue, and some yellow. How shall we separate reality from mental construction in such simple observations? The best way to do this separation is to accept the following as useful assumptions:

1. We live in an orderly universe that behaves consistently over time. Events take place in the universe that are independent of mind. Those events pro-

duce perturbations that are constant for every object, animate or inanimate, with which they interact. I will call those objects and events that are independent of mind *reality*.[4]

2. When any perturbation interacts with an object, the effect is a function of *both*. Neither the perturbation that is external to the object nor the nature of the object itself fully determines the result.[5] *Knowledge* is the internal construction that results when an intelligent being is a participant in such interactions.

Consider my statement concerning the red, blue, and yellow books in light of these assumptions. My statement implies that a perturbation exists in the universe (reality) that we call light (knowledge). When this perturbation interacts with the books on my shelf, the result differs from one book to another because the books differ.[6]

What happens when light strikes each book depends on the nature of the light and on the nature of the books. Although (my knowledge assumes) the light striking all books is the same, the light reflecting from them is not. Thus, as a result of the interaction between light and books, a new, altered perturbation (a new reality) exists. When this new light strikes the retina of my eye (or the retina of any other eye, my knowledge presumes), the interaction results in a stimulus being passed to the brain (a third reality), but that stimulus is a function of both the nature of the light and the nature of the eye it enters. Eyes of some animals appear not to detect color at all (knowledge constructed as a result of social discourse and numerous interactions with reality), and human perception of color may vary with retinal pigmentation (*see* Cole and Scribner's discussion of Pollack's work in the section in Chapter 4 entitled *Heritability of IQ*). Thus, although the initial reality is the same for every eye, the perception of that reality is not. When the signal from the retina reaches the brain, it undergoes additional processing that is a function of the schemas already there (more knowledge).

My son and I lived together for 18 years. During that time we undoubtedly talked about colors many times. Throughout that period nothing suggested that the words we used implied different perceptions. We assumed that we *saw* the same thing. But when my son wanted to become a Navy Seal, a carefully designed interaction with reality (called a test for color blindness) provided evidence that our common understanding about color was not common at all. My son did not see in those multicolored blobs the same figures that I see. Consequently, people in the Navy who apparently see what I see said my son is color-blind.

If I talk about what I see and others talk about what they see, our words may be the same even though our perception is not (my color-blind son and I, for example). Conversely, our words may be different even though our perception is the same (a non-English-speaking person and I, for example).

> Over the very long run . . . ideas which once were the most firmly established part of scientific knowledge can be replaced or even come to be seen as wrong.

---

[4]Radical constructivists appear to reject this proposition. Their position is that *object* and *event* have no meaning except as they are constructed by a mind. Thus, radical constructivists reject the notion of an independent reality. Whether an external reality exists is inherently unknowable, but discussing learning is far easier if we assume that external reality exists.

[5]An interaction could not be detected if the interaction was not a product of both objects and events that interact.

[6]A more detailed analysis might call attention to the fact that *book* and *shelf* are specific constructions of mind (i.e., knowledge as outlined in proposition two), but the words denote objects that proposition one grants an existence apart from mind (reality).

Whether this is interpreted as meaning that there is no such thing as "truth" or no "out-there-ness", as the relativists would have us believe . . . makes very little difference to sociologists studying the behavior of scientists [or, I would interject, to educators studying science learning]. What should make a difference . . . is that scientists [and students] *believe* that there is a nature and that some facts conform with it and others do not. As W. I. Thomas said, "If men define situations as real, they are real in their consequences." (Cole, 1992, p 55; my words in brackets).

Whether real by definition or fact, we adapt to an external reality through the complementary processes of assimilation and accommodation. Even though we do not receive knowledge intact from the external world, we are constrained by information in it, and knowledge constructions that are inconsistent with information apprehended from external reality are not viable.

SOCIAL INFLUENCE ON SCHEMA CONSTRUCTION

Now let us return to the paradox of how we sense that information is transferred intact from person to person, even though we know that this transfer is impossible. The appearance is there for two reasons: First, we are all subjected to the same reality. The universe is not capricious, and perturbations in the universe are common to all people. Second, what is perceived as a result of interactions with these perturbations and the knowledge derived from those interactions is *not* common because it is a function of both the perturbations and ourselves.

When we think of the external world, we generally think about the inanimate objects and events in it, but the animate objects are just as real and just as important— although some people (e.g., Doise and Mugny, 1984; Vygotsky, 1986) would argue that the inanimate objects are more important.

We are social animals who continually negotiate the meaning of events in our lives so that we can benefit from the experiences of others as well as our own. Through our interactions, we come to what we believe to be common understandings. In most cases we are reasonably successful. If our understanding is not the same, our interactions do not reveal the discrepancy. However, discrepancies in understanding often remain, and we are surprised when later interactions reveal them (as when I learned that my son is color-blind).

Because we are social animals, individual survival is enhanced by common understanding. When I look out the window and see cars, I am confident that others would see cars too. Nobody would believe me if I said that I dropped my pen and it fell up rather than down. Anyone who insists on walking through walls because they are not there is likely to be sent for counseling; they may be declared mentally ill (in our society) or possessed by spirits (in other societies) and committed to an institution. Saying that we each construct our own idiosyncratic meanings does not imply that it is not shaped by other people. It is. Our survival depends on shared constructions.

CONSTRUCTIVISM III: SURVIVAL

Our cognitive constructions are not random. From the beginning of life, cognitive development has direction; it proceeds toward adaptation and survival.

Survival is enhanced by conserving energy. Thus, when given a choice, the need for survival dictates that we make the choice that requires the least expenditure of energy over time. In the cognitive domain, I think of this as *the Principle of Least Cognitive Effort*. Simply put, the Principle of Least Cognitive Effort dictates that, faced with a cognitive decision, we make the choice that would appear to require the least effort over our lifetime. I was proud when I discovered the Principle of Least Cognitive Effort; I was humbled when I learned how old it is (c.f., Zipf, 1949)!

The Principle of Least Cognitive Effort comes into play in several situations. When we are asked to do something for which we see no need, we do not do it. When huge gaps exist between the information in our environment and what our existing schemas can assimilate, and all attempts to accommodate fail, we are confused and sometimes frightened. If we must respond, as we often must, we invoke other schemas—defense mechanisms—which allow us to survive in the face of puzzlement. We retreat into a private shell. We claim that the information is not worth knowing. We invoke mystical explanations. We memorize what we see or are told and assume that we can do nothing else.

SUMMARY

The discussion of learning in this book is organized around schema theory. Schema theory says nothing about what actually takes place in the brain, but its imagery describes knowledge at an appropriate level for teachers. Schemas are large, complex bodies of related knowledge that guide the way we think and act. Schemas include sights and sounds as well as verbal information; they often contain misinformation and gaps as well as useful knowledge.

In describing information in schemas, Piaget made a distinction between information we can get directly through observation (physical knowledge) and generalized information about relationships that we abstract from those observations (logicomathematical knowledge). Logicomathematical knowledge includes such things as number relationships and sequences, the permanence of unseen objects, and perspective taking. Perspective taking is looking at events from a different perspective—usually that of another person.

Knowledge is not transmitted from one person to another; it is constructed by each learner as a result of interactions with reality and negotiations of meaning with other people. Radical constructivists insist that all knowledge exists only in peoples' minds, that external reality is not a viable construct. In this book I talk about an external reality and assume that it exists, whether it does or not. However, I try to make clear that we have no way of knowing exactly what that reality is. We must realize that our ideas of atoms, molecules, light, electricity, and the like are, in a very real sense, figments of our imagination. This realization is not to say that no knowledge exists "out there", but instead that the way we think about what is "out there" is only in our heads.

We construct knowledge through an adaptive process consisting of two complementary parts: assimilation and accommodation. The technical meaning of these terms is not important. What is important is the realization that what we learn is dictated by two influences: the schemas that we already have in our heads and the information contained in the external stimuli to which we respond.

Schemas often contain errors because we are forced to interpret situations by using schemas that are inadequate to the task. For the most part, we eventually realize that discrepancies exist within our store of knowledge, and we reconstruct schemas as necessary to sort them out. We develop intellectually.

Intellectual development is subject to the same kind of pressures that underlie evolution: We need to survive. That development is difficult to do in the face of contradictory information. However, if the energy required to resolve inconsistencies is greater than the energy that we are likely to save by doing so, we may invoke one of several defense mechanisms rather than doing the hard intellectual work of reconstructing existing schemas. I refer to this as the Principle of Least Cognitive Effort.

6

■

Implications for Teaching

The view of learning outlined in Chapter 5 is very different from one that accounts for cognitive development in terms of associations, connections, and conditioning. This second view, which has been most influential in education for almost 50 years, focuses on what the teacher is doing to the learner; the view outlined in Chapter 5 focuses on what the learner is doing to what is to be learned. The implications of this change in view are substantial. (For an elaboration of this point, *see* Inhelder, 1976, and Renner, 1982.)

The newer views presented in Chapters 4 and 5 are too general to be of much practical use. As with atomic theory, Newton's laws of motion, and thermodynamics, a great deal of engineering is required before we can expect "better things for better living". This chapter and Chapters 7–16 are about engineering. In this chapter, I focus on *general* models of instruction that follow from a constructivist view of learning; in Chapters 7–16, I deal specifically with teaching problem solving, concepts, language, and higher order intellectual skills.

EXPOSITORY TEACHING

We must be careful not to misunderstand what constructivism implies about learning. Although each of us responds to experience in unique ways, our responses have a great deal in common and we constantly negotiate meaning to reach consensus (Doise and Mugny, 1984; Vygotsky, 1986). To stop writing books or giving lectures or making films or producing paintings because we "know" that knowledge is not transmitted intact from one head to another would be insane. Exposition in its various forms is useful. But equally insane is the practice of operating schools as though telling is synonymous with teaching—as though the idea we try to convey is virtually certain to be the idea implanted in the minds of students. We can continue to use traditional tools to communicate ideas, but we must alter instruction so that we gain more control over results.

For constructivists to build useful models for instruction, they need to understand the limits of exposition, be aware of differences in students' understanding of expository material, and know how to interact with students so that the constructions arrived at are consistent with those accepted by society at large. This approach usually means that teachers must listen more and talk less, and they must provide greater opportunities for students to negotiate meaning among themselves.

COOPERATIVE LEARNING

In Chapter 5 I pointed out that we construct meaning on the basis of interactions with our environment, including the people in it. In normal conversation we are inclined to accept at face value what we are told until we hear something that does not fit. The statement may be something preposterous such as, "The sparrow weighed 200 pounds", something at odds with what we already believe such as, "Heavy and light objects fall at the same rate", or something we do not want to believe such as, "Your son is on drugs." When such incongruities occur, we may respond in several ways. We may get angry and stalk away. We may ignore what was said. We may ask for clarification or argue about whether the statement is true. Only when we interact in an effort to come to some common understanding are we likely to develop intellectually (Doise and Mugny, 1984; Johnson, 1981; Johnson and Johnson, 1992; Kubli, 1983; Perret-Clermoat, 1980). The primary rationale for cooperative learning is to provide the opportunity for this development to occur.

RESEARCH ON STUDENT ACHIEVEMENT

A great deal of research has been done on cooperative learning (Aronson, 1978; Johnson, 1981; Johnson and Johnson, 1985; Johnson, Johnson, and Maruyama, 1983; Patterson, Dansereau, and Newbern, 1992; Sharan, 1992; Sharan, Kussell, Hertz-Lazarowitz, Bejarano, Raviv, Sharan, Brosh, and Peleg, 1984; Slavin, 1990; Slavin, Sharan, Kagan, Lazarowitz, Webb, and Schmuck, 1985). This research clearly shows that, used appropriately, cooperative learning enhances learning. Johnson and Johnson (1992) reported a meta-analysis of 323 studies that indicated the following:

> The average cooperator performed at about two-thirds of a standard deviation above average competitors (effect size = 0.67) and three-quarters of a standard deviation above the average person working within an individualistic situation (effect size = 0.75). . . . When only the high-quality studies were included in the analysis, students at the 50th percentile of the cooperative learning situation performed at the 81st percentile of the competitive and individualistic learning situations (effect sizes = 0.86 and 0.88, respectively). Further analyses revealed that the results held constant when group measures of productivity were included as well as individual measures, for short-term as well as long-term studies, and when symbolic as well as tangible rewards were used (p 24).

These results do not say that placing students in groups automatically leads to learning. Students can sit in groups without interacting, or they can interact in negative ways. Some students do not know how to interact in a positive manner. They try to enhance their own self-esteem by winning arguments, putting down less able group members, ignoring other group members, or engaging in other negative behavior rather than listening to what others have to say and considering whether it makes sense. Cooperative groups must *cooperate* if they are to make progress (Johnson and Johnson, 1992, p 27).

Knight and Bohlmeyer (1992, pp 14–15) cited a number of factors that may account for the favorable effect of cooperative learning.

COGNITIVE PROCESSING. Most cooperative learning strategies require students to teach other members of the group. When students explain something to others, try to understand what others are explaining to them, and try to fit pieces of information together, they think differently than when they study alone. They use elaborative and metacognitive strategies more frequently, and they use more higher level reasoning (Johnson and Johnson, 1992, p 25).

ACADEMIC TASK STRUCTURE. The structure of cooperative learning differs from that of traditional learning. Students spend more time on the learning tasks, less able students have more opportunity to learn from the more able, students talk more, and more repetition of ideas is likely.

REWARD STRUCTURES. More opportunity for recognition from peers occurs in a cooperative learning environment, as does more opportunity to share ideas that seem to contribute to group learning.

ROLE DIFFERENCES. Roles played by teachers and students are different. Teachers act as mentors, facilitators, and resource persons rather than dispensers of information. They distribute attention more equitably, and they give students more responsibility for learning (Hertz-Lazarowitz and Shachar, 1992, p 86). In cooperative learning situations, teachers discipline and interrupt students less, and they do not hurry pupils when they work (pp 87–88).

CONTROVERSY IN COOPERATIVE LEARNING

Controversy generally enhances cognitive development, but much of the benefit depends on how controversy is handled. Johnson's (1981) research summary points out that controversy is more effective and information is communicated more accurately and completely in a cooperative environment than in a competitive one. Controversy is more beneficial if disagreement is valued and people feel free to express feelings as well as ideas. Also, if students are skilled at identifying similarities and differences among positions, are good at summarizing, have lots of relevant information, and have learned how to disagree agreeably rather than putting one another down, cooperative learning will be more successful. As common sense would suggest, the more heterogeneous the group, the more likely controversy will occur.

GROUPING CONSIDERATIONS

Many people advocate cooperative learning specifically to promote better understanding between different racial, ethnic, religious, age, and gender groups. Research indicates that the potential exists, but whether that potential is realized depends on several factors.

Miller and Harrington (1992) have done a careful analysis of factors that determine whether interactions among culturally different groups will help or hurt. Their conclusions are virtually identical to the arguments made in the Social Science Statement attached to the *Brown v. Board of Education* briefs of the U.S. Supreme Court in 1952. As stated, improvement in race relations depends on the circumstances under which members of previously segregated groups first come in contact with others in unsegregated situations. . . . [A]vailable evidence . . . indicates . . . the importance of such factors as: the *absence of competition* for a limited number of facilities or benefits; the possibility of *contacts which permit individuals to learn about one another as individuals*; and the possibility of *equivalence of positions and functions among all of the participants* (cited in Sharan et al., 1984, pp 437–438; my emphasis).

Miller and Harrington (1992) provided detailed suggestions that are worth reading by anyone forming cooperative groups for the purpose of promoting intercultural understanding. However, these suggestions boil down to this statement: Blind yourself to group differences and treat each student as an individual. Gordon Allport's (1954) recommendations are similar.

The extensive body of research on cooperative learning is summed up succinctly by Wells, Chang, and Maher (1992): "Our conclusion . . . is that to achieve most effectively the educational goal of knowledge construction, schools and classrooms need to become communities of literate thinkers engaged in collaborative inquiries" (p 104).

CONSTRUCTIVIST MODELS FOR TEACHING

As important as the contributions of Piaget and other constructivists may be in helping us to understand how learning takes place, these contributions provide only a hint of what we must do to teach.

Several people who are concerned about teaching rather than developmental psychology have extended and built upon Piaget's theory in various ways. Among these are Abraham and Renner (1985), Arons (1984), Barnes (1976), Cantu and Herron (1978), Case (1972, 1975, 1978a, 1978b, 1979), Erickson (1979), Karplus (1977), Karplus and Thier (1967), Karplus et al. (1977), Lawson (1975, 1979a, 1979b, 1980, 1982), Nussbaum and Novick (1981, 1982), Osborne and Freyberg (1985), Pascual-Leone (1970), Pascual-Leone and Smith (1969), Renner (1982), Renner and Stafford (1972), Rowell and Dawson (1977, 1983), Shayer and Adey (1981), and Skemp (1979). Each of these extensions has merit, and many contain common elements.

THE LEARNING CYCLE

One of the earliest and, for Americans, most familiar models for applying Piaget's ideas to teaching is the *learning cycle*. The idea behind the learning cycle was introduced by Atkin and Karplus (1962) and was developed by Karplus and others associated with the Science Curriculum Improvement Study (SCIS). The idea was formally introduced by Karplus and Thier in 1967. Several other science educators played important roles in the development and application of the idea. Lawson, Abraham, and Renner (1989) provided an excellent summary of learning cycle development and its use in instruction.

EXPLORATION. The learning cycle begins with an *exploration* phase, in which students become familiar with phenomena that embed the idea to be learned. Students receive little guidance during this phase as they explore new materials and new ideas, but experiences are selected and planned with the expectation that students will encounter phenomena that they cannot explain or totally understand.

INVENTION. Exploration is followed by an *invention* phase, in which the idea is given a name and its specific attributes are explored. Because few teachers found *invention* descriptive of the activities that take place during this phase of the cycle, this phase was later called *concept introduction* (Karplus et al., 1977) or *term introduction* (Lawson et al., 1989). Glasson and Lalik (1993), who reinterpret the learning cycle from a social constructivist perspective, call this phase *clarification*.

The idea behind the invention phase of the learning cycle grew out of the discovery learning movement, which was popular during the late 1950s and 1960s. Many teachers seemed to understand discovery learning to imply that students should discover everything through free experimentation. Because concept names and theoretical explanations of natural phenomena are social constructions that are *invented* by those who study the phenomena, those names and theories cannot be discovered. As Atkin and Karplus (1962) pointed out, students can discover natural phenomena (as they are expected to do during the exploration phase of the learning cycle), but the terms, theories, and accepted explanations of natural phenomena must be introduced by the teacher in some manner.

APPLICATION. The third phase of the learning cycle, originally called *discovery*, is now called *concept application* or just *application*. (Glasson and Lalik call it *elaboration*.) During the application phase, the idea is consolidated and elaborated by investigating a variety of applications. (Detailed teaching suggestions are found in Abraham and Renner, 1985; Karplus et al., 1977; and Lawson et al., 1989.)

GENERATIVE LEARNING MODEL

Mark Cosgrove and Roger Osborne (1985) reviewed several instructional models, including the learning cycle, before describing their own. As they pointed out, similarities among the models are far more salient than differences (p 105). However, they believe that certain preconditions need to be taken into account. First, they argue, "the teacher needs to understand the *scientists' views, the children's views,* and *his or her own views* in relation to the topic" to be studied (p 105). Second, these authors believe that pupils must have an opportunity "to explore the context of the concept, preferably within a real (everyday) situation" (p 106). A third precondition is that students have an opportunity to clarify their own views early in the learning process (p 106). Finally, opportunities for consolidation and elaboration of the idea being taught must occur.

Those familiar with the learning cycle will be aware that the last three preconditions outlined by Cosgrove and Osborne are addressed, but the first may not be. Osborne and Freyberg (1985) focused on children's science and gave a great deal of attention to common misconceptions and how these misconceptions affect the learning of canonical science. Thus, in their four-phase generative learning model, they added a preliminary phase in which the teacher assesses students' preexisting views concerning the concept or principle to be taught. Otherwise, the model is very similar to the learning cycle and other models that they review.[1]

THE LIMITATION OF MODELS

The learning cycle and other models for organizing instruction within the constructivist paradigm certainly have merit, and they are undoubtedly of considerable value to teachers whose current teaching practice is incompatible with the constructivist view. However, such models are no panacea, and they can become something of a fetish.

After the learning cycle was popularized by Karplus, Lawson, Renner, and others who were influenced by Piaget's constructivist ideas, many teachers thought it would be helpful to share lesson plans based on the learning cycle. These plans were collected and published in a newsletter for others to use. The newsletter contained many excellent ideas, but some lessons were forced into a learning cycle format when they could have been taught more sensibly without it. Like most teaching methods, the learning cycle and other models provide a useful guide for instructional planning, but teaching and learning are complex. Success requires that knowledgeable teachers exercise professional judgment during on-the-spot assessment of *all* variables that affect instruction. Teachers must do what makes sense. In general, this approach will follow a pattern similar to the one described as the learning cycle. However, exceptions will occur for which no apology should be made.

THE TEACHER'S ROLE

Teaching in a manner that is consistent with the constructivist paradigm calls for a reorientation in the teacher's thinking: a reconceptualization of what a teacher is and what she or he does. Two ideas, *mediating learning* and *transcendent messages,* are developed here.

MEDIATED LEARNING

Schools are founded on the assumption that learning takes place more efficiently if it is guided by someone (a parent or teacher) who has already developed elaborate and effective schemas for processing information. By placing children in schools, teachers can mediate the learning environment and speed the process of learning as well as direct it toward development of schemas that are valued in the culture.

Most instruction is not based on mediation at all; it is based on transmission. As I have pointed out, transmission of information intact is impossible. Knowledge is constructed in the minds of each learner. Outsiders can affect what is learned by mediating the environment, but they cannot transmit knowledge. The key to effective teaching, then, is being able to detect when a student is having difficulty constructing knowledge, making informed guesses about what might be standing in the way, and then manipulating the environment so that the student has a reasonable

---

[1]Two of the models that Cosgrove and Osborne review are actually the same. John Renner's 1982 article, "The Power of Purpose", was his presidential address before the National Association for Research in Science Teaching. In contrasting instruction based on the learning cycle with traditional science instruction, he judiciously avoided using terms normally associated with either teaching style. However, in his research reports he uses the normal terminology.

chance of getting out of this state of confusion. The things that the teacher can manipulate to mediate learning are outlined in Figure 2.1 (*see* Chapter 2).

The purpose of mediated learning (or instruction) is to produce independent learners. If this goal is to be realized, students must develop schemas that can be used later to obtain meaning from the environment. In other words, teaching chemistry is not enough. We must teach students *how to learn* chemistry as well.

TRANSCENDENT MESSAGES

Most of what we teach students about how to learn is done implicitly through transcendent messages rather than explicitly through words and actions. What is meant by a transcendent message is best seen in an example.

When I visit classrooms I frequently hear teachers make statements such as, "You had better pay attention to this because you will need to know how to calculate density for the test next week." The teacher is, in effect, mediating learning by pointing out that density is an important concept. It should be understood. But a transcendent message occurs that focuses on the wrong goal. The teacher's words not only communicate that density is important, but they communicate that the purpose of school is to pass tests and earn high grades. Given the real goals of education, this message is wrong.

A more appropriate remark might be: "Density can be useful in several ways. You may need to know the volume of an irregular object and not be able to measure it directly. If you know the density, you can calculate the volume after you weigh the object on a balance. Or you may want to know if a transparent solid is plastic or glass. Density can be an important clue." The transcendent message of this second communication is that the course content has value in attacking practical problems and that our purpose in teaching the course is to prepare students to solve such problems.

To achieve the goal of helping students develop attitudes, habits, and other managerial schemas that we describe as learning how to learn, we must attend to the transcendent messages conveyed by what we do and say. We must communicate the purpose behind schooling and the content we teach. Such communication is as much a goal of mediated learning activities as is chemistry content.

Teachers must keep these general goals of mediated learning in mind as they teach. Teachers should

1. Identify and correct deficiencies in students' general thought processes.
2. Teach specific concepts, operations, and vocabulary required by the course.
3. Develop an intrinsic need for sound thinking and the spontaneous use of operational thinking by the production of crystallized schemas and habit formation.
4. Produce insight and understanding of the teacher's own thought processes, in particular those processes that produce success and those responsible for failure.
5. Produce task-intrinsic motivation that is reinforced by the meaning of the curriculum in a broader social context.
6. Change the students' orientation toward the teacher from passive recipient and reproducer to active generator of information (Adapted from Feuerstein et al., 1981, p 275).

SUMMARY

This chapter has moved the discussion of how students learn one step closer to teaching practice. It began by pointing out that the constructivist view does not imply that there is never a place for expository teaching. It does caution, however, that overemphasizing expository teaching is easy.

The constructivist view of learning suggests that learning is enhanced by interactions (with nature and with other people) that cause cognitive disequilibrium and then provide opportunities to regain equilibrium by constructing new schemas. Cooperative learning provides such opportunities, and strong evidence exists that, when properly managed, cooperative learning in small groups is more effective than the more common whole-class instruction. The caveat "when properly managed" is necessary because many variables modify the effectiveness of cooperative learning. Unless teachers attend to these variables, the implementation of cooperative learning is doomed to failure.

A great deal of research has gone into efforts to develop instructional models that are consistent with a constructivist view of learning, and some of these efforts have been successful. Two popular models, the learning cycle and the generative learning model, have produced significant improvements in science learning in several controlled studies.

Although models like the learning cycle and generative learning model are useful guides, teachers cannot take leave of their senses and blindly force every lesson into any given pattern. If teachers are confident that another pattern serves the same function as one of the steps outlined in the model, they should exercise their professional judgment in deciding the best way to organize instruction for a given group of students studying a particular lesson.

Teachers are not dispensers of information; they are mediators of learning. They provide students with messages about issues such as what is important and what is not, how to organize their time, and how to approach a problem. Teachers also provide transcendent messages that go beyond their mediating words and actions. Sometimes those transcendent messages are quite significant, and we need to pay attention to them.

Chapters 1–6 of this book provided two kinds of overview, one informal and the other theoretical, for specific instructional suggestions taken up in Chapters 7–16.

How To Teach Chemistry

7

Problem Solving

> Note: Early in this chapter you will need a tape recorder to do a suggested exercise. You may want to find one now so that your reading will not be interrupted when you reach that point.

Chapters 1–6 of this book gave an overview of learning and included these major ideas:

1. Students differ. They learn in different ways and they fail to learn for different reasons.
2. The learning process is influenced by many variables, such as characteristics of the learner, what is to be learned, what the learner does to learn, and what is taken as evidence that learning has taken place.
3. Intelligence differs from one person to another, and it is partially determined by genetics. Differences in schemas with which we are born produce differences in the manner and speed with which stimuli are processed. However, no evidence suggests that normal people lack the equipment required to learn science.
4. Knowledge is constructed in our heads under the control of schemas that were either present at birth or constructed at a later time through assimilation and accommodation.

We now turn to more specific considerations of learning that are important in chemistry. We begin with problem solving because it represents the ultimate goal of chemistry education. Individuals who can address novel situations and arrive at a suitable course of action are valued in society. Such behavior is what we mean by problem solving.

WHAT IS A PROBLEM?

Problems exist in many forms: We cannot balance our checkbooks, we cannot get our cars started, or we cannot get students to be quiet. All of these are problems, and they appear to have little in common except that we are temporarily prevented from reaching some goal.

We say that we have solved a problem when we overcome all impediments and proceed to the goal. Problem solving, then, is the process of overcoming some apparent or real impediment and proceeding to a goal. Said another way, "Problem solving is what you do when you do not know what to do."

Problems exist along a continuum, from well-defined to ill-defined, and along another continuum, from routine to nonroutine. *Routine problems* are those that we frequently encounter, and we develop well-defined procedures for solving them. In this book we refer to routine, well-defined problems as *exercises,* and we refer to nonroutine, ill-defined problems as *problems.*

Most adults balance their checkbooks routinely and have well-practiced procedures for doing it. When they follow their normal procedures to balance their checkbooks, they are completing an exercise. If a person comes to the end of the exercise and finds that the checkbook does not balance, he or she has a problem. The person must abandon the well-practiced procedure and do something else. What the person does is ill-defined and may vary from one occasion to another. This latter procedure is called problem solving.

Whether a task is an exercise or a problem depends on the person, the task, and the conditions under which it is done. A task may require problem solving for one individual and not for another. Take, for example, the following stoichiometric problem:

> How many moles of water will be produced when 12.23 g of methane burn?

This question is an exercise for advanced chemistry students and chemistry teachers. They know what to do. However, for beginning students this question can be a difficult problem. Beginners do not know (or do not recognize that they know) any procedure that will lead to a solution. What the beginning student does when faced with such tasks and what the experienced chemist does are quite different. The beginning student is solving a problem; the experienced chemist is completing an exercise.

COMPLETING EXERCISES AND SOLVING PROBLEMS

ALGORITHMS

Algorithms are carefully developed procedures for getting right answers to exercises and routine tasks within problems with a minimum of effort. This ability is important, but learning algorithms does not prepare students to deal with novel tasks. Too much emphasis on exercises can actually impede solution to problems such as this one:

> The chloride of an unknown metal is believed to have the formula MCl_3. A 2.395-g sample of the chloride is dissolved in water and treated with excess silver nitrate solution. The mass of the AgCl precipitate formed is found to be 5.168 g. Find the atomic mass of M, the unknown metal (Boikess and Edelson, 1978).

Unless you have taught general chemistry recently, you will need to think how to solve this problem, and you may make a few false starts before you succeed.

Two additional tasks that are likely to involve problem solving are the following problems that I recently faced:

The furnace at our church was not getting enough air for combustion, and an outside door into the furnace room was left open to supply the needed air. In the winter, pipes freeze. I needed to cut an opening that would provide the necessary air. The oil burner on the furnace is rated at 18 gallons per hour. How much air will be required, and how large must the opening be to supply that amount of air?

My wife and I wanted to cut a pattern for a Christmas tree skirt that we had seen on television. The skirt had a circular opening 8 inches in diameter. The diameter of the outer circle making up the skirt was 48 inches, but the 16 pie-shaped pieces making up the skirt were cut, and the wide end was folded and sewn across the end so that a point was formed at the outer end of each piece. This point extended about 4 inches beyond the 48-inch diameter of the outer circle. Of course, allowances had to be made for half-inch seams to hold the 16 pieces together. Our problem was to cut a pattern from which the 16 pie-shaped pieces could be cut and sewn together to make a finished skirt with an 8-inch hole in the center, an overall diameter of 48 inches for the outside circle, and 4-inch points extending beyond the 48-inch circle. The final pattern would look something like Figure 7.1.

PROBLEM SOLVING: AN EXERCISE

Research on problem solving will mean much more if you consider it after you have analyzed what *you* do when you solve problems. You are good at problem solving, and much of what you do is done subconsciously. You need to become aware of the many things you do when you solve problems. Following this suggestion will help you do that:

Use a tape recorder to record what you say as you solve one of the problems given previously or some other problem that you do not immediately know how to solve. Try to express what you are thinking and doing in your head as you solve the problem. Imagine that someone is inside the recorder and you are explaining what you are doing. Nobody need hear the recording other than you, so try to forget about how foolish you sound and express your every thought.

Keep any scratch paper that you write on as you solve the problem. When you finish, the tape recording and scratch paper will constitute the raw data from which you will analyze how you solve problems.

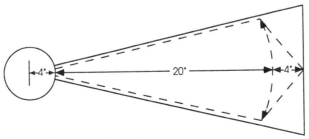

Figure 7.1. Pattern for making a tree skirt.

The kind of problem you solve makes no difference, but the discussions that follow are related to the unknown metal chloride problem given previously or the cryptarithmetic task that follows. If either of those problems seems interesting, solve one of them as you think aloud and record your thoughts.[1]

> What digits must be substituted for each letter if DONALD + GERALD = ROBERT and D=5? Each letter represents a single, unique digit.

> *Please stop reading now, find a tape recorder, and solve one of the problems while recording what you do in the process.*

TYPES OF PROBLEMS

ROUTINE VERSUS NONROUTINE

Problems may be classified in many ways. In distinguishing exercises (routine problems) from problems (nonroutine problems), we have indicated one classification dimension.

WELL-DEFINED VERSUS ILL-DEFINED

Another classification dimension separates well-defined problems such as, "What is the molar mass of NaCl?" from ill-defined problems such as, "How can recycling be improved? What would cars of the future look like? How can creativity be enhanced?" (Finke, Ward, and Smith, 1992, p 168). Well-defined problems have one solution and a limited number of solution paths; ill-defined problems may have no solution at all and many ways to address the problem.

ADVERSARIAL VERSUS NONADVERSARIAL

A major classification dimension in the research literature is adversarial versus nonadversarial problems (Mayer, 1989). An adversarial problem is one in which one person competes against another, and one of the best examples is the game of chess. Most problems in chemistry education are nonadversarial and have no winners and losers.

Problems may also be classified according to subject matter, such as economic problems, social problems, political problems, managerial problems, and judicial problems. What constitutes a problem and a solution in each of these domains varies greatly. As you might expect, the literature on problem solving is vast, and only that portion that is closely related to chemistry is discussed in this book.

GENERAL PROCESSES IN PROBLEM SOLVING

Problem solving involves many activities, and in some sense every problem solving activity is unique. Still, procedures can be identified that are common to most prob-

---

[1]Episodes 15 and 16 of SourceView (1992), the videotape portion of the ChemSource Project, provide excellent examples of algorithmic and more open approaches to problem solving in a high school chemistry class. Group discussion of these episodes can provide useful insights into problem solving.

lem solving behavior. We will speak of four: understanding the problem, representing the problem, executing a plan for solution, and verifying. We will discuss each of these in turn.

UNDERSTANDING THE PROBLEM

UNDERSTANDING THE GOAL. Understanding the problem means understanding the task. One aspect is identification of the *goal*. The goal in the cryptarithmetic problem presented previously is to assign a digit to each letter. In the stoichiometric problem, the goal is to find the atomic mass of the metal designated by M. These problems are well-defined, and their goals are transparent.

Even when the goal of a problem is transparent to the originator of the problem, it may not be transparent to others. For example, when elementary schoolchildren were shown pictures of several coins and asked whether they had enough money to purchase an 87-cent item, many students worked at counting the number of each kind of coin or counting the total amount of money represented for a considerable length of time before realizing that they had missed the goal of the problem (Wheatley, personal communication). In our research with college chemistry students solving the unknown metal chloride problem, many students reported the molecular mass of the metal chloride as their answer rather than the atomic mass of the unknown metal. In many cases, the goal of the task was correctly identified in early stages of the solution process, but the student shifted to an alternative goal during solution and never realized the mistake.

This behavior of elementary mathematics students and college chemistry students is not unusual. Mayer (1989) pointed out that systematic distortion of problems to be consistent with prior knowledge is one of six common constraints on problem solving (p 45). The point is that accurately identifying the goal of a problem task and keeping that goal in mind are not trivial matters.

Neither is it clear that identification of the problem goal at the beginning of the task is always helpful, particularly in the case of ill-defined problems. Clearly articulating a goal at the beginning may limit the solutions that will be considered. For example, I posed the problem of insufficient air getting to the furnace at my church in terms of "cutting an opening to provide the needed air." However, cutting an opening is not the only option, and it may not be the best.[2] By visualizing the problem in terms of calculating the size of the needed opening, other options are not likely to be considered. As Finke, Ward, and Smith (1992) pointed out, "Reconceptualizations and redefinitions of problems often lead to important insights; thus, adherence to the original goals of a problem can block creative solutions" (p 178).

UNDERSTANDING THE CONDITIONS. A second part of understanding a problem is understanding the *conditions* placed on the problem. Sometimes these conditions are stated in the problem, and sometimes they are only implied. In DONALD + GERALD = ROBERT, you are told that D = 5 and that each letter represents a unique digit. In the stoichiometric problem, the atomic masses of elements are assumed and one must assume that the laws of chemical combination pertain. These conditions are not stated in the problem, but presumably they are available to the problem solver.

---

[2]Edward de Bono has written extensively about this problem and has developed curriculum materials aimed at developing what he calls "lateral thinking"—thinking that produces new, creative ideas. *See* de Bono (1970) for an introduction to his ideas.

In some cases, understanding the problem is trivial—the unbalanced checkbook, for example—but in other cases, it is not. In particular, understanding just what one is permitted to do is not trivial for anyone operating in an unfamiliar area—the beginning chemistry student tackling stoichiometry, for example. Even experienced problem solvers may make unwarranted assumptions about conditions imposed on a problem. A well-known example is the nine-dot problem:

> Connect the nine dots shown in the following array with four straight lines drawn without lifting your pencil from the paper or retracing over any portion of a line greater than a single point.
>
> • • •
> • • •
> • • •

Most people are unable to solve this problem because they assume that the lines must not extend beyond the field defined by the nine dots, even though no such condition is stated in the problem. Once the condition that the lines may extend beyond the field is understood, most people find a solution.

Understanding that the lines may extend beyond the field may be necessary to solve the nine-dot problem, but that knowledge is not sufficient. When subjects performed the nine-dot problem and were told that they had to go outside the square to solve the problem, they still were unable to solve it immediately. "Solving these and other types of insight problems occurs only when problem-specific knowledge is drawn from memory" (cited in Finke, Ward, and Smith, 1992, p 146). In case you are unable to dredge up such knowledge from your memory, a solution is given in **SOLUTIONS TO PROBLEMS** at the end of this chapter.

UNNECESSARY ASSUMPTIONS IN CHEMISTRY. The nine-dot problem may seem like an interesting psychological puzzle that has little bearing on problem solving in chemistry, but it is not. Students frequently make assumptions that seem perfectly foolish to us but not to them. For example, in some of our early work on problem solving in chemistry, a high school student struggled for a long time with a stoichiometry problem and gave up because he did not realize that he could use the periodic table. His high school teacher did not allow it, and he assumed that it was not allowed for the problems that we presented either.

A somewhat more subtle way that such assumptions affect chemistry teaching appears when we prepare problems for an assignment or test. Given our advanced knowledge of the subject or our familiarity with problems of a particular type, we fail to state conditions that are important but not obvious to the student being asked to solve the problem. For example, when I first wrote the DONALD + GERALD = ROBERT problem, I forgot to include the stipulation that each letter represents a unique digit. Being familiar with cryptarithmetic problems, I assumed this condition without stating it. A person who is unfamiliar with the task would have no reason to assume this condition, and I added the condition before finishing the chapter.

Many difficulties that students have in understanding problems are related to vocabulary and word meaning, but these difficulties are often subtle. For example, one student working the unknown metal chloride problem read "treated with excess silver nitrate" several times and then asked, "What do you mean by 'treated'?" and "Does 'excess' mean that there is more than needed?" "Treated" and "excess" are common words that this student almost certainly uses in other contexts. However, he was not familiar with their use in this context and had difficulty understanding the problem as a result. These difficulties are described in more detail in Chapter 12, which deals with language.

Often students are quite certain that they understand when they do not. One student translated the phrase, "the chloride is dissolved in water" by writing this partial equation:

$$Cl_2 + H_2O \rightarrow$$

Whether the student misread "chloride" as "chlorine" or simply does not understand that "chloride" is a general label for a class of compounds of which the MCl_3 is an example is not clear. What *is* clear is that the student did not understand what the problem said.

READING ERRORS. A surprisingly common impediment to understanding problems is misreading the problem. For example, in reading the stoichiometric problem, a student may read "MCl_2" rather than "MCl_3" or read "kg" rather than "g" or "5.392" rather than "2.395". Such errors are distressingly common, and little progress is made until deliberate attention is given to the difficulty. Techniques for addressing this and other "mechanical" errors in problem solving are discussed in Chapter 8.

REPRESENTING THE PROBLEM

Understanding the problem involves those aspects of the problem that are *out there* in actuality. Representing the problem pertains to how we understand the problem *in here*, in inner reality. The way a problem is represented dictates how, and often whether, it will be solved.

At first blush, problem understanding and problem representation may seem to be the same; they are not. When we encounter a problem, information incorporated in memory is influenced by our existing schemas as much as by external information. New information is juxtaposed to selected bits retrieved from memory. As a result, the problem as it exists inside our heads is substantially different from the problem as it exists outside our heads.

When I first encountered the unknown metal chloride problem, my thoughts went something like this:

1. Some (2.395 g) compound (MCl_3) is **placed in water,** where it dissolves. **Because the problem does not say anything about it reacting with the water, I assume that it dissolves without reacting**.
2. Next silver nitrate is **added to this solution**. The MCl_3 and $AgNO_3$ **react to form a white precipitate** of silver chloride.
3. **The silver nitrate is added until no more chloride precipitates. Because silver chloride has a very low solubility in water, I may assume that all of the chlorine originally in the unknown compound now resides in the silver chloride precipitate**.
4. **Somehow the silver chloride is separated, dried, and weighed.**
5. The mass of the resulting silver chloride is 5.168 g.

Notice that this is *not* what the problem says. It is a particular meaning that I attached to the problem when I first encountered it. My representation incorporates information that is not given in the problem (the portions in bold face), and it describes an actual physical procedure that I imagined taking place. It even hints that keeping track of the amount of chlorine may play an important part in the solution of the problem.

Others working on the metal chloride problem represent it quite differently. Some people work with mathematical relationships and do not think about real chemicals at all. Others cast the problem in terms of chemical equations and mole relationships.

PROBLEM SPACE. Newell and Simon (1972) refer to the elements included in a representation of a problem as the *problem space*. The problem space consists of states of knowledge and operators that can be applied to elements in the space to produce new states of knowledge.

You might think of my representation of the metal chloride problem as a *chemical reaction* problem space. The states of knowledge and operations that I visualized were all related to the process of chemical change. In contrast, a person who represents this problem in terms of mole relationships might be said to be working in a *mole problem space*.

The initial problem space that I described for the metal chloride problem includes information given in the problem, but it also includes mental images and factual information that were not given. The problem space in which I represented the problem influenced the way I went about searching for a solution.

We continue to augment the problem space as we solve a problem. As Newell put it, "In essence, problem spaces are always exponentially growing trees" (Newell, 1977, p 48). After thinking about the metal chloride problem in terms of actual laboratory procedures, I focused on the mass of silver chloride and wondered how many moles of chlorine it contained. My representation grew to include mathematics and proportional relationships implied by formulas and equations. By the time I returned to the metal chloride that I imagined in the beginning, it was more than an imagined white solid dissolving in water. It was now a number of metal ions and chloride ions, and this number was related to the mass of silver chloride produced in the end. Our representation of a problem changes over the course of solution.

In discussing the DONALD + GERALD cryptarithmetic problem, Newell and Simon suggested that the original problem space is augmented by information concerning the various carries to columns and partial information such as "R must be greater than 5", "E is not equal to 0", or "R is either 7 or 9" (1972, p 154). This problem space is where most of Newell and Simon's subjects represented the DONALD + GERALD task, but Newell and Simon pointed out that instead of representing the problem as a task of individually assigning digits to letters, one could work in an *algebraic* problem space where the information in each column is represented by an algebraic equation (1972, p154):

$$2D = T + 10c_2$$
$$c_2 + 2L = R + 10c_3$$
$$c_3 + 2A = E + 10c_4$$
$$c_4 + N + R = B + 10c_5$$
$$c_5 + O + E = O + 10c_6$$
$$c_6 + D + G = R$$

[c_2, c_3, c_4 ... represent the carries to the numbered column.]

One of Newell and Simon's subjects spent 25 minutes at the beginning of his first cryptarithmetic task searching different ways to represent the problem. He generated at least seven problem spaces including one in which an attempt was made to associate the words in the problem with specific numbers, another in which a single rule was used to assign numbers (e.g., A = 1, B = 2, and C = 3), and another in which surface features of letters were associated with surface features of numbers (E looks like 3, S looks like 2, and lower-case L looks like 1). You may have considered such representations before hitting on the representation that led to solution of the problem.

Clearly, familiarity with a particular type of problem facilitates solution because we quickly represent the problem by using a problem space that has worked in the past for similar problems. However, such shortcuts occasionally lead to trouble. During exploratory research on problem solving in chemistry, we presented a problem designed for an undergraduate organic chemistry class to a professor whose research involves similar problems. The professor took longer to solve the problem than some undergraduate students, apparently because he immediately began to work in the problem space he had found so effective as he pursued his research, but one that would not work on the somewhat artificial problem presented in our example. Student subjects in our research often read a problem, focused on cues that suggested a particular algorithm that they had learned, and forced the problem into the problem space defined by the algorithm.

DIFFERENCES BETWEEN EXPERTS AND NOVICES. Perhaps a more interesting example of differences in problem representation is seen in research contrasting the problem solving behavior of experts and novices. Larkin (1981) discussed the general conclusions of this research in relation to a problem involving a toboggan moving down a slippery hill.

The novice's representation is in terms of the surface features of the problem: the toboggan, the top of the hill, relative speed at various points, and so forth. By contrast, the expert's representation of the problem is in terms of derived concepts abstracted from the surface features: potential and kinetic energy, forces, acceleration, and conservation laws (*Also see* Chi, Feltovich, and Glaser, 1981). Such differences in representation of the problem have nontrivial implications for solution of the problem.

Greenbowe (1984) found that many students had difficulty with problems such as this one:

A 1.00 gram mixture of cuprous oxide, Cu_2O, and cupric oxide, CuO, was quantitatively reduced to 0.839 grams of metallic copper by passing hydrogen gas over the hot mixture. What was the mass of CuO in the original sample? (Mahan, 1975, p 29)

Much of the difficulty that students had with the problem can be traced to early steps taken to represent the problem. Many of them wrote the following equation and proceeded to solve for the amount of CuO that reacted:

$$Cu_2O + CuO + 2H_2 \rightarrow 3Cu + 2H_2O$$

As written and balanced, this equation presumes that the original mixture consisted of a 1:1 molar ratio of the two oxides, even though the problem statement clearly indicates that the composition is unknown; indeed, the goal of the problem is to find the composition of the mixture.

Research on problem representation is relatively new, and we still have much to learn. What we do know is the following:

1. Considerable variation occurs in the way that problem solvers represent problems; in particular, major differences often exist in the way that problems are represented by novices and experts.
2. The solutions given by authors in textbooks bear little resemblance to what experts do when they work *unfamiliar* problems. (Textbook solutions are generally algorithmic. They describe the most efficient pathway to a solution and probably represent how an expert who solves such problems routinely would approach the task.)
3. Poor representations of problems often lead to incorrect solutions to problems, whereas good representations frequently make the solution pathway transparent.

EXECUTING A PLAN FOR SOLUTION

Once one has an internal representation of a problem, a way to reach the goal must be planned. Given the variety of problems that one may encounter, a single kind of plan seems unlikely. Indeed, observation of several individuals solving the same problem reveals considerable variation in strategies. (To verify this, share your tape of the problem solving task completed earlier in this chapter with others who solved the same problem. Look for similarities and differences in the various procedures.) Still, observation of similarities is possible.

First of all, similarities exist in the decisions one must make during the problem solving process. Newell and Simon (1972, p 826) suggested the following:

1. One must decide what operation to carry out next.
2. At the completion of the operation, one must decide whether to continue from the new state of knowledge or, not seeing where it will lead, to back up to some previously determined state of knowledge.
3. Before backing up or going on, a decision must be made about the value of remembering the present state of knowledge.
4. Once the decision is made to back up, one must decide where to back up to.

In discussing problem representation, Winograd (1977, p 67) described slightly different but related decisions that must be made during problem solving:

1. What should I do next?
2. What knowledge might I try using?
3. Does it apply? How?
4. What can I conclude from it?

> As an exercise, review the tape of your problem solving process and see if you can identify these decisions.

That decisions of the kind outlined above are made seems fairly clear. What is not clear is the basis for making the decisions. However, here again, we are not totally ignorant.

A great deal of information may come from the task environment. To the extent that the problem one faces is (or appears to be) similar to previous problems, that previous experience is an important source of knowledge.

The knowledge that we bring to a problem from previous experience is of several types. Various authors use different terms for such knowledge, but *content knowledge*,[3] *logical operations*, and *strategic knowledge* are three labels that suggest what I have in mind, and a few examples will clarify.

In the stoichiometric problem, the atomic mass of chlorine, the meaning of a chemical formula, and the relationship between atomic mass and moles are examples of content knowledge that would be required to solve the problem. It is knowledge that is peculiar to the discipline and represents facts, commonly accepted meanings, or accepted states and relationships.

Logical operations[4] represent schemas that operate on one set of content knowledge and transform it in some way to generate new facts or relationships that were not previously known. I like to say that it is what we use to make sense out of the information we have. Examples of such knowledge important to the stoichiometry problem includes *if, then, therefore* reasoning, the systematic consideration of possibilities, proportional reasoning, and such rudimentary operations as the comparison of two results to see if they differ.

Strategic knowledge is knowledge that controls the sequence of steps taken in searching for a solution; that is, knowledge of what answers to the questions posed by Newell and Simon and by Winograd were useful in the past.

Research suggests that we gain a vast array of such knowledge as a result of experience and that we apply it subconsciously. For example, students learn a great deal about taking tests, and their scores on the tests reflect both their knowledge of the subject being tested and their knowledge of the test procedure itself.

In addition to the various kinds of knowledge that we bring to the problem from past experience, problem solving itself is a source of knowledge. Relationships that were not evident in advance of the task may become evident during solution, and this new knowledge may direct the course of the solution process. However, this progression can only happen if students have developed the operational knowledge that enables them to profit from the information generated. If students blindly follow an algorithm or model solution that was learned by rote, the initial solution path has little chance of being altered as a result of information generated in the process. Unfortunately, this limitation is precisely what is happening when beginning chemistry students work textbook problems (Gabel, 1981; Greenbowe, 1984; Nurrenbern, 1980).

VERIFYING

All teachers know that good students check their work. Teachers also know that admonitions to do so fall on deaf ears. Research on problem solving also reveals that successful problem solvers spontaneously engage in many strategies that effectively verify that procedures and conclusions are logically sound and accurately executed. So why do we have so much trouble getting students to check their work?

One of David Frank's (1986) observations seems to suggest part of the problem. His study was conducted with students in a chemistry course for science and

---

[3]Cognitive scientists call this domain-specific knowledge.

[4]This terminology is Piaget's. Cognitive scientists operating from an information processing frame of reference refer to procedural knowledge.

engineering students at Purdue University. These students were probably among the more successful problem solvers within their age group. Frank observed that these students engaged in several verification strategies when they were working less difficult problems used as warm-up exercises, but they did little checking when trying to solve more difficult problems presented during the interviews. Frank suggested that the students realized the value of verification, but that they were so unsure of their ground when working on the difficult problems that appropriate verification strategies were seldom apparent. Spending a great deal of time verifying that the operation had been performed correctly seemed somewhat pointless when the student had little confidence that the operation itself was valid.

Many verification strategies depend on familiarity of content. When an experienced chemist makes a measurement in metric units or calculates a molecular mass, judging whether the result is reasonable is easy because of expectations that have grown out of experience. The beginner may be unable to do this. In fact, previous experience on the part of students may even lead them to question the validity of correct results.

A chemistry teacher recently reported this experience: Students had plotted data and calculated the slope of the resulting line. Several students were sure that they had done something wrong because the calculated slopes were far too large. Their previous experience in math classes had produced values for the slope that were less than 10; the data plotted in this graph produced values several hundred times larger. They did not understand that the value of the slope depends entirely on the scale of values plotted along the axes of the graph, and their previous experience led them to question their unusual result. The students were using a valuable verification strategy, but their limited understanding of the mathematical operation led them to the wrong conclusion.

However, many verification strategies can and should be used in unfamiliar contexts. One is rereading the problem statement.

Problems are frequently misread or incorrect inferences are drawn from the reading. Transcription errors—writing down the wrong number or the wrong units—are also common. In protocols for student solutions of the metal chloride problem we saw many examples of such errors. For example, one student read the problem and wrote MCl_2 rather than MCl_3 and assigned the mass, 2.395 g, to silver nitrate rather than to the metal chloride.

The exact reason for such errors is not clear, and it may not be very important. What is important is that such errors are made, and they are made by both successful and unsuccessful problem solvers. Although protocols for experts solving problems do not reveal as many instances of such errors, these errors are still common. Students should be aware of this fact. We do a disservice when we present ourselves as infallible, and we divert attention from the very thing that must take place for improvement. Students are led to believe that competent people do not make errors, and neither should they. *The truth is that competent people make many errors, but they detect and correct them before the errors cause much damage.*

Recall the Bruner and Postman (1949) study described on pp 44–45, about identifying playing cards. In that study, experimental subjects were asked to identify playing cards that were presented to them. Most of the cards were normal, but some, such as a *red* six of spades and a *black* four of hearts, were not. At first subjects in the experiment made many errors, but once they realized that some cards were anomalous, they were able to correct their "card-naming schema" so that they quickly detected anomalies in other cards as well. In other words, they were able to detect and correct their perceptual errors before misnaming the card.

The student who made the errors cited in the unknown metal chloride problem was not a good problem solver, although he did eventually solve the problem. I find the portions of his protocol where he detected his errors instructive:

S: "I'm trying to think what to do next. Find the atomic weight of this . . . of this [*indicating M*]. Have to find the atomic weight of M, MCl$_2$, **MCl$_3$**! [*S changes formula*] O.K., find the atomic weight of MCl$_3$.

S: "Oh boy. O.K., hmmm, I do not know what to do . . . Hmmm. Seems to me you need to know how many moles of the chloride are given. . . . Let's see . . . of the chloride . . . Wait a minute here. . . . O.K., you've got grams of the silver precipitate, silver nitrate. **No**, let's see, **this** [*the metal chloride*] is 2.395 grams . . . That's wrong." [*S crosses out the 2.395 g under the silver nitrate and writes 2.395 g under the metal chloride.*]

The student made other corrections in the protocol, but in each case the correction resulted from an accidental discovery. The student was stuck and reviewed his work to look for a new start, and he discovered the error in the process.

This example is in sharp contrast to protocols for graduate students and professors working similar problems. The protocols for these *experts* are riddled with statements such as: "Let me check that"; "Let me read that again"; "This does not seem right"; "That does not seem sensible"; "Let's back up a bit here"; "Yes, **that** seems OK"; and "**That** expression seems OK". These and similar comments suggest continuous, subconscious checking of facts, interim results, and procedures.

Time and again, we find that poor problem solvers read a problem statement once, apply a familiar algorithm, and report an answer. Good problem solvers read the problem statement (or portions of it) several times, review procedures, check units, look at the magnitude of their answer, compare the answer with other information that they know, and apply other verification strategies to be certain that their answer and procedure are sensible.

> Listen to the tape recording of your own problem solving again and see how often you applied some kind of verification strategy.

WHERE STUDENTS GO WRONG

Several specific difficulties that students have solving problems were already mentioned. Mayer (1989) summarized many general pitfalls in six constraints on problem solving:

1. *Humans systematically distort the problem to be consistent with prior knowledge* (p 45). In this category, Mayer discussed such things as reinterpreting a problem so that it is easier to do or makes more sense.
2. *Humans focus on inappropriate aspects of the problem* (p 47). Problems can be represented in many ways. Focusing on certain aspects of a problem make solving the problem difficult, but focusing on other aspects (i.e., generating a different representation) can make the solution transparent.

3. *Humans change the problem representation during problem solving* (p 49). Some of these changes represent insights gained as a result of early problem solving steps; others represent losing sight of the original goal of the problem: finding the molecular mass of the unknown metal chloride rather than the atomic mass of the unknown metal, for example.

4. *Humans apply procedures rigidly and inappropriately* (p 50). An example described in this chapter was students forcing into a stoichiometry algorithm problems that do not fit the algorithm. (*See* Nakhleh, 1993, and Nakhleh and Mitchell, 1993.)

5. *Humans are intuitive and insightful and creative* (p 51). If you were able to listen to tape recordings of others solving problems presented in this chapter, you probably observed several instances of such insight and creativity.

6. *Humans let their beliefs guide their approach to problem solving* (p 53). Among the beliefs that impede problem solving in chemistry are that only one way exists to solve chemistry problems, and only very bright people can solve these problems.

STUDENT BELIEFS[5]

Carolyn Carter (1987) analyzed the beliefs of undergraduate chemistry students and the influence of belief on problem solving. Carter interviewed nine students enrolled in the first semester of a year-long general chemistry course for science and engineering majors. Students were presented with a variety of traditional and nontraditional problems in chemistry and nonchemistry contexts in a series of interviews throughout the semester.

Carter described the beliefs held by four students, Billie, Sid, Ron, and Jo, to illustrate the range of beliefs observed in her study. Billie and Ron see chemistry as abstract and alien. Their job is to absorb and reproduce knowledge presented by the teacher, an authority from another world. Problems are tasks that require calculations and an answer but not a question, and the only purpose of solving problems is to get an answer that *they* want. The way to do problems is to reproduce algorithms and recognize problem types; creativity has no role in chemistry. The way to succeed is to work the same problems over and over until they are memorized.

Jo and Sid hold quite different beliefs. To them, chemistry is a creative way of understanding concepts and problems. They see themselves as the source of knowledge, and they see their role as putting concepts together and applying them to solve problems. The teacher is there to motivate, answer questions, and explain when necessary. Problems are tasks in which one must think creatively and synthesize ideas; problems are not algorithmic. The goal of problem solving is to understand ideas and apply them to new contexts, and the way to get good at it is to work problems, think about the concepts involved, and relate ideas to previous knowledge.

It would be comforting to believe that students like [Billie] and [Ron] are just "dumb", or "lazy", and that their difficulties in learning chemistry are not related to their formal instruction in chemistry. [Billie] and [Ron] are neither dull nor

---

[5]Material in this section is adapted from Herron, 1990, pp 32–33. The student names have been changed from those used by Carter to names that are not exclusively used by males or females in order to reduce sex stereotyping.

lazy, however. They [are] bright, hardworking students trying as best they [can] to make sense out of chemistry as taught in their high school and college courses. They put in many hours on what they believed was productive study. They tried to get good grades—which they believed measured understanding and success. The idea that chemistry may be understood in ways other than their instrumental methods would have astonished [Billie] and [Ron] (Carter, 1987, p 315).

Students such as Billie and Ron are estranged from chemistry as a discipline. Chemistry is *out there* and the people who do it are some undefined *they* who want students to perform in mysterious ways that do not make much sense. The student's job is to figure out what *they* want done, to recall how *they* say a problem should be solved, and to apply *their* procedure to generate an answer *they* expect. There is little consideration of whether the answer to a problem or the problem itself makes sense. It is not supposed to.

This impression that chemistry is largely arbitrary and meaningless is not confined to students in college courses. A group of 13- to 16-year-olds in Australia seemed to share the same impression. They were asked by Ellerton and Ellerton (1987) to make up questions that would be difficult to answer and to indicate how they would respond if the question were on a test. Most of the questions emphasized memory, facts, and formal treatment of symbols, presumably what the students thought chemistry is about. But the most interesting result was the students' indication of how they would respond to the questions they had written. They frequently said that they would not or could not respond, that they would skip the question, or that they would simply panic.

EFFECT OF PROBLEM COMPLEXITY

Student beliefs are only one of many factors that influence student performance in problem solving. The complexity of the problem plays a role as well. When operating in a relatively familiar domain, such as stoichiometric calculations, students are able to solve one-step problems such as finding the mole mass of a compound or calculating the number of moles corresponding to a given mass of the compound. However, they are quite unsuccessful when a problem requires the stringing together of such steps to solve a more complex task. Frazer and Sleet (1984) gave tests consisting of three complex problems and separate items in addition to the subproblems involved to 76 sixth-form students (age 17–18) in four schools in England. For the first problem, 77% of the students were successful on all of the subproblems, but only 37% succeeded on the main problem; for problem two, the respective percentages were 83 and 52%; for problem three, they were 73 and 57%. Lazonby, Morris, and Waddington (1985) report similar findings.

Several possibilities may explain this kind of result. One might argue that in one-step problems students are not solving problems at all but are simply applying well-practiced algorithms. When the more complex problem is presented, deeper understanding of the relationships among concepts and skills is required to string steps together to produce a sensible result. Alternatively, failure on the more complex tasks may be due to poorly developed metacognitive skills that govern the organizing of work, sequencing of tasks, and checking of results. It is not certain what is going on. It *is* certain that the number of elements to be considered during problem solving plays a role in success.

WORKING MEMORY AND M-DEMAND

LIMITATIONS ON WORKING MEMORY

In all stages of problem solving we rely on information in memory. The amount of information that we can store appears to be limitless. However, we are not able to attend consciously to everything that we know at once, and this inability has implications for problem solving.

To explain the difference between what we are able to store and what we are able to attend to at any point in time, theories of cognition postulate a *long-term memory* of unlimited capacity and a *short-term* or *working memory* of limited capacity.

On the basis of Miller's (1956) classic paper, the capacity of working memory is generally described as "seven bits" plus or minus two. Just what constitutes a bit is open to interpretation, and other theories attribute different capacities to working memory (Case, 1978b; Pascual-Leone and Smith, 1969). In addition, Eysenck (1989, pp 290–291) pointed out certain limitations in the theory outlined here. However, no disagreement exists over the basic ideas that we have a limited capacity for conscious attention and that capacity is relatively small.

EFFECT OF M-DEMAND AND M-SPACE ON PROBLEM SOLVING

The neo-Piagetian theories of Pascual-Leone and Smith (1969) and Case (1978b) describe the capacity of working memory as M-space, and that term is commonly used by other researchers as well. According to neo-Piagetian theories, intellectual performance is influenced by both the intellectual level of the task and the M-demand of the task.

Niaz and Lawson (1985) examined the influence of intellectual development, M-space, and field dependence–field independence on success at balancing chemical equations by trial and error. They concluded that balancing equations by trial and error requires formal thought and that success on more difficult equations (those that require several steps) depends on M-space. (Field dependence had little effect.) In a subsequent study Niaz (1987) analyzed chemistry test items for their M-demand (i.e., the number of *bits* of information that must be held in working memory to solve the problem) and measured the M-space of students. Niaz observed that the greater the M-demand of the items, the lower the success rate; and the greater the M-space of the students, the higher the success rate. In an earlier study, Johnstone and El-Banna (1986) obtained similar results. They found a sharp drop in the proportion of students getting test items correct when the M-demand exceeds five or six.

So what shall we make of this research showing that problems with a high M-demand are not solved by our students? Such problems remain, and experts in chemistry have somehow learned to solve them. Have experts enlarged their M-space or have they managed in other ways to deal with complex tasks?

Although work by Pascual-Leone (1970), Case (1972), and Scardamalia (1977) indicates that M-space increases with age, M-space seems to level off in the mid-teens. Little evidence exists to show that the M-space of experts is appreciably greater than that of high school or college chemistry students. However, experts organize information in chunks that can be manipulated with less demand on M-space than would be the case for the novice.

Working memory can be used to hold information or to process information. The more processing that we do, the fewer items of information we can process. The more information we need to handle simultaneously, the less processing we can do. It is little wonder that chemistry students are unable to solve complex problems involving several steps, even when they can complete each step presented separately.

By necessity, we must take one thing at a time, break complex tasks into subtasks, solve these smaller tasks independently, and integrate the solutions to the subtasks at a later time. If we cannot keep everything in working memory, we must compensate in some way to deal with complex tasks (*see* Roth, 1990). Strategies used to compensate for limitations in working memory are discussed in Chapter 8.

SUMMARY

Problem solving is an extremely broad subject, and the material in this chapter is limited to problem solving as it pertains to chemistry teaching and learning. Several problems were presented, and you were urged to solve one of them while talking into a tape recorder to record your thought processes.

Problems range from routine to nonroutine, from well-defined to ill-defined, and from one domain to another. The major distinction made in this book is between routine, well-defined problems that experts solve by using well-practiced algorithms (*exercises*) and tasks that are so novel to the solver that no algorithmic solution is apparent (*problems*).

No set of problem solving steps can be taken to solve all problems, but most problem solving includes, at some point, efforts to *understand the problem, represent the problem, plan and carry out some kind of solution,* and *verify that the procedures followed or the solution reached are valid.* If you thought aloud into a tape recorder as you solved one of the problems suggested early in the chapter, you should have observed these processes in your own problem solving.

SOLUTIONS TO PROBLEMS

THE NINE-DOT PROBLEM

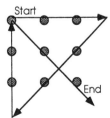

Figure 7.2. A solution to the nine-dot problem.

THE METAL CHLORIDE PROBLEM

The textbook solution to this problem is given in Chapter 8, **Modeling**, but none of the subjects in our research approached the problem in such a straightforward manner.

As suggested by the description of my representation on p 69, thinking about the problem in a very concrete manner seemed to be important in my solution of the problem. I realized quickly that the chlorine in the metal chloride ended up in the silver chloride precipitate. Most successful subjects in our research did too. I used the information given in the problem, along with the mole mass of AgCl, to calculate the number of moles of chlorine present. Others found a percent composition of AgCl and used that to calculate the mass of chlorine. By using either of these pieces of information and the formula, MCl_3, the atomic mass of M can be found.

THE CRYPTARITHMETIC PROBLEM

There are many ways to solve the problem. Mine was a combination of trial-and-error and deductive reasoning. It went something like this:

> DONALD
> +GERALD
> ―――――
> ROBERT
>
> D=5 (Given)
> Then T=0 (5 + 5 = 10) and R must be an odd number other than 1 or 5 (We carry a 1 when the D's are added, and R is the sum of 2L + 1. L cannot be 0 because T=0, so R must be greater than 1 and odd. It is not 5 because D=5)
> R = 3, 7, or 9
> G is less than 5 because the sum of D (5) and G gives a one digit result (R). G is not 0 (T=0).
> G = 1, 2, 3, or 4.

At this point in my solution, I saw no obvious next step, but I did write down other relationships that had to be true.

> All carries are either 0 or 1. (Because only two digits are added, the largest possible sum in a column is 19: 9 + 9 + 1.)
> I see that O + E = O or O + E + 1 = O. (Second column from the left.) How can this be? E is not zero because T = 0. The sum of O and E must be greater than 10, so there is a carry of one to the left column, and G must be less than 4. G = 1, 2, or 3.

I now resorted to trial and error:

> Trial 1. Assume that R = 9.
> Then L = 4 and E is even.
> G = 3 or 4, but we have already established that G cannot be 4 and a 1 is carried to the left column. Under this assumption, G must = 3.
> Trial 2. Assume that R = 7.
> Then L = 3 and E is even, or L = 8, E is odd, and G = 1.
> Trial 3. Assume that R = 3.
> Then L = 1 and E is even, or L = 6, E is odd, and G = ...
> Impossible. Because D = 5, G must be less than zero in order for R to be as small as 3. I do not consider negative numbers to be digits, so I think the conditions of the problem would be violated.
> R cannot be 3.
> R = 7 or 9.

Because the conditions of Trial 1 are easier to test, I will try those conditions in the problem and see what relationships are revealed. First, let me list what I know under this assumption in systematic order.

TRIAL 1 TEST

0 = T; 1 = ?; 2 = ?; 3 = G; 4 = L;
5 = D; 6 = ?; 7 = ?; 8 = ?; 9 = R

I also know that E must be even. What could A be?

A = 1, 2, 6, 7, or 8.

N could not be 1. (When N is added to R, the result is B. Because R = 9, B would have to be either 0 or 1 if N = 1, but T = 0 and if N = 1, B cannot be 1.)

N could be 2, but if it is, there can be no carry to the column containing N.

N could not be 6. (When N is added to R, the result is B. Because R is 9, B would have to be 5 or 6, but D is 5 and if N is 6, B cannot be 6.)

N could be 7, but if it is, there can be no carry to the column.

N could be 8, but if it is, there can be no carry to the column.

N = 2, 7, or 8 and there can be no carry to the column containing N.

Then A = 1 or 2 because any other possibility requires a carry to the column containing N.

Let me again resort to some trial and error:

Trial 1a. Assume that A = 2.
Then E = 4 and N = 7 or 8.
Trial 1a1. Assume that N = 7.
Then B = 6, and a 1 is carried to the O column.
O + E(4) + 1 = O.

Wait a minute! The only way I see to get the sum of O in this column is for E + the 1 carried to be equal to 10! That means that E = 9! I should have seen this earlier, but I did not! I can quit this mess of assuming that R = 9 right now.

Let me summarize again:

TRIAL 2 TEST

0 = T; 1 = G; 2 = ?; 3 = ?; 4 = ?;
5 = D; 6 = ?; 7 = R; 8 = ?; 9 = E

From the earlier Trial 2, I know that L is 3 or 8.

Trial 2a. Assume that L = 3.
Then E is even, but I know that E = 9, so L is not 3.
L = 8 and a 1 is carried to the A column.

0 = T; 1 = G; 2 = ?; 3 = ?; 4 = ?;
5 = D; 6 = ?; 7 = R; 8 = L; 9 = E

A must be 4 and there is no carry to the N column.

0 = T; 1 = G; 2 = ?; 3 = ?; 4 = A;
5 = D; 6 = ?; 7 = R; 8 = L; 9 = E

The only letters left are N, O, and B; the only digits left are 2, 3, and 6. I see something that leads me to work on O first. I know that . . . Nope! It does not work, but I remember that I previously established that there is a carry to the O column. Then R + N must be greater than 10 so N must be 6.

This makes B = 3 and O must be 2. I think I've got it! I'll substitute the numbers for the letters and add to see if it checks:

```
0 = T; I = G; 2 = O; 3 = B; 4 = A;
5 = D; 6 = N; 7 = R; 8 = L; 9 = E.
  D O N A L D
  5 2 6 4 8 5
+G E R A L D
  I 9 7 4 8 5
  R O B E R T
  7 2 3 9 7 0
```

It checks!

8

Teaching Problem Solving

One can argue that there is no need to teach problem solving because we do it every day. All learning involves problem solving. However, some people are better at dealing with novel tasks than are others, and those who take on novel problems and produce creative solutions are highly valued. If we can make people better problem solvers, we should.

The discussion of problem solving in Chapter 7 was organized around understanding the problem, representing the problem, executing a plan for solution, and verifying. These processes are not steps to solving problems, and they are seldom carried out sequentially. Rather, problem solvers vacillate from one process to another and back again. However, separating the processes makes it easier to focus on important aspects of problem solving. We consider here several tools that help students understand and represent problems; then we turn to teaching the solution process and verification strategies. This discussion is followed by suggestions that pertain to the process as a whole.

TEACHING STUDENTS TO UNDERSTAND PROBLEMS

FOCUS QUESTIONS

Common classroom tasks are exercises for teachers, and these tasks are often expected to become exercises for students. Still, these tasks are problems when seen for the first time, and a great deal can be taught about problem solving if the tasks are treated as problems. Many teachers do. Rather than presenting students with an algorithm to copy, they ask questions that help students understand the tasks and develop sensible strategies for tackling these tasks. In the process, teachers convey transcendent messages about what is important in problem solving. Questions such as these are common:

Question: What are you asked to do in this problem? (Problem goal.)
Transcendent message: When solving any problem, an important step is identifying the goal.

Question: What are you told about conditions? (Stated conditions.)
Transcendent message: When solving any problem, it is important to understand constraints that must be considered.

Question: Is there anything that you can assume is true, even though you aren't told? (Unstated conditions.)
Transcendent message: Solving problems may require information that is not explicitly stated.

Question: Is there anything that you know *can't* be true under the conditions stated? (Unstated conditions.)
Transcendent message: Negative information may be useful in solving a problem.

Question: Do you see anything inconsistent in the problem, e.g., any apparent contradictions? (Possible misinterpretations, transcription errors, etc.)
Transcendent message: It is unlikely that you will solve a problem that doesn't make sense. If the problem doesn't make sense, try to find out why.

During the discussion of these questions, it is useful to ask several students the same question to see if they understand the goals and conditions differently. They probably will, and students need to be aware that various interpretations are possible. When one is stuck on a problem, consciously looking for alternative interpretations or asking others how they see the problem may lead to new insight and eventual solution.

METACOGNITIVE QUESTIONS

Questions such as those previously listed focus student attention on important factors to consider in understanding problems, but as long as the teacher is asking the questions, students will not become independent problem solvers. Students must develop the metacognitive habit of posing questions to themselves in addition to learning what questions to ask.

After students become adept at answering questions such as those in the previous section, teachers should shift to more open questions such as these:

Question: What is the first thing that you will do (or did) in solving this problem?
Transcendent message: You should have some general procedures that you follow when you solve problems.

Question: Then what?
Transcendent message: You should be able to trace the logical steps that you have taken in solving a problem.

Question: Can you see anything else you could have done to help you understand the problem?
Transcendent message: Even when following a general procedure that has worked in the past, you should consider new possibilities for improving that procedure.

The purpose of this line of questioning is to encourage students to incorporate the previous questions into their problem-solving schemas and to use them automatically to manage problem-solving activities.

The most common error in teaching students to understand problems is to ignore the process altogether. Teachers assume that routine problem statements are as clear to students as they are to themselves. This assumption is almost never true, even with college students. Research to elucidate the problem-solving behavior of experts and novices reveals that experts spend more time than novices on the concomitant processes of understanding and representation (Reimann and Chi, 1989, pp 165–171), and anecdotal evidence suggests that forcing students to spend more time on understanding and representation is more efficacious than using that time for additional practice (Davison, M. A., Purdue University, personal communication, 1993).

PAIR PROBLEM SOLVING

Whimbey and Lochhead (1986) reported improvement on standard textbook problems with a paired problem-solving strategy. One student acts as the problem solver while the other acts as a checker. The problem solver reads the problem aloud and continues to talk while solving the problem. The checker monitors what is said and may stop the solver and ask for clarification when a procedure is not clear. This process is particularly effective in revealing errors such as these listed in Whimbey and Lochhead's (1986) checklist of errors:

1. Student read the material without concentrating strongly on its meaning. . . . He did not constantly ask himself, "Do I understand that completely?"
2. Student read the material too rapidly, at the expense of full comprehension.
3. Student missed one or more words (or misread one or more words).
4. Student missed or lost one or more facts or ideas.
5. Student did not spend enough time rereading a difficult section to clarify its meaning.

. . .

16. Student skipped unfamiliar words or phrases, or was satisfied with only a vague understanding (pp 18–19).

Whimbey and Lochhead emphasized that when such errors are detected, the checker should say something such as, "You read that wrong", or "You wrote that down wrong", rather than saying what the error is. I have used this technique with students who have difficulty solving routine problems, and I am convinced that this advice is sound. Students often read a problem in the same incorrect manner (reading 2.54 g rather than 4.52 g, for example) three or four times before realizing their mistake and correcting it. Only when students are able to detect their own errors and correct them are they developing true problem-solving skills.

Although pair problem solving is a potentially useful technique, it requires skills that are missing in some populations. Hutchinson (1985) found that the disadvantaged adults he worked with had difficulty assuming the roles of problem solver or listener. When challenged on an issue, they felt personally attacked and responded defensively. In addition, they lacked the communication skills required by the technique (pp 507–508).

TEACHING STUDENTS TO REPRESENT PROBLEMS

DIFFERENCES IN EXPERT AND NOVICE REPRESENTATIONS

A great deal of research has compared the way that experts and novices represent problems, and the contrasts are striking (*see* Reimann and Chi, 1989, pp 165–171 for a review; Kumar, White, and Helgeson, 1993). These differences in representation appear to be the key to differences in problem-solving success. Unfortunately, we know little about how to teach novices to represent problems as experts do.

The ability to represent problems depends on finding connections between the task at hand and information in memory that might be used to solve the problem. Expertise means that a great deal of information is carefully organized in a semantic network that connects many ideas. Such expertise is not readily taught. Still, problem representation can be enhanced in ways other than building expertise in the content domain.

The ability to make connections is related to the context in which we experience a problem. Classical work on "functional fixedness" illustrates the difficulty that most people have in viewing problem tasks with sufficient flexibility to develop novel solutions. Functional fixedness is easily illustrated in the context of the Maier two-cord problem, a task used in many psychological studies of problem solving. In this task, two cords are hung from the ceiling far enough apart so that a subject cannot hold one cord and move close enough to the other cord to grasp it and tie the cords together, which is the problem goal. The problem can be solved by tying a heavy object to one of the cords, setting the resulting pendulum in motion, grasping the other cord, and waiting for the pendulum to swing close enough to be grasped. A pliers, stapler, hammer, and other objects that could be used as a pendulum bob but are normally used for other functions are available in the room, but few subjects think to use them. If a plumb bob is available, or if, before presenting the problem, subjects are told "a pliers could be used as a pendulum bob", success on the problem is greatly increased. However, subjects seldom have the insight to use the available materials for a bob without prompts, just as few individuals think to solve the nine-dot problem (*see* p 79) by extending lines beyond the field defined by the dots.

Similar behavior is frequently observed in school settings. Virtually every chemistry teacher has observed students who do well in mathematics classes but fail to transfer their math skills to problems in science. These students fail to see the applicability of those skills within the new context.

An existing context may also prompt students to attempt inappropriate representations of problems. In research at Purdue, we gave chemistry students the following task during a unit on writing chemical formulas (Niaz, Herron, and Phelps, 1991).

> Write an equation using the variables S and P to represent the following statement: "There are six times as many students as professors at this university." Use S for the number of students and P for the number of professors.

A frequent response of students was to write S_6P, an expression like the chemical formulas they had been writing but certainly not the mathematical equation that was expected.

Many chemistry students attempt to represent problems almost exclusively by recalling an algorithm from memory and then applying that algorithm blindly to the problem at hand (Gabel, 1981; Gabel and Sherwood, 1984; Gabel, Sherwood, and Enochs, 1984; Greenbowe, 1984; Nurrenbern, 1980). These students do not seem to consider alternative representations for problems, nor do they seem to question the validity of the representation that they make.

Normal chemistry instruction provides a context that encourages such behavior. More often than not, model solutions are presented as *the* way to solve particular problems. This presentation is usually done before students are asked to consider how *they* might represent the problem. This model solution is then followed by a number of practice problems, which students are asked to do using the model solution. The focus of instruction, then, is on the presentation of an efficient solution pathway, which students are asked to accept and use.

CONTEXTUAL INFLUENCES ON REPRESENTATION

Examples given thus far show how the context of instruction can interfere with good problem representation. Bransford, Sherwood, et al. (1985) suggested that appropriate instructional contexts might aid students in representing problems.

In early work, Bransford and his associates used video recordings of segments from popular movies such as *Raiders of the Lost Ark* and *Swiss Family Robinson* to provide a meaningful context within which problems are presented. For example, in one segment from *Raiders*, Indiana Jones must replace a golden idol by something of equal mass to prevent a trap from being sprung. In the movie, a bag of sand is used to replace the idol. After being presented with information concerning the density of gold, sand, and other materials, students view this segment and are asked to consider whether the maneuver is actually possible. When presented in this manner, students are quite successful in using available information concerning density to solve the problem.

This early work was extended in the Jasper Woodbury Problem Solving Series (Cognition and Technology Group at Vanderbilt, 1992a, 1992b, 1992c, 1992d, 1992e, 1993). Videos in the Jasper series end with one of the characters in the 14–18 minute story posing a problem as a challenge to students. The presented problem is complex and requires students to generate several subproblems to generate a solution, but all of the data required to solve the problem can be found in the video (Cognition and Technology Group at Vanderbilt, 1992a, p 70). The Vanderbilt group is still exploring various implementations of the Jasper series, but early research indicates that this environment is a rich one for teaching students to represent and solve complex problems.

This work is part of a substantial body of recent research suggesting that thinking is situated in or tied to particular contexts (Brown, Collins, and Duguid, 1989; Cognition and Technology Group at Vanderbilt, 1990; Collins, Brown, and Newman, 1989; Qin and Simon, 1990; Resnick, 1989). Other research shows that problems encountered in the context of meaningful, real-life activities are more easily solved than "academic" problems presented in the absence of contextual cues (Robertson, 1990, p 254).

In a study of high school students' abilities to solve molarity problems, Gabel and Samuel (1986) administered tests that included common molarity problems and analogous problems involving making lemonade. The results suggest that learning analogous problems might help students understand molarity. (For a discussion of research on everyday problem solving, *see* Sinnott, 1989.)

All of these studies suggest that we might be far more successful in teaching students to represent novel problems if we begin by presenting those problems in a context rich in contextual cues. Tech Prep, a program meant to prepare general education students for technical careers, provides contextual cues by presenting science and mathematics in the context of real problems encountered in technology courses (Dornsife, 1992; Grubb et al., 1991; Hull and Parnell, 1991; Indiana Department of Education, 1992). Preliminary research suggests that the approach has merit.

The more familiar the context in which the problem is presented and the closer the problem is to the everyday experience of students, the more likely students are to make the necessary connections and arrive at an appropriate representation of the problem. However, if we want the procedures used to solve problems in the familiar context to transfer to less familiar problems, more time must be spent looking back at the solution and focusing attention on what was done that was helpful (Chi and Bassok, 1989; Polya, 1957). By following this "looking back" activity with less familiar problems that can be solved by using similar strategies, students are likely to elaborate the procedural schemas and begin to see how these schemas can be used in broader contexts.

CONNECTING SYMBOLS WITH REALITY

Before leaving the issue of problem representation, let us consider an issue that is particularly relevant to mathematics and chemistry. Both fields use an elaborate representational system that greatly facilitates the solution of problems by experts in the respective fields. Chemists write chemical equations to represent events taking place in both the macroscopic and microscopic domains. Mathematicians use letters to represent variables and arrange the letters in equations that can be manipulated to reveal relationships that are not readily apparent in the absence of such symbolic representations. The power of such symbolic systems in solving problems is great, but the symbolic representations must be understood and must be connected to the situations and events they are meant to represent.

Students often manipulate both mathematics and chemistry symbols with apparent facility but arrive at conclusions that are utter nonsense. The nonsense is not apparent unless the connection between the symbolic representation and the reality represented by the symbols is clearly established.

Consider Gayle[1], one of the students in Greenbowe's study (Herron and Greenbowe, 1986). The following problem was difficult for most of the students in the study (only 32% solved it), and Gayle's errors were typical.

> A 1.00-g mixture of cuprous oxide, Cu_2O, and cupric oxide, CuO, was quantitatively reduced to 0.839 g of metallic copper by passing hydrogen gas over the hot mixture. What was the mass of CuO in the original sample (Mahan, 1975, p 29)?

Gayle wrote the following equations and then applied the factor-label algorithm to find the mass of CuO in the original mixture:

$$CuO + Cu_2O \rightarrow Cu_3O_2$$
$$Cu_3O_2 + 2H_2 \rightarrow 3Cu + 2H_2O$$

---

[1]In keeping with the practice adopted for this book of using names that are used for males and females, Gayle has been substituted for the name used by Greenbowe.

Not only did Gayle assume a 1:1 molar ratio for the two oxides in writing the equations, but Gayle reported that there was 1.05 g of CuO in the 1.00-g mixture. Gayle was oblivious to both errors and seemed confident in the answer. Gayle clearly did not connect the symbolic representations (equations) with the physical reality they were meant to represent. Anamuah-Mensah (1986) reported similar behavior among high school students.

Chemists habitually operate in three worlds: the macroscopic world of everyday substances, the microscopic world of atoms and molecules, and the symbolic world of formulas and equations. Representations in one of these worlds can reveal information about a problem that is not obvious when the problem is represented another way. By habitually representing problems in several problem spaces, selection of the best representation for the problem at hand is facilitated, and judging the validity of particular symbolic representations of the problem is easier.

I encourage students to think about what chemical equations mean in terms of macroscopic and microscopic observations. I ask questions such as, "What would you see if you were doing this in the laboratory?" and, "If you were the size of atoms and molecules and were watching what is described by this equation, what would you actually see?" When performing demonstrations or conducting experiments, I ask students to describe what is taking place by using equations, as described on p 93.

TEACHING THE SOLUTION PROCESS

Research on problem solution processes is of two kinds:

1. In the information processing field, computer programs have been written to simulate human behavior, and the behavior has been described in terms of production systems or flow charts.
2. Those who make recommendations from the perspective of classroom practice normally couch their recommendations in terms of heuristics.

HEURISTICS AS SOLUTION AIDS

Heuristic reasoning is described by Polya as "reasoning not regarded as final and strict but as provisional and plausible only, whose purpose is to discover the solution of the present problem" (Polya, 1957, p 113). A heuristic is a general strategy that may be applied to many (but not all) problems to provide insight into a more systematic plan for solution.

Polya's (1957) work in mathematics, though old, provides some of the best suggestions for use of heuristics in mathematics. Anyone teaching mathematics would do well to study his book and follow his suggestions in teaching the subject. In particular, his insistence that a problem is not finished when an answer is obtained should be heeded by any teacher who believes that it is important to teach problem solving rather than generate answers to routine exercises.

Polya insists that the solver should look back over the problem-solving procedure and consciously search for other problems that could be solved by the same method. A similar suggestion was made by Hayes (1981) in his sixth problem-solving step, consolidate gains, which involves looking back at the problem-solving process to see what was learned about problem solving as opposed to the content dealt with in the problem. Chi and Bassok (1989) indicated that good problem solvers do something like this when they learn from examples.

Research aimed at deliberately teaching students to use heuristics of the type described by Polya was conducted by several people (Lee, 1982; Lucas, 1974; Pereira-Mendoza, 1980; Schoenfeld, 1979; Wheatley and Wheatley, 1982). David Frank summarized this research with the following statement:

> A summary of this mathematical research on heuristics shows that heuristics can indeed be taught to students in mathematics classes. Not only do students make use of the heuristics, but students who use them have higher test scores. However, it is not entirely clear which heuristics can be most easily and effectively taught. Nor is it clear how much time should be devoted to the teaching of heuristics (Frank, D., 1986).

Although some research of this kind was done in chemistry, it is not as extensive nor does it appear to be as carefully done as the work reported in mathematics. Heuristics are commonly employed in unstructured approaches to problem solving.

UNSTRUCTURED APPROACHES TO PROBLEM SOLUTION

Approaches used to teach problem solving vary from highly structured approaches, where students are given an explicit set of steps to follow, to approaches that emphasize general strategies (heuristics) that may help in solving some problems but afford little help in solving others.

David Frank's (1986) study in an introductory chemistry course for science and engineering majors is typical of unstructured approaches to problem solving. In Frank's study, each of two graduate students taught two recitation sections. Problem solving was stressed in one section, but not in the other. The instruction was loosely based on Polya's problem-solving model:

- understanding the problem
- designing the plan
- carrying out the plan
- looking back (Polya, 1957)

However, the first two stages were combined into one, called "planning the solution", partly because Yackel's (1984) work had shown that these stages are not done separately or sequentially.

Students were encouraged to ask questions such as the following: What is the unknown? What are the conditions? What do these substances look like? How do the atoms and molecules involved here interact? How could you symbolize what you see? Instructors asked similar questions when they modeled problem solutions, and they pointed out when a student had neglected one of the steps, such as quitting a problem without judging whether the answer made sense.

Much of the class time was spent with students working on novel problems in small groups. The instructors monitored the progress of the small groups and gave help only when it was needed. The class period concluded with a discussion of heuristics used and what was learned about problem solving.

Students in the treatment group outperformed students in a control group on the four major exams in the course, and think-aloud interviews conducted at the end of the semester showed that the experimental students made more generalizable representations of problems, had fewer uncorrected math errors, were more persistent, and evaluated their work more frequently than did students from the control group.

According to Frank, the most likely explanation for the modest improvements that he observed was that when groups were stuck on a problem, they shared ideas, argued about the validity of suggested strategies, and otherwise interacted in ways that led students to think carefully about problems and how they could be solved. This interpretation is consistent with research on the effects of social interaction on cognitive development (*see* discussion on pp 190–192).

Frank's study involved only the recitation portion of a college chemistry course. I believe that greater improvement would require intervention in all parts of the course. This kind of intervention is what Fasching and Erickson (1985) did in their study. One half of each lecture period was spent in unstructured, small-group problem-solving activity. In addition, a research project was required, and students engaged in open-ended laboratory activities.

Fasching and Erickson apparently did not conduct class discussions of general problem-solving strategies, an important component of Frank's intervention, but otherwise their procedures appear to be similar. Unfortunately, Fasching and Erickson were unable to conduct a controlled evaluation of their treatment, but their qualitative analysis of student performance suggests noticeable improvement in problem solving by about 40% of their students.

STRUCTURED APPROACHES TO PROBLEM SOLUTION

One of the best examples of the structured approach to problem solving is that employed by Mettes and his associates in the Netherlands (Kramers-Pals, Lambrechts, and Wolff, 1983; Mettes, Pilot, and Roossink, 1981; Mettes, Pilot, Roossink, and Kramers-Pals, 1980, 1981). In their approach, students are given a general Program of Actions and Methods (PAM), which is divided into four phases:

1. read and analyze the problem
2. transform the problem into a standard problem
3. carry out the routine operations of the standard problem
4. review the result

Although these phases are similar to recommendations made in less structured approaches, here the steps are explicit and sequential.

The PAM has been translated into a Systematic Approach to Solving Problems chart (SAP), a flowchart that spells out in detail the steps students must take in solving a thermodynamics problem. Students solve problems individually or in small groups by using the SAP worksheets and are expected to design their own charts of key relations listing such things as important laws, definitions, and equations. Students taught to use the SAP charts in a university thermodynamics course outperformed other students on course examinations.

Bunce and Heikkinen (1986) were less successful with their structured approach to problem-solving instruction. Bunce used a worksheet when solving mathematical chemistry problems in college lectures. The worksheet contained the following:

- a sketch of the situation described in the problem
- all information (such as rules, equations, and definitions) needed to solve the problem
- labels for information given or requested
- a qualitative statement regarding what the problem asked
- a breakdown of the problem into subproblems
- a list of steps needed to solve the problem
- the solution

This technique had little effect, partly because students did not use it. According to students' reports, it was too time-consuming.

Although highly structured approaches may be effective in getting students to obtain right answers to routine problems (exercises), research suggests that they will not help students solve novel problems. In a study involving fifth- and sixth-grade students, de Leeuw (1978) found that, when tested immediately after instruction with problems used in instruction, students who had been taught specific algorithms performed better than students who had been taught general heuristics. However, when tested later and on transfer tasks, those who had learned heuristics did better.

TEACHING VERIFICATION STRATEGIES

Although we know that verification is an important part of successful problem solving, we know little about how to teach it. If David Frank is correct in his suggestion that even good problem solvers do little verification when they are working with difficult problems in unfamiliar domains, we may be wasting time and frustrating students when we insist that they check their work when they are attempting to solve difficult, novel problems. Our goals might be better served if instruction in this area is tied to exercises rather than problems.

Listed here are some of the strategies that are commonly used to increase confidence that a correct or useful solution to an exercise has been found.

1. *We have increased confidence in a result if we know that others reached the same result independently.* This tactic is commonly used in real-world situations, and students know and use it. They look on their neighbors' papers. Rather than discourage the practice, we should point out that this verification strategy is not always available—during a test, for example—and that they need to develop others.

2. *We have increased confidence in a result if it is reached repeatedly by retracing the solution path.* Some students use this tactic, and it is the one most commonly taught in school. However, it is not a very powerful tactic because an incorrect piece of information or an error in reasoning is likely to be repeated.

3. *We have increased confidence in a result if it is reached repeatedly by* different *paths.* This tactic is a far more powerful check than retracing the same path because the different paths normally use different information and different reasoning. The more dissimilar the procedures, the more powerful the test.

   Examples of this procedure in the history of science are abundant—using independent means to estimate Avogadro's number, for example. Estimation procedures for numerical problems or working the problem with units only are simple examples of the procedure.

4. *We have increased confidence in a result if it is consistent with other information that we have, thus making it a* reasonable *answer.* A simple example of such a procedure is what might be called a magnitude check: "I calculate the mass of a single molecule as 8 g. I know that a single molecule is too small to see or feel. I know that I can see and feel a mass of 8 g. My answer is not reasonable." "I estimate the age of my friend as 47, and I know that she is older than her brother who is 45, so my answer is reasonable."

By systematically asking students how they know their answer is correct, other verification strategies are likely to emerge.

ADVICE TO PROBLEM SOLVERS

After reviewing research on problem solving in chemistry, Frazer (1982, p 180) listed the following points as "general strategies or advice to problem solvers". Although Frazer based his advice to problem solvers on research evidence, some of this advice is questionable, and most of it requires elaboration if it is to be interpreted properly. Still, it affords a convenient framework for discussing many important points about problem solving.

> 1. Work backward from the goal, not forward from the given information.
> 2. Break down the problem into subgoals and work at each separately. Do not try to cope with too much information at any one time.
> 3. Convert an unfamiliar problem into a familiar problem and then apply an already learnt procedure.
> 4. Make a guess at the solution and work backward to see if the guessed solution is consistent with all the information available.
> 5. Check that all information stated in the problem has been used and that all other sources of information (memory, literature, experts, and experiment) have been exhausted.
> 6. Check that all the stages of problem solving ... have been used.
> 7. Check whether there are any guidelines ... or algorithms ... applicable to this problem.
> 8. Try to see the problem as a whole.
> 9. Draw diagrams, verbalize the problem, convert a statement into a question, and convert statements into mathematical expressions.
> 10. *Brainstorm*; that is, write down all the ideas that come to you, however foolish or irrelevant they seem.
> 11. Rest to allow time for "incubation" of the problem. (p 180)

The following material elaborates on the points in Frazer's list, but not in their original order.

POINT 1: WORK BACKWARD FROM THE GOAL

Research on problem-solving behavior of experts and novices reveals that experts normally work forward, whereas novices work backward. This research does not necessarily conflict with the advice in Frazer's list. The problems used in the expert–novice research most likely were easily represented and understood by the experts but not by the novices. Consequently, experts were able to devise a plan working forward, whereas novices lacked sufficient information to do so. Working backward can be a useful strategy in some instances, and students should recognize it as a potentially useful heuristic; however, it is no panacea.

Alex Johnstone claims that students must be encouraged to work backward. Almost certainly, they require instruction on just what one means by working backward. For the problem concerning the copper oxide mixture (p 88), working backward might mean guessing that the mass of CuO in the original mixture is 0.5 g, calculating the mass of copper in 0.5 g of CuO and 0.5 g of Cu_2O, comparing that with 0.839 g, guessing again, and continuing until the structure of the problem is clear or the iterative procedure closes in on 0.839 g as the mass of copper. But what does working backward mean in the context of the cryptarithmetic task (p 66) or the nine-dot problem (p 68)?

POINT 4: GUESS THE SOLUTION AND WORK BACKWARD

This suggestion, as outlined in the previous section, is consistent with the common heuristic of "guess and test". It is a useful strategy. Unfortunately, work at Purdue suggested that students view guessing as inappropriate in problem solving. David Frank (1986) reported that even when students successfully used the strategy to solve a problem, they apologized, saying: "I really should not guess. I should know how to do the problem." The students apparently failed to see guess and test as a legitimate strategy to use.

POINT 2: BREAK THE PROBLEM INTO PARTS

This suggestion is clearly consistent with what is known about problem solving and the limitations of working memory. We are incapable of keeping all of the information related to a complex problem in conscious memory while we solve the problem. We must break the problem into parts. However, unless instruction elaborates on the suggestion, the suggestion is of little value.

Many strategies for the process of breaking a problem into parts can be taught. Several authors have suggested constructing *road maps* outlining the logical steps in solving a problem. Figure 8.1 illustrates a road map that might be used in solving the unknown metal chloride problem on page 64.

Once students learn to construct and interpret such diagrams, these diagrams serve several purposes:

1. They provide a convenient external record of relationships that may be referred to when solving the problem.
2. If the steps indicated in the map are actually used, the diagram serves as an external record of the logical steps executed in solving the problem.
3. If the student works on a subgoal and then forgets why the work was done, the map will refresh his or her memory.
4. If the student thinks to employ the strategy of working backward, the map suggests how that might be done.
5. The map may also suggest where particular algorithms may be used as part of the problem-solving process (Point 7 in Frazer's list).

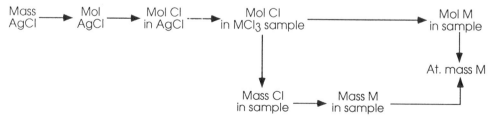

Figure 8.1. Road map showing important relationships in the metal chloride problem.

6. The map may also assist students in seeing the problem as a whole (Point 8 in Frazer's list).

However, unless the student sees the problem as a whole, a complete map cannot be constructed! Students' inability to see the problem as a whole is the greatest limitation of the road map strategy.

POINT 8: VIEW THE PROBLEM AS A WHOLE

As indicated above, this suggestion is undoubtedly sound, but the advice is useless unless we elaborate on the procedure and illustrate it repeatedly. Viewing the problem as a whole requires considerable understanding of the problem for the strategy to be employed effectively (*see* Chi and Bassok, 1989, pp 257–258). The same thing can be said about Points 3 and 9 in Frazer's list.

POINT 3: CONVERT TO A FAMILIAR PROBLEM

If one is to convert a problem into a familiar one, the problem must be well understood or the familiar problem that results will be a misrepresentation that leads to erroneous results. Chi and Bassok cite a study by Van Lehn that provides indirect evidence that 85% of systematic arithmetic errors derive from incorrect learning from examples (p 259). Many errors observed in chemistry problem solving at Purdue can be explained in a similar manner (Carter, 1988; D. Frank, 1986; Greenbowe, 1984).

POINT 9: DRAW DIAGRAMS, VERBALIZE THE PROBLEM

Similarly, diagrams are not always effective. Yackel (1984) found that most students drew diagrams to represent a problem about lakes that she used in her research. However, the utility of the diagrams varied considerably and depended on how closely the diagram modeled salient facts given in the problem. Diagrams that looked very much like lakes were less helpful than diagrams that looked nothing like lakes but did illustrate the differences in water level of the lakes, which was the salient information. The student had to recognize that salience to draw the diagram.

The focus of instruction should not be on drawing diagrams per se. Rather, instruction should focus on the *kind* of diagram (or other transformation of data) that was useful in solving a particular problem, why that diagram was useful but others were not, and how the generator of the diagram thought to construct the useful diagram.

One of the places that I have effectively used the strategy of "thinking of a similar problem" is in dealing with exponential numbers. Students who understand the rules of arithmetic when applied to familiar, whole-number problems often make mistakes when dealing with exponential numbers or algebraic expressions that have variables replacing numbers. I suggest that students see if they can make up an analogous problem by using simple, whole numbers and use the simple problem to figure out how the mathematical operations will work in the more complex problem. Take for example, the following:

$$2.5 \times 10^{23} + 1.7 \times 10^{24} =$$

When students have difficulty, I suggest that they solve the following problem:

$$2 \times 2^2 + 3 \times 2^3 =$$

I then ask them to compare the operations that they must perform with the exponents to get the correct answer and return to the first problem. Although not immediately helpful, students can eventually learn to use this strategy to verify that a particular mathematical procedure is valid.

POINT 6: CHECK THAT ALL STAGES OF PROBLEM SOLVING WERE USED

Because I have never found a set of problem-solving stages that generally apply, I have never given students a list to check. I do, however, suggest that students make a list of questions that they will want to answer habitually when they are solving problems in chemistry. I suggest that they keep the list handy when they are doing homework, and I suggest that they go over the list *after* they have finished a problem and *before* they check their answer against the answers provided in the text. Some of the items that I suggest for the list are as follows:

1. Are the numbers you used in the calculation the ones given in the problem; that is, did you make a transcription error?
2. Did you record the answer to the proper number of significant digits?
3. Did you include appropriate units?
4. Do the units that you obtained in a factor-label solution make sense; that is, are they appropriate units to describe what you were attempting to calculate?
5. Did you reread the problem to be sure that you calculated what you were asked to find?

I suggest that students add any items to the list that they think may keep them from forgetting to do an important operation. Although students are encouraged to use the written list when doing homework, they are not allowed to use the list on an exam. The suggested steps should become automatic.

Translating problem statements into chemical equations, mathematical equations, and other symbolic representations (Point 9) is not a trivial task. Much more attention should be given to teaching this skill. In my course, asking students to translate chemical equations into ordinary English and vice versa is standard practice. I do the same with mathematical sentences (equations). Several times during the course I ask students to write chemical (or mathematical) equations to describe an event that is being demonstrated (e.g., write the equation to describe the synthesis of antimony triiodide, based on what is seen as I do the synthesis as a demonstration). Afterward we discuss information that is left out of the equation as well as information implied by the equation but not made explicit by what is observed. Such exercises are instructive.

POINT 10: BRAINSTORM

Students are reluctant to share ideas because they fear that their ideas will seem foolish. They have been rewarded for getting it right and punished for getting it wrong too often to feel comfortable sharing ideas before they check their validity. This fear is counterproductive in problem solving. Thus, the brainstorming suggestion is probably one that should be followed during instruction on problem solving. If brainstorming is to be used successfully, care must be taken so that *foolish* ideas are evaluated but not scorned.[2] Students need to be rewarded for making suggestions. They will discover for themselves which ideas are helpful and which are not.

POINT 5: CHECK THAT ALL INFORMATION WAS USED

This recommendation is of dubious value. A common error made by poor problem solvers is to force information into the solution of the problem when it does not belong (Muth, 1991). The nature of exercises included in textbooks is likely to teach students that every problem will have all information needed to solve the problem and no more. This characteristic is unfortunate because most real-world problems are either overloaded with irrelevant information or deficient in information. One of the more difficult problem-solving tasks is to sort out what is needed from what is not.

After (but not before) students have developed some initial skill at problem solving in a new content area, exercises should be assigned that have too much information or too little, and students should be told that discovery of superfluous information or the lack of information needed to solve the problem is part of their task.

Frazer's advice to problem solvers says nothing about verification strategies, and our work suggests that checking habits often spell the difference between success and failure at problem solving.

GENERAL SUGGESTIONS

Several suggestions for teaching problem solving have been made in connection with understanding problems, representing problems, solving problems, and verifying. We now turn to more general suggestions.

INCREASE MOTIVATION

Two of the most apparent differences between successful and unsuccessful problem solvers is that successful problem solvers believe that they can solve problems, and they persist for a long time before giving up. Unsuccessful problem solvers do not believe that they can solve problems, and they give up easily.

The most important factor in overcoming such attitudes is success on the part of the student.

> Any new problem-solving activity should be carefully planned to maximize the chance of early success. Initial problems should be reduced in complexity and level of abstraction and should capitalize on automatized skills of the learner.

Detailed suggestions for the development of instructional materials that increase in complexity and abstraction are found in Feuerstein's discussions of mediated

---

[2]Finke, Ward, and Smith (1992) recommended that students work individually to generate ideas before subjecting them to evaluation in group settings because research shows that group influences reduce the number of ideas generated in group settings (pp 185–186).

learning (Feuerstein, Rand, and Hoffman, 1979; Feuerstein, Rand, Hoffman, and Miller, 1980) and Case's discussions of instructional design (Case, 1978b, p 208).

Problems that are related to issues that are important in the life of the student or that are related to familiar events are likely to be more interesting and command attention long enough for skills to develop. The question of how to make instructional materials more interesting is taken up in detail in Chapter 19.

SILLY ERRORS

Both good and bad problem solvers misread problems, but good problem solvers normally catch their errors. However, Feuerstein et al. (1981) pointed out that many students have not developed basic attentional skills. They are unable to decode information inherent in written sentences or diagrams. Feuerstein developed a program to overcome such difficulties, and he claims that the program is successful. Whimbey and Lochhead (1986) provided specific suggestions for improving attending skills in the context of normal classroom instruction. Bransford, Arbitman-Smith, and co-workers (1985) wrote an excellent review of these programs. Although the research evidence supporting the two programs is limited, what does exist is generally favorable. (For a review of Feuerstein's work, *see* pp 35 and 36.) Both programs are worth knowing about. Even though few teachers are likely to use the materials exactly as they were developed, every teacher will find suggestions that can be incorporated into existing teaching patterns.

Suggestions by both Feuerstein and Whimbey and Lochhead are time-consuming and require commitment on the part of the student. The activities must be inherently interesting (as Feuerstein claims his materials are) or students must be sufficiently mature and self-disciplined to stick with the procedure long enough to change behavior.

The work of the Cognition and Technology Group at Vanderbilt described on p 87 is particularly interesting because of its high potential for motivation. By connecting video disks with computers, considerable flexibility can be achieved, and students can explore problems in a variety of ways. Other research groups are working on computer software that may have some of the motivational appeal of video games and will keep students involved long enough to develop important problem-solving skills. However, this work is in its infancy, and much of the existing software is both dull and ineffective in developing problem-solving skills.

MODELING

Most problem-solving instruction is likely to take place in small group settings, but many aspects of problem solving can be modeled by the instructor or by students in large-group environments. These suggestions indicate what can be done and why it is important:

1. Think aloud as you solve problems in class. Even though you may want to work out examples in advance of presentation to students, make your presentation as though you had not. Ask aloud such questions as, "How shall I think about this problem?" and, "What is it really about?" In answering these questions, you might occasionally suggest a representation and reject it, showing students why it is not a good way to represent the problem (*see* Whimbey and Lochhead, 1986).

2. Ask students to suggest representations for problems you work in class. When a suggestion is received, ask for additional representations. Ask students which representation seems most sensible and why. (This provides information about students' thinking and gives students practice in representation.)
3. If you have students work problems at the chalkboard, ask them to begin discussion of their solutions by describing how they visualize the problems and why they see their solutions as sensible representations. Ask other students if they visualize the problem differently. Ask students what led them to a particular representation.

Katona's (1940) work on problem solving has shown that the least effective strategy for teaching students to solve problems is working examples for the learner—the strategy typically used in introductory texts and classroom presentations. Focusing the learner's attention on various principles that might be followed to solve problems of a given type is considerably more effective. Still more effective is what Katona called "learning by help". In this strategy, the instructor uses illustrations or examples to give cues to the student. The student is not told what to do to solve the problem, but the cues clearly focus attention on principles that are to be considered to solve the problem. The student has to *discover* the solution to the problem, albeit, not without help.

The contrast between the suggestions given here and the presentation normally found in textbooks and presented in lectures is important. Here is the solution to the unknown metal chloride problem (*see* p 64) given in the textbook:

$MCl_3 + 3Ag^+ \rightarrow 3AgCl(s) + M^{3+}$
Let x = the atomic weight of M.
Then the mol wt of $MCl_3 = x + 3(35.453) = x + 106.36$.
$2.395 \text{ g } MCl_3 \times [1 \text{ mol } MCl_3/(x + 106.36)\text{g}] \times$
$\qquad 3 \text{ mol } AgCl/1 \text{ mol } MCl_3 \times (143.32 \text{ g } AgCl/1 \text{ mol } AgCl)$
$\qquad = 5.168 \text{ g } AgCl$
$x = 92.90$

Although this summary is an efficient way to solve the problem, it is not what the author would have done if he were solving this as a novel task, which is what beginning students are doing. If we want to teach students the skills that experts use when they face novel tasks, we must somehow reveal what experts do when they encounter novel tasks. In other words, we must occasionally demonstrate what happens in real problem solving. This demonstration is the intent of the suggestions given previously.

These comments should not be interpreted to mean that example problems and worked-out solutions are bad. Chi and Bassok's (1989) research on learning from examples in physics indicated that good problem solvers as well as poor ones want examples and find them useful. How the examples are presented and used is what makes a difference. Stated as simplistically as possible, good problem solvers use examples in an effort to understand what is going on; poor problem solvers use them as templates to copy.[3]

---

[3]If you recorded your own problem solving as suggested in Chapter 7, review the tape now and consider what you would do if you were modeling problem solving for your students. What points evident from your tape would you try to make to students through modeling? What precautions would you take to ensure that you accurately model problem solving without confusing students with endless wandering?

GROUP PROBLEM SOLVING

Both theoretical arguments and empirical evidence support having students work on problems in pairs or small groups. However, group work in itself does not enhance problem-solving skills. The right kind of interactions must take place if cognitive gains are to occur.

Dimant and Bearison (1991) described several studies involving pairs of students working on problems used by Piaget to identify levels of intellectual development. In interactions where disagreements, contradictions, and contrary solutions were expressed in a balanced fashion between partners, cognitive gains were more evident than when such interactions were absent. According to Dimant and Bearison, "What is most revealing in the findings . . . is that it is not how many experimental tasks the dyads completed nor how difficult these tasks were that accounted for individual cognitive gains, but the quality of subjects' social interactions independent of whether they were paired with same-level or more competent partners" (p 283; also *see* Johnson, 1981, p 8; Johnson and Johnson, 1992, p 26).

In a typical implementation of group problem solving, students are asked to work on problems in groups of two or three and to share their ideas about what the problem means, how it can be represented, and how to go about solving it. After the groups work for 20–30 minutes, results are shared and discussed by the class as a whole. During this discussion, attention is called to the various strategies that were used. The focus is on understanding the reasoning and strategies used rather than on the *right* answer. The teacher does not correct wrong answers or indicate whether a particular answer is correct. Rather, students are encouraged to develop their own strategies for determining when a result is generally accepted.

Such "problem-centered instruction" has been used successfully with students from grade two (Yackel, Cobb, Wood, Wheatley, and Merkel, 1990) through college (D. Frank, 1986). Depending on the age of the students and the goals of instruction, the discussion may extend to consideration of how the same strategy might be applied to less familiar problems, and additional problems may be assigned and discussed.

When the problematic task is one for which a standard procedure is normally used by an expert, the expert's procedure can be introduced by the teacher after the presentations by students. However, it should be presented as one more way to represent the problem rather than *the* procedure that must be followed. If students understand the procedure and it truly is more efficient than those developed by the students, they will adopt it. If students do not understand the procedure, encouraging its use is counterproductive, regardless of how efficient the procedure may be.

Two purposes are served by introducing standard algorithms after students have solved problems rather than before:

- Students have an opportunity to experience actual problem solving and to apply previously learned knowledge in a novel context.
- This exploration makes students aware of what must be done to solve the problem so that the expert's algorithm is more likely to make sense when it is introduced.

OVERCOMING LIMITATIONS OF WORKING MEMORY

In Chapter 7 we saw that humans are limited in the amount of information to which they can consciously attend. This limitation has important consequences for problem solving. As problems become complex, one must consciously attend to many things, and overloading working memory is easy. Similarly, as things become more abstract, more transformations of information are required, bringing into play more logical operations. These operations also require conscious attention. Most problems encountered in an introductory chemistry course probably require attention to more than the "seven bits, plus or minus two" that we are capable of holding in working memory. Then how do we manage?

At least three processes are used to reduce the load on working memory: automatization, chunking, and using an external memory.

AUTOMATIZATION. Automatization refers to overlearning to the point that performance of a task requires little or no conscious attention. I am presently typing this chapter, and in doing so I am using several schemas that are automatized. First of all, a great deal of information concerning the English language—vocabulary, spelling, and rules for sentence construction—is applied to my thoughts without conscious effort. In addition, my fingers are striking the keys on the keyboard without any conscious thought about where they go. (In fact, I am unable to tell you where any letter is located on the keyboard without *typing* it out.) I apply all of this knowledge automatically, but I did not always do so.

Similarly, when I solve problems in chemistry, most math skills and a great deal of the chemistry knowledge that I use to solve the problem is applied automatically. Little conscious attention must be given to these processes, and little load is placed on working memory. This automatization is *not* the case for beginners.

CHUNKING. Chunking is another way that we conserve space in working memory. Chunking subsumes several pieces of information under a single *idea*. Alex Johnstone's example of the functional group in organic chemistry is an excellent example of chunking (Johnstone, 1980). Before one learns the carboxylic acid group, the information in the formula

$$R–C=O$$
$$|$$
$$O–H$$

must be stored as "R connected to a carbon atom bonded to two oxygen atoms, one of the oxygens connected by a double bond and the other oxygen connected to the carbon and a hydrogen atom by single bonds". However, once the schema for the carboxylic acid group has been constructed, the information in the formula is simply encoded as "R connected to a carboxylic acid group".

Similarly, when we are working in a familiar problem area, much of the information that we use is likely to exist in chunks, and more *space* is available in working memory to deal with logical relationships and conditions that are unique to the problem at hand. When we work in an unfamiliar area, less information is chunked, and the demands on working memory are increased.

EXTERNAL MEMORY. External memory is a frequently overlooked but important means of overcoming limitations on working memory. The simplest external memory is some written record: a list, a diagram, a calculation, a graph, a table, or a sentence. For some problem-solving tasks, three-dimensional constructions (e.g., a molecular model of an enzyme or a scale model of a building) constitute an important kind of external memory. Molecular-modeling computer programs serve such functions in addition to aiding the visualization of spatial relationships.

INSTRUCTION ON WORKING MEMORY

I discuss working memory as a strategy for convincing students that several of my recommendations are worth following. Students are reluctant to practice procedures or rehearse important information enough to automatize the knowledge. Undoubtedly their reluctance grows out of a healthy skepticism about the real importance of what we tell them to remember. Too often their experience has been that the knowledge had no inherent value and that they suffered no real loss when they forgot the material after the examination. Still, not all of what we teach is useless, and much of the knowledge is used so frequently that it should be automatic.

I call attention to basic math skills, knowledge of metric units, and the various relationships involving the mole as examples of knowledge that will be used so frequently in the study of chemistry that it *must* become automatic.[4] I use an example from their experience (shoe tying when first learned compared to the same skill at age 18) to illustrate that a skill that required undivided attention when first learned requires virtually no attention once it becomes automatic. I then point out that most chemistry problems are sufficiently complex to overload the working memory of anybody, including the experts in the field. I insist that some automatization of frequently used information is absolutely essential in chemistry or any other complex field.

For similar reasons, I teach efficient algorithms for such routine tasks as balancing chemical equations (after I am convinced that the student knows what a balanced equation is and why we want one) and encourage students to use them. I emphasize the point that the algorithm should be sensible (i.e., we know what the product of the procedure means) but should not require them to think any more than necessary. Indeed, the purpose of an algorithm is to reduce the load on working memory and save time.

I also discuss working memory when I teach students the factor-label algorithm. I point out that it is not a foolproof procedure for getting right answers. However, I show the students that once you learn to *read* units, it can be a very efficient external memory for the logical steps taken to solve a problem. However, the importance of this external logical record is seldom seen until students recognize the importance of verification strategies in solving problems.

The relationship between limitations in working memory and the need for effective verification strategies is seen in the context of external records that we make in the laboratory or when we solve complex problems. We constantly admonish students to include units when recording quantities, but they seldom do. The problem, I believe, is twofold. First, we do not make the notion clear to students that limitations in working memory make it impossible for *any* person to remember all of the information that they will need to make sense of an investigation or solve other problems. When we record units and other labels along with numbers, these elements of the external record aid us when we attempt to recall what was previously done. The second problem is that the records that are meaningful and useful for experts may not be so for beginners.

When I record "Mass of beaker = 124 g", I know that I weighed the beaker and that it has the given mass. But many of my students do not understand mass, and they do not recognize "g" as the symbol for a unit that describes the mass of an

---

[4]Insisting that certain information should be overlearned so that its use is automatic is not an argument for rote drill. Most of what we use automatically was not learned by rote. True, practice is required, but practice in meaningful contexts is more effective than repetitive drill outside of any meaningful context.

object. Units of cubic centimeter tell me that I have measured a volume and recorded the amount of space that is occupied by an object. But many of my students do not recognize cubic centimeters as a description of volume, nor do they know that volume denotes the space occupied. When this lack of understanding occurs, the kind of external record that we admonish students to make does not serve the intended function. Because students are unable to use the information as intended, they attach little importance to producing the record.

Our limited working memory suggests why learning is facilitated by using familiar examples and why logical arguments that are transparent to an expert are hopelessly opaque to a novice. The expert (by definition) is dealing with knowledge that is largely automatized; the novice is not. The expert has used the logical operations inherent in the development many times before; the novice has not. The expert has much of the knowledge in the field chunked; the novice does not. The expert has a large repertoire of familiar strategies for recording information in external memory; the novice does not. (For example, a factor-label representation of a complex stoichiometric problem can serve as an efficient external memory for the logical steps that were taken in solving a problem, but it cannot serve that function for the novice who is just learning to use the factor-label algorithm.)

SUMMARY

Because problem solving is complex and problems take on myriad forms, there is no single way to teach problem solving. Essentially any activity that increases conceptual knowledge, encourages persistence, increases motivation, and helps students to see connections among ideas, to reflect on and check what was done, to consider alternative interpretations, and to try different strategies is likely to improve problem solving. Teaching step-by-step procedures probably will not.

This chapter describes a variety of activities that can be used to develop aspects of problem solving; perhaps the most potent activity is group problem solving that includes thoughtful argumentation and negotiation of meaning.

The chapter does not come close to covering all of the work in the field. Wickelgren (1974), Hayes (1981), Mettes, Pilot, and Roosink (1981), and Mettes, Pilot, Roosink, and Kramers-Pals (1981) approached problem solving from an information-processing frame of reference, and their recommendations are not always consistent with those given here. You may wish to refer to their work as you think about how you will teach problem solving.

Perhaps the best way to summarize this chapter is to say that a great deal of work has been done on the development of procedures to teach problem solving. That work is encouraging, but it is far too early to predict what the best procedures will be. What can be said is that deliberate attention to teaching problem solving is required if we expect to improve student performance. At the present time, nearly all chemistry teachers do little more than teach by example, and even our example is often inconsistent with what research indicates that knowledgeable people do when they confront novel tasks.

Although alluded to at several points, this chapter has not addressed the most important factor in improving problem-solving success by chemistry students. Repeatedly, research on problem solving has shown that students cannot solve problems because they do not understand the concepts mentioned in the problem (Carter, 1988; Gabel and Sherwood, 1980; Greenbowe, 1984; Nurrenbern, 1980; Nurrenbern, 1982; Robertson, 1990). *If students are having trouble solving problems, the first thing to check is their understanding of the concepts in the problem.* In making this check, care must be taken to be sure that concept understanding of the kind described in the following chapters has been obtained rather than the simple ability to mouth words.

9

Concept Learning

I started the section called How To Teach Chemistry with problem solving because problem solving is the ultimate goal of chemistry education. In addition, problem solving reminds us of just how complex learning can be. Even though we realize that the complexities of teaching and learning are important, we cannot deal with everything at once. If we try, we overload working memory in the process. Until we look more carefully at learning that is applied in the process of solving problems, we will have little impact on problem solving itself.

Chapter 8 ended with the reminder that the most common cause of failure in solving problems in chemistry is the lack of conceptual understanding. Why is this? How can concepts be presented so that students build in their heads the same schemas that we have in ours? How can we check to see whether this happens? These questions are addressed in this chapter and in Chapters 10 and 11. This chapter deals with the meaning of *concept*, differences in formal and informal concepts, and how concepts are learned. Chapter 10 develops a tool called *concept analysis* and shows how it can provide insights about difficult concepts. Chapter 11 considers difficulties in teaching concepts and provides suggestions for testing to see how concepts are understood.

One goal of this chapter is to lead you to construct in your head a schema labeled *concept* that is like the schema in my head. This kind of goal characterizes most chemical education.

FORMAL CONCEPTS

Cognitive scientists have developed an explicit definition of *concept*, just as physical scientists have developed an explicit definition of *work*. Both ideas originated in everyday experience, but they have been altered to formulate an internally consistent system of concepts and principles that make up cognitive science, on the one hand, and physical science on the other.

CONCEPT DEFINITIONS

Markle and Tiemann (1970) described a concept as "a class of entities". Merrill and Tennyson (1977) went further: "A concept is a set of specific objects, symbols, or events which are grouped together on the basis of shared characteristics and which can be referenced by a particular name or symbol" (p 3). Others give slightly different definitions, but all use *concept* to refer to a *set* or *class* rather than a single entity. Given this definition of concept, *man* is a concept because it names a class of entities, but *Dudley Herron* is not a concept because it labels a particular man. *Dudley Herron* lacks an important attribute or characteristic of concepts: It does not name a class, and *concept* is the name given to a *class* of entities.

But consider what was just said in another light: Admittedly, *Dudley Herron* can be viewed as an example of the concept *man*, but *Dudley Herron* is not the same as *man*. *Dudley Herron* labels an *idea* that some people have. That idea has characteristics that are not common to other persons on Earth. You might even say that *Dudley Herron* labels a set, but the set contains a single member. Reasoning in this manner, *Dudley Herron could* be called a concept.

This argument does not redefine the concept of *concept*, but it does reconsider the interpretation of one of its attributes: Specifically, may a class (set) have a single member or even no members at all (e.g., an ideal gas)? The point of this diversion is that even when we are very specific about the meaning of concepts, the meaning is subject to interpretation. People holding the same concept may make different classifications as a result of different interpretation of attributes. When concepts are taught, we must concern ourselves with both their meaning and the manner in which that meaning is interpreted.

CRITICAL AND VARIABLE ATTRIBUTES

Clarifying any concept involves identifying attributes that must be present for an entity to be a member of the class and identifying attributes that may vary across examples and nonexamples. Those attributes that must be present are called *critical attributes*, and those attributes that may vary are called *variable attributes*. The following are critical attributes of *concept*:

C1. A class or set is named.
C2. Each member of the set shares common characteristics.
C3. The class can be referenced by a name or symbol.

Man is a concept. Man names a class (C1) that includes Dudley Herron, Ronald Reagan, and John Adams, all of whom share common attributes (C2)—attributes that will remain hidden for the sake of propriety! Furthermore, all three men can be referenced by the name, *man* (C3). Thus, *man* has all the critical attributes of *concept*.

Just as certain critical attributes are shared by all instances of *concept*, other attributes vary from one *concept* to another. Some, but certainly not all, variable attributes of *concept* are the following:

V1. The class may name objects, symbols, or events.
V2. The number of shared characteristics does not matter.
V3. The reference to the class does not matter.
V4. Members of the class may be real or imagined.
V5. Each set can have two or more members.

Man, punctuation mark, and *disaster* are all concepts; however, the first is a class of objects, the second is a class of symbols, and the third is a class of events (V1). Men have many characteristics in common, but the only characteristic that seems to be shared by disasters such as floods, hurricanes, train wrecks, and stock market crashes is that damage results. Some concepts hold many attributes in common; others do not (V2).

The list of variable attributes for *concept* is incomplete. Normally one lists only those characteristics that might cause confusion. For example, V4 makes clear the fact that ideal gas may be a concept, even though it is only imagined. This fact is not clear from the definition or from the list of critical attributes. Similarly, V5 addresses the issue of whether *Dudley Herron* should be a concept. If we want to include ideas such as Dudley Herron, V5 must be rewritten to say, "The number of members in the set can be any number, including zero and one."

The meaning of any concept can be made explicit by carefully delineating its critical and variable attributes. However, until students apply the definition and attributes to distinguish examples from nonexamples, the concept will probably not be fully understood. As we often do in instruction, I will leave you in this partial state of understanding of *concept* and move to a discussion of how concepts are learned informally.

WHAT MAKES A CONCEPT DIFFICULT?

Most chemistry teachers find *concept*, as presented here, difficult to understand. Asking the question, "Why?" can be instructive. Discuss the issue in groups of three or four. If some members of the group understand better than others, try to determine why. (Do they, for example, have more background in psychology?) How many words were new or used in an unfamiliar way? Would different examples help? If so, what kind? How abstract is *concept*? How explicit is the meaning sought? Finally, do similarities exist between your learning of *concept* and your students' learning of categories such as element, compound, and chemical change?

NATURALLY LEARNED CONCEPTS

All students have a concept for *work* before they encounter work in a science class. However, important differences exist between the schema that the student calls *work* and the one that scientists call *work*. The student's concept developed naturally as a result of informal experience. As you will soon see, many concepts are of this kind. What we do to teach such concepts is alter the student's schema so that the concept label is associated with the explicit, formal concept that we have in mind.

What I am trying to do in this chapter is alter your concept of *concept*. Each of you has some schema that is accessed when you hear *concept*. Unless you have had a formal course in concept learning, your concept of *concept* developed as a result of informal experience. You probably use *concept* in a global sense, as another name for *idea*. This understanding is the commonly held one for *concept*, just as "expending effort" is the common understanding of *work*. By contrast, the understanding of *concept* presented in this chapter is formal. It grows out of research in cognitive science.

FORMATION OF NATURAL CONCEPTS

Most concepts are learned informally. We were not formally taught what a chair is and what it is not, the characteristics that distinguish a cat from a dog, or the difference between a mountain and a hill. We simply developed the schemas for these concepts through experience and interaction with others. These concepts changed (and continue to change) over time as schemas accommodate new information.

Saying that most concepts are learned informally does not imply that we learn them without instruction or mediated learning. The child who says *dog* upon viewing a cat or a squirrel is likely to be laughed at, corrected, or given other information that affects the way the schema for *dog* is ultimately constructed. These interactions may even include transcendent messages that characterize mediated learning and affect the way other schemas will develop: "Listen to the sounds animals make. Those are important clues to identification," or "Watch what they do. Squirrels often climb trees, cats occasionally climb trees, but dogs never climb trees."

With or without mediation, we learn by obtaining information through our senses and processing it under the influence of existing schemas. We appear to be born with (or develop soon after birth) schemas that guide perception through a process of feature analysis. We recognize things that have particular features and place them into categories that were previously formed. These categories of objects or events with common features are what we call concepts.

Concept formation is a pervasive feature of perception. We are bombarded by stimuli. We ignore the majority; the rest we organize to bring meaning to experience. We organize the stimuli into classes with common features; we form concepts.

Even though the evidence is strong that we perceive by feature analysis and then classify stimuli into categories of common features, assuming that this process is a conscious one would be a mistake. Although you had a schema for *concept* before you began this chapter, you probably would have had trouble defining it, and, had you been asked to list the critical and variable attributes of concept, you would have found this task difficult to do. This obfuscation is characteristic of naturally learned concepts. We believe that "I would know one if I saw one", but we cannot explain *how* we would know.

PROTOTYPES

Prototypes play an important role in naturally learned concepts. What we mean by a prototype is illustrated by the concept *bird*:

"Is a robin a bird?"
"Certainly!"
"A chicken?"
"Well, I *suppose* so."
"A penguin?"
"Well, yes, but—"

Most science teachers have a *formal* concept of birds as well as a *natural* concept formed through experience. Consequently, you may feel equally comfortable with all three examples. However, research on naturally learned concepts reveals that most adults are far more comfortable calling a robin a bird than calling a penguin a bird.

Many features characterize birds: Birds have feathers and wings; they fly, chirp, and eat worms. Robins do all of these things, but chickens and penguins do only some of them. Some birds are better birds than others! Our "best of all possible birds" is our *prototype* for the concept bird. For naturally learned concepts, we appear to hold information in memory in terms of such prototypes (Rosch, 1977).

Science teachers often face conflicts between naturally learned concepts and formal concepts that have a more explicit meaning, such as *work* in physics, *burning* or *combustion* in chemistry, or *plant* in biology. During instruction, the naturally learned concepts are elaborated or modified so that students will classify according to the specific attribute rules that characterize formal concepts.

When classification is done on the basis of naturally learned concepts, the stimulus is compared with prototypes held in memory. If a "strong family resemblance" occurs between the new stimulus and a prototype, the new stimulus is included in that category, and the prototype may change slightly if such accommodation is required. However, the rules of classification are seldom explicit.

When we categorize on the basis of naturally learned concepts, we have little difficulty with stimuli that closely resemble a prototype, but we have considerable difficulty with stimuli that do not resemble a prototype or that resemble two or more prototypes equally well. Without decision rules based on the critical and variable attributes that characterize formally taught concepts, classification errors are common. In everyday experience, these errors may be unimportant, but they are often critical in science. Students must understand that poorly differentiated concepts may be quite adequate for everyday purposes but not for the precise work that characterizes any science.

CLOSE IN AND FAR OUT EXAMPLES

Figure 9.1 represents information about birds that might exist in someone's memory. For this person, robin would probably be the first word recalled when told, "Name a bird". It is the prototypic example of bird for this person. However, sparrow, dove, and canary might be named quickly because they also have many features of the prototype. They are *close in* examples. If pushed, hummingbird, chicken, wild turkey, or ostrich might be named, but they are unlike the prototype in several ways. They are *far out* examples. They are birds, but they are not good birds. If this person is told about an extinct animal called a dodo and is asked whether a dodo was a bird, the person may have trouble deciding. How many features of the prototype, robin, must something have before it is called a bird?

The problem is complicated further if two closely related concepts can be chosen. Figure 9.2 illustrates the problem in a generalized form, where circles represent

Figure 9.1. Schematic representation of the prototype, *robin*, in relation to other possible examples of the concept, *bird*.

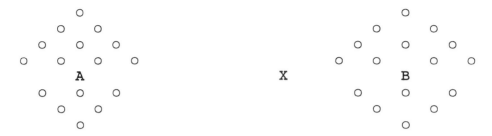

Figure 9.2. Generalized representation of a cognitive space consisting of all possible examples of two closely related concepts.

the location of examples of a concept relative to the prototypic examples of the concepts represented by the letters, A and B. X represents some stimulus with features of both concepts.

Because the new stimulus, X, has features of A as well as B, classification is uncertain. If some resolution is forced, the new stimulus is likely to be incorporated as an example of the concept identified by the prototype it most closely resembles, B. If told that X is an example of B, the prototype for the concept may change as a result of new features associated with X.

MISCONCEPTIONS[1]

The condition represented in Figure 9.2, where several concepts are represented in memory and one concept fades into another, is easy to imagine. Stimuli at the periphery of one concept may intersect the periphery of another and lead to errors of classification or misconceptions.

A phenomenon such as this may explain many of the findings concerning misconceptions in science. Pfundt's work with early adolescents illustrates the kind of misconception commonly seen by teachers (Pfundt, 1982).

Pfundt had students heat copper sulfate pentahydrate, reproduce the pentahydrate from the anhydrous product, burn alcohol, evaporate water, and reduce lead(IV) oxide to lead. She then conducted clinical interviews in which she asked students to describe what happened.

"Of the 10 pupils interviewed, most [discussed] the burning of alcohol not only in connection with familiar burning processes but with the evaporation of water" (Pfundt, 1982, p 11). Why might students mistake this example of *burning* for an example of *evaporation*?

Conjure up your best image of burning and describe what you see in your mind's eye. (Really do it! I'll be here when you finish!)

---

[1]Many people prefer *alternative conception, naive conception,* or some other term in place of misconception. They argue that conceptions that are at variance with commonly accepted scientific conceptions are arrived at logically and make just as much sense in light of information held by children and naive adults as do the conceptions accepted by the scientific community. To call these misconceptions has a negative connotation that should be avoided. Such arguments have merit, but they ignore the question of why misconception has a negative connotation for some. Until that question is addressed, it is a matter of time before other labels are tainted. Learning is impossible in the absence of discrepancies between current understanding and the understanding of others. Misconception is a necessary part of intellectual growth. The term should carry no negative connotations.

I see a bright yellow-orange flame, black and white smoke, and a solid being consumed and replaced by black and white ash. I feel heat, and I hear an occasional crackling sound. I smell a distinctive odor. These images are my prototype of burning.

Now think about evaporation. What is your prototype of that process?

I see water in a shallow pool on hot concrete by a swimming pool. The sun is strong, and I can see the water gradually disappear, first at the edges, and then throughout the pool of water. It disappears completely and without a sound.

Now place alcohol in a shallow dish and light it. (If you have done it before, you may do this in your mind; if you have not, you really should do the observation as you consider the question raised here.) Compare what you see with your prototype for burning and your prototype for evaporation. It is not exactly like either one. Which does it appear closest to? How many of the cues of sight, sound, and smell are common to your burning schema; how many are common to your evaporation schema? Should we be surprised to find that students who have not formally studied burning and evaporation and used the critical attributes of those concepts to classify events may relate the burning of alcohol to evaporation?

Many misconceptions encountered in science are no more than a reflection of poorly differentiated concepts that have developed naturally rather than through carefully planned instruction. Research on such misconceptions is important because it pinpoints confusion, but it should not surprise us. Because many—perhaps most—of the concepts that we use in science developed informally, one of our tasks is always to identify the nature of the concepts already held by students and to provide instruction that shapes these natural concepts into the formalized schemas required for science.

BASIC CATEGORIES

Clarification of naturally learned concepts is not the only task that we face. Concepts fall into categories at various levels of abstraction or generality (Figure 9.3). When we see an object and classify it, we do so at a particular level of abstraction or generality that represents our *basic level*.

What do you see in Figure 9.4? You could say that you see *clothing* and *furniture*, but you probably did not. You could say that you see a *dress shirt* and a *coffee* (or cocktail) *table*, but you probably did not. Most likely, you see a *shirt* and a *table*. For most adults, shirt and table represent the basic level for the objects shown.

Shirt is the supraordinate category for dress shirt, T-shirt, knit shirt, and other kinds of shirts. Shirt is also a subordinate exemplar of the supraordinate category *clothing*. (Depending on the focus, it can also be subordinate to *sewing*, *manufactured product*, and other categories.)

Research by Rosch and her associates (Rosch, 1977; Rosch et al., 1976) indicates that we store information at a particular level of abstraction, which Rosch (1977) refers to as the "basic level". The way information is stored is not arbitrary because "real-world activities . . . do not occur independently of one another. Creatures with feathers are more likely also to have wings than creatures with fur, and objects with the visual appearance of chairs are more likely to have functional [sit-on-ableness] than objects with the appearance of cats" (p 213).

Basic-level concepts can change as a function of how knowledge is used. Iron and rust are likely to represent basic level concepts to the average person dealing with matter in everyday life. However, as the chemist considers matter, element and

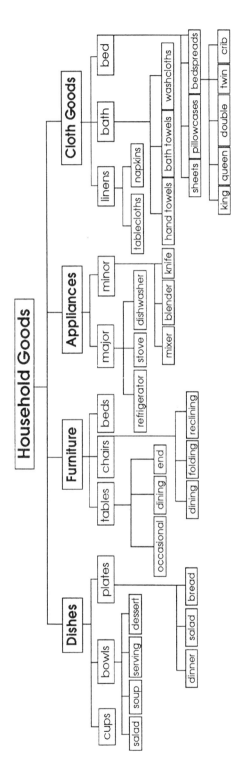

Figure 9.3. This partially developed hierarchy shows increasingly more specific categories of things found in typical homes.

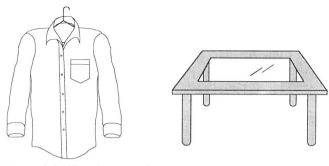

Figure 9.4. What do you see?

compound are far more basic than iron and iron oxide. *In teaching science, we teach new concepts and shift students' basic-level concepts from one level of abstraction to another.*

The idea that we are changing students' basic level of cognitive attention to a higher level of abstraction is consistent with differences in the way novices and experts represent problems in science (Larkin, 1981; Chi, Feltovich, and Glaser, 1981). Novices tend to represent problems involving toboggans rolling down a hill in terms of the surface (real-world) elements such as *toboggans* and *hills*. Experts, on the other hand, represent the problem in terms of concepts such as *energy* and *forces*.

In practice, the basic-level concepts for experts are at a higher level of abstraction. These concepts became the basic level of representation because experts learned that the problems encountered in their world can be solved with less effort by using more abstract concepts. The novice has yet to learn this lesson.

In teaching concepts in science, we not only need to teach what the concepts are, but we need to show students that thinking about events at one level of abstraction may reveal relationships that are obscured when those events are considered at some other level. We need to encourage students to change their basic level of conceptual thought.

NATURALLY LEARNED CONCEPTS: A SUMMARY

We might summarize what is known about the development of natural categories or natural concepts in this way: Because we are faced with the multitude of stimuli coming from the environment, we are unable to incorporate all of the information into schemas. We attend to those stimuli that seem most important, and we impose meaning on what we perceive by organizing information into categories. This organization is far from arbitrary because the world in which we live is not arbitrary. Certain features tend to go together. As Rosch pointed out, "It is an empirical fact 'out there' that wings co-occur with feathers more than with fur" (1977, p 222).

The principle of minimum cognitive effort appears to operate as we form natural concepts. We code information in forms that allow the greatest use with the least cognitive effort. Two important results occur. First, we store information about categories in terms of prototypes that incorporate the salient features with which we are familiar. Second, the prototypes selected are compromises between a need for maximum information about the world and a need to conserve cognitive space by storing few prototypes.

RETURN TO SCHEMAS

Perhaps this is a time to reemphasize the idea that schemas consist of more than words or visual images. Various kinds of content exist in schemas. My prototype for burning includes visual images, odors, the feeling of warmth, crackling sounds, and probably other kinesthetic features of which I am unaware. Rosch's work included investigations of motor movements associated with concepts such as chair and claimed that "humans have consistent motor programs with which virtually all humans interact with virtually all objects of a certain class" (Rosch, 1977, p 216).

We have a strong tendency, particularly in secondary and tertiary schools, to think of learning exclusively in terms of verbal knowledge. This assumption is a mistake. Strong evidence states that nonverbal information (including emotional reaction) is an important component of schemas we construct.

SCHEMAS CONSTRUCTED IN THE LABORATORY

The implications of what has just been said are particularly important in regard to the laboratory. Science educators have insisted that laboratory experiences are an irreplaceable component of science instruction, and yet studies seldom reveal the worth that we are certain exists (Hofstein and Lunetta, 1982; Blosser, 1980). These studies typically compare test scores for students who have had laboratory experiences with the scores of students who have not. The tests that are given generally focus on verbal knowledge such as definitions or the solution of *textbook* problems.

The worth of laboratory work probably derives from opportunities provided for *nonverbal* components of schemas to develop. I am not sure that we know how to measure such learning, and I am confident that we have not designed laboratory experience to maximize such learning. However, in view of Rosch's research suggesting that categories are maintained as prototypes of the most characteristic members of the category, concrete experiences such as those gained in a laboratory seem to be important. Much of the knowledge incorporated into schemas may be kinesthetic, and that knowledge cannot be obtained in the absence of kinesthetic experience.

HOW SCHEMAS CHANGE

Now that we have seen how schemas containing conceptual knowledge develop, let us consider how they are altered through learning. Rumelhart (1980) described three kinds of learning, *accretion*, *tuning*, and *restructuring*. The differences among these three kinds of learning are seen when they are related to assimilation and accommodation (*see* Chapter 5).

Schemas are constructed through the complementary processes of assimilation and accommodation. Assimilation is the taking in of information through the senses. Information can be assimilated only if a schema already exists that can accommodate that information. If the information can be accommodated by some existing schema and is assimilated, the incorporation of that information will lead to a modification of the existing schema so that it has a more elaborate or an entirely new form. In effect, the schema accommodates to the information that is assimilated, just as the new information must accommodate to the existing schema to be assimilated in the first place.

The differences among learning by accretion, tuning, or restructuring are related to the degree to which an existing schema must be modified to accommodate new information.

ACCRETION. *Accretion* involves assimilation of factual information as presented accompanied by little or no modification of the existing schema. Acquisition of factual information such as "the weather is warm today", "he is 46 years old", or "Israel became an independent state in 1948" is unlikely to result in any major change in the schemas to which this information is assimilated. Rumelhart suggested that no change in schemas is involved in such learning. Much of the learning of adolescents and adults occurs in this manner.

TUNING. *Tuning* involves minor changes in schemas as concepts are clarified. The clarification may be of several types. Each time we encounter a new instance of a concept, we have opportunities to *fine tune* our schema for that concept. Each time we relate the concept to some other concept so that the relationship between the two is clear, we add *interiority*—new information about the concept—to our concept schema. As we use the concept, we sometimes discover that classification would be easier if we relaxed a previously held constraint or imposed additional ones. The historical development of the concept *base* affords an example.

As the concepts of acid and base developed, bases were first limited to substances that produce hydroxide ions, but later the notion that the concept is more useful if it is expanded to include other proton acceptors became obvious. Still later, extending the concept further to include any substance that acts as an electron-pair donor became useful.

Such learning appears to involve relatively minor change in existing schemas, it is relatively easy to accomplish, and little conflict exists between the schema that results from such tuning and the schema that existed before.

RESTRUCTURING THROUGH PATTERNED GENERATION. According to Rumelhart, *restructuring* involves the creation of new schemas, whereas tuning represents minor adjustments in existing schemas to accommodate new conditions. Two forms of restructuring exist: patterned generation and schema induction.

In *patterned generation*, a well-established schema is copied with a few modifications. For example, a person who already understands molarity can develop a separate schema for molality by copying the molarity schema with a minor modification: "mass of solvent" is substituted for "volume of solution" as the quantity to which moles of solute are compared.

A former student of mine likened restructuring to the way patterns are changed in a kaleidoscope. In his analogy, patterned generation would be like replacing one of the colored chips that form the pattern in the kaleidoscope with a chip of a different color or shape. The new pattern seen in the kaleidoscope would be similar to the old one, but it would have new features produced by the new chip.

Whether the development of new concepts of acid and base represents tuning or restructuring through patterned generation depends on whether one maintains a single concept of acid that changes over time (tuning) or one maintains three distinct acid concepts (Arrhenius, Brönsted, and Lewis) that are related but are distinct (restructuring through patterned generation). The label placed on this kind of learning is not important; understanding that new conceptual insights are gained through minor, but unambiguous, modifications in existing schemas *is* important.

RESTRUCTURING THROUGH SCHEMA INDUCTION. The other restructuring process by which new schemas are acquired is called *schema induction*. Schema induction takes place when a pattern of events is recognized to occur repeatedly, and no existing schema can accommodate that pattern. Furthermore, the pattern that is recognized is sufficiently different from any existing schema that the new schema cannot be formed through patterned generation.

Using the kaleidoscope analogy, schema induction is like rotating or shaking the kaleidoscope to produce a pattern that is totally unlike the one previously seen. The elements (colored chips) that cause the pattern may be the same ones present in other patterns, but those elements are arranged so differently that the final result is very different.

LEARNING DIFFICULTY AS A FUNCTION OF SCHEMA MODIFICATION

Figure 9.5 places the learning processes that we have just described along a continuum showing the degree to which existing schemas must change for the new information to be assimilated.

RELATIONSHIP OF SCHEMA CHANGE TO PRINCIPLE OF LEAST EFFORT

The continuum in Figure 9.5 is a continuum of effort required for learning to occur. All learning is influenced by the principle of least cognitive effort. Our basic-level concepts are those that provide the most information at the least intellectual cost, and we are understandably reluctant to abandon useful ideas.

Learning formal concepts such as *atom* and *molecule* involves schema induction. Consequently, the principle of least effort opposes the construction of such concepts. In the absence of some indication that an adequate return will occur on the effort invested, students hold on to alternative frameworks that are perfectly satisfactory for explaining their real-world discoveries: Heat destroys water; electricity dries up water; and thunder makes water.

When confronted with the task of learning any new idea (atomic theory, for example), *not* learning the idea requires less effort. The principle of least cognitive effort predicts that a person will choose to not learn if the tasks commonly encountered can be handled with the schemas already in place or if no value is seen in the tasks that the new schema prepares one to perform. Conversely, the principle predicts that a person *will* learn a new idea if the new schema prepares one to handle valuable tasks that cannot be done now, or if the new schema enables one to handle current tasks with less effort.

Motivating students is a major concern of all teachers, and motivation is largely a matter of convincing students that the effort required for the learning at hand will pay off by reducing the effort required to perform important tasks that they face.

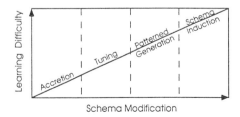

Figure 9.5. Learning becomes more difficult as schemas are increasingly modified to accommodate new information.

Most experienced teachers realize that students are more motivated to learn when instruction is related to the world of the student. One reason is quite simple. When instruction is related to the existing concerns of the student, the student can perceive the potential value of learning. Only then can the student see that learning a new idea or developing a new procedure has the potential for reducing cognitive effort.

Motivating students to learn theoretical concepts is particularly difficult. It is made more difficult when those concepts are presented before the student has any need for them. Until chemistry students encounter some facts about matter that cannot be explained without the idea that matter is made up of discrete particles, they have no need for *atoms*. Convincing students that atomic theory is worthwhile is a difficult task. However, when students try to explain chemical events without the idea of atoms and realize how difficult this explanation can be, they welcome the concept of atoms and its explanatory power because of the principle of least cognitive effort.

Theories and formal concepts are best taught when experiences faced by the student make clear that some new idea is needed to make sense of those experiences. When those theories and concepts require the induction of totally new schemas, they are often learned more effectively in stages rather than all at once. For example, if atomic theory is introduced by inducing a simple schema that views atoms as particles that hook together, that schema can be used to explain a great deal of chemistry, consolidating the schema in the process. As experience is gained, this schema can be modified gradually through accretion, tuning, or patterned generation, processes that require far less cognitive effort than the original schema induction.

Keep in mind that development of a particular concept may be favorable for one student and not for another. The potential use of the concept varies with the student's educational goals. In addition, the effort required to learn a useful concept is not the same for all students. Some students have previously developed intellectual tools that enable them to learn particular concepts (abstract concepts such as *atom*, for example) with relatively little effort; students who lack those intellectual tools learn the same concept only with major effort, if at all.

THE IMPORTANCE OF INSTRUCTION

The processes of accretion, tuning, and restructuring can occur naturally, as a result of informal experience, or they can be precipitated by instruction. However, many of the most useful science concepts are not perceptible from interactions with the environment in the absence of mediating experience. For example, the concepts of *atom* and *molecule* enable us to make sense of the world with less effort than is required without these concepts. The principle of least cognitive effort argues that we are better off with such concepts. However, atoms and molecules are not apparent from direct observation. Consequently, the concepts are not likely to develop naturally during any single lifetime, and the cognitive effort required to develop such concepts through mediated learning is likely to be greater than the effort required to use less powerful concepts that are acquired naturally.

The importance of mediated learning should be clear. Much of what we are and what we are able to do is the result of schemas developed over several generations and reconstructed in succeeding generations in the form of cultural traditions. Such information is unlikely to be discovered without mediated learning of some kind.

Much of what we value is not discoverable. Under the right conditions, students can discover that water open to the atmosphere gradually disappears, that passing an electric current through water in a closed tube causes the water to disappear and large volumes of gas to appear in its place, or that an electric spark in this resultant gas (more correctly, mixture of gases) produces a loud noise and a small amount of water. However, the student cannot discover that water is made up of molecules formed from two hydrogen atoms and one oxygen atom. This mental model to explain the observations concerning water (and thousands of other observations) is *invented*. Such models have been constructed over several generations as powerful schemas that enable us to fit our own discoveries into a self-consistent organization of knowledge. Without some form of mediated learning, we would not arrive at these models. (*See* pp 56–58.)

SUMMARY

Knowledge of the way natural categories develop is important to science teachers. Many (perhaps most) concepts are constructed in this way. However, we cannot expect all science concepts to evolve naturally. For one thing, many concepts that we teach are abstract and have few features that are easily derived from direct experience. Second, we frequently need to make exact classifications that differ appreciably from the prototype for the concept. Consequently, one must be consciously aware of the features used to classify. Thirdly, many uses for science concepts are at a higher level of abstraction than the basic level categories that develop naturally. Teachers need to think about the concept of *concept* in a formal way for the same reasons. If they do not, they may be unaware of what must be done to teach concepts to students.

Chapter 10 introduces a procedure for formal analysis of concepts. Like balancing equations or performing equilibrium calculations, this procedure requires effort, but the payoff in understanding is substantial. This chapter focused on how students develop understanding and misunderstanding of science concepts. The next chapter gives reasons that misunderstanding so easily takes place, and Chapter 11 suggests what can be done about it.

EXAM ON CONCEPT

In preparation for a discussion on testing at the end of Chapter 11, answer the following questions about the concept of *concept* without going back to review any of the material in this chapter. Keep your responses for the discussion at the end of Chapter 11.

1. Write a definition for *concept*.
2. Give five examples for *concept*. Do not use any of the examples discussed in this chapter.
3. Name five ideas that do not represent concepts. Do not use any of the examples discussed in this chapter.
4. Circle each of the following that names a concept.

a. animal	f. Chicago	k. Hydrocarbons burn.
b. man	g. the contents of my office	l. hydrocarbon
c. Dudley Herron	h. love	m. mass
d. $2X + 4 = 10$	i. equilibrium	n. oxidation
e. city	j. I love my wife.	o. Na

5. For each item that you **did not circle** in question 4, state why you did not consider it to be a concept.
6. For each item in question 4 that **you did circle** but **are not confident** that you should have, state why you are uncertain.

10

Analyzing Concepts To Clarify
Their Meaning

The previous chapter described *concept* and how concepts are learned and mis-learned. In this chapter, we will scrutinize chemistry concepts to show how they differ and how those differences affect learning. This careful scrutiny requires a tool, and the tool I have selected is appropriately called *concept analysis*.

CONCEPT ANALYSIS

ELEMENTS

I illustrate the elements that are commonly included in a concept analysis by using an artificial concept invented by Smoke (1932) and used in some of our research several years ago (Herron et al., 1976). *Mib* is the name Smoke gave to the concept selected for analysis. The complete concept analysis for mib is presented as Appendix A.

THE CONCEPT NAME. A concept analysis begins by naming the concept under analysis, in this case, *mib*.

A DEFINITION. The analysis usually continues by defining the concept. The definition of the concept names the critical attributes and states the relationship among them. The definition of *mib* is as follows: *A mib is a plane geometric figure made by attaching a segment to the short leg of a right triangle so that it is an external, perpendicular bisector of the short leg.*

The definition that I have given for *mib* is not the first one written. Seldom is the first definition for a concept the one finally adopted. As concept analysis proceeds, it becomes obvious that the definition is faulty and must be revised. The definition for *mib* that is written above was derived by analyzing examples and nonexamples presented by Smoke (1932), using the resulting definition for instruction and finding it faulty, and eventually doing a concept analysis. Some of the difficulties in defining *mib* are discussed in Herron et al. (1976).

CRITICAL (DEFINING) ATTRIBUTES. Critical attributes are those features that must be present for an instance to be considered an example of the concept. In the case of *mib* we said that

C1 the figure must contain a right triangle
C2 a segment must be attached to the short leg
C3 this segment must be external
C4 this segment must be perpendicular to the short leg
C5 this segment must bisect the short leg

An adequate definition of a concept always includes all critical attributes.

VARIABLE ATTRIBUTES. Variable attributes are those features that really do not matter; they may be present or absent in examples or nonexamples. From the instances for *mib* presented in Smoke's original paper, one can infer that the size of the figure, its orientation, the length of the segment, and whether additional segments exist are variables that have no bearing on whether a figure is or is not a mib.

Any concept has an infinite number of variable attributes, and only those attributes that might cause confusion are usually mentioned. What to include in a list of variable attributes depends on the purpose. In preparing a lesson for elementary schoolchildren, you might include items that would be omitted in a lesson for adults.

SUPRAORDINATE, COORDINATE, AND SUBORDINATE CONCEPTS.Virtually all concepts are related to other concepts in a hierarchical fashion, and an understanding of the concept often involves knowing the supraordinate, coordinate, and subordinate concepts to which it is related.

Supraordinate concepts are at a higher level of generality than the concept under analysis. Plane figure and geometric figure are supraordinate concepts for *mib*. The target concept is always an example of a supraordinate concept; *mib*, the target concept, is an example of plane figure, one of the supraordinate concepts in the analysis.

Concepts at the same level of generality are *coordinate* concepts. Because *mib, square, rectangle*, and *right triangle* are all examples of the supraordinate concepts *plane figure* and *geometric figure*, they are at the same level of generality. *Mib, square, rectangle*, and *right triangle* are coordinate concepts.

Subordinate concepts are at a lower level of generality than the target concept. Because *mib* is an artificial concept, no subordinate concepts exist, but we could invent some by subdividing mibs into classes. We could, for example, separate mibs whose shortest leg is less than a given length and call those *small mibs*. Similarly, we could identify mibs with more than one segment as *complex mibs*, and so on.

Concept analyses could never include exhaustive lists of supraordinate, coordinate, and subordinate concepts, and no attempt is made to do so. However, a few examples are commonly included to place the target concept within a larger context. This practice is particularly true when concept analysis is done as part of instructional planning.

EXAMPLES. A concept analysis further clarifies the target concept by listing examples. Examples in a concept analysis should be as divergent as possible. Examples set the boundaries for the class under discussion. If all examples are similar, the boundaries are not clear. Usually, at least one example is given for every variable attribute that might cause confusion. Examples of *mib* are shown in Appendix A. The variable attributes clarified by each example are identified. In some cases, the collection of examples rather than a specific example demonstrates that an attribute varies across examples and nonexamples.

NONEXAMPLES. Examples of a concept are selected to clarify the *variable* attributes of a concept. Variations in the list of examples make clear that those variations are not critical in determining whether an instance is an example of the concept. Attributes that *are* critical are clarified through nonexamples. Nonexamples lack one or more of the critical attributes of the concept. At least one nonexample must be given for each critical attribute to provide the needed clarity. Nonexamples are often matched with examples so that the nonexample is as near in likeness to the example as possible but is still lacking some attribute that is critical.

ADDITIONAL FEATURES

Some concept analyses include additional features that make the analysis more useful, but the features mentioned here are usually included and provide the essential elements for understanding a concept. Understanding a concept always implies

- an ability to classify instances as examples or nonexamples,
- knowledge of the critical attributes that must be present for an instance to be an example, and
- knowledge of attributes that may vary across examples.

Although the following conditions are not always required as evidence of understanding, they are common.

- Name the concept and define it. (This condition is probably necessary for communication about the concept, but it may not be necessary for understanding as described by the previous three examples.)
- Place the concept in relation to supraordinate, coordinate, and subordinate concepts.

CONCEPT ANALYSIS OF *CONCEPT*

Before I ask you to do a concept analysis of your own, consider one more concept analysis: my analysis of the concept *concept*. An examination of it may clarify how *concept* is used in this book. The analysis is shown as Appendix B, to which you will need to refer as you read the following discussion.

DEFINITION

This definition for *concept* is taken from Merrill and Tennyson (1977): A *concept* is a set of specific objects, symbols, or events that are grouped together on the basis of shared characteristics and that can be referenced by a particular name or symbol.

If you accept this definition for *concept*, what ideas that you previously called concepts would be excluded? What ideas that you previously called concepts would be included?

CRITICAL ATTRIBUTES

Critical attributes of *concept* are that

- it represents a class or set,
- members of the set share common characteristics, and
- the set is referenced by a particular label.

Using these attributes, is *Statue of Liberty* a concept? Why?

VARIABLE ATTRIBUTES

Some variable attributes of concept are as follows:

- The set can be almost anything: objects, symbols, events, attributes, and others.
- The reference label may be a name, a symbol, or any other kind of label.
- The set can have two or more members.

Given these variable attributes, can abstract concepts such as *psychopath* be called concepts?

SUPRAORDINATE, COORDINATE, AND SUBORDINATE CONCEPTS

Concepts are a kind of knowledge. Therefore, *knowledge* is a concept that is *supraordinate* to concept. Conversely, concept is *subordinate* to knowledge.

Other concepts that are examples of knowledge are *principle*, *skill*, and *rule*. Because principle, skill, rule, and concept are all examples of knowledge, they are at the same level of generality. They are *coordinate* concepts.

Much of this chapter is devoted to various kinds of concepts and the learning difficulties they precipitate. The names listed as subordinate concepts of *concept* represent a few of the categories described later in the chapter.

EXAMPLES

Examples of concept that I included in the analysis are:

1. sentence
2. man
3. hydrocarbon
4. furniture
5. mass
6. love
7. pH

These examples are not actual ones but *names* for examples. This distinction may seem trivial now, but when difficulties associated with concept learning are discussed later, the importance of this distinction will be apparent.

Concepts exist in people's heads. Presumably each of us has in his or her head some schema that is referenced by *sentence*. *The schema is the concept*; what we have listed as *sentence* is the *label* for that schema. Students (and apparently, teachers) easily confuse the two, and an important aspect of teaching concepts is designing instruction to minimize such confusion.

NONEXAMPLES

The following nonexamples of concept were selected to contrast with the examples having the same number in the list above.

1. $2X + 4 = 10$
2. Dudley Herron
3. Hydrocarbons burn.
4. the contents of my office
5. 72 kg

6. I love my wife.
7. *

The rationale for the examples and nonexamples is contained in the section called Comments at the end of Appendix B.

PRACTICE EXERCISES
GROUP CONCEPT ANALYSIS

Before continuing with this chapter, you should do at least one concept analysis. This exercise can be done independently, but it is more instructive if it is done with group interaction.

If possible, form a group of three to six people and agree on a chemistry concept to analyze. Because the nature of the analysis will depend on the level at which the concept is taught, agree on that as well. For example, you may agree to analyze the concept of acid as it might be taught in an introductory college chemistry course. Do the analysis individually. After you are satisfied with your individual analyses, come together as a group to discuss your respective products.

GROUP DISCUSSION

No right analysis exists for any concept, but you should expect disagreements about whether a particular attribute is critical or whether a given instance is an example of the target concept. *The purpose of this exercise is to reveal how much knowledgeable, intelligent adults may disagree about concepts that they have used for years!*

Your discussion should focus on resolving as many differences as possible, but be adult about it. Who has it right and who has it wrong is beside the point. Rather, the point is what the differences are, their origin, and how they affect what you teach.

Differences are likely to fall into two categories. First, you will disagree about the precise meaning of the concept—whether A is or is not a variable attribute, whether B is or is not an example, and so forth. Second, you will disagree about whether the concept, as you describe it, is appropriate for a particular group of students. Try to resolve all disagreements of the first kind before you attempt to resolve those of the second kind.

INDIVIDUAL ANALYSES

After you are satisfied with this first analysis, do several others. The second time around, each group member may analyze different concepts. Share your analyses in writing, and then come together to discuss them. After you have discussed several analyses, answer the following questions:

1. Did you disagree about the exact meaning of some concepts? If so, why?
2. Were some concepts harder to analyze than others? If so, why?
3. Did you discover characteristics of concepts that might make them difficult to teach? If so, what were those characteristics?

WHAT CONCEPT ANALYSIS CAN REVEAL

If you did the concept analyses as recommended in the previous section and shared them with your peers, several points should be clear.

1. Much of what we teach in chemistry falls in the category of concept learning.
2. Knowledgeable, intelligent chemists differ in their understanding of the concepts taught.
3. We have difficulty separating concepts from other important ideas that are related to them.
4. The kind of clear, explicit description of a concept that is called for in a concept analysis is difficult to construct.
5. A concept can be *mis*understood in several ways.

Concept analysis can shed light on the nature of concepts we wish to teach, suggest reasons the concepts may be difficult, and provide clues to successful instructional strategies. The rest of this chapter deals with a classification of chemistry concepts that is based on insights derived from concept analysis. Several of the analyses that led to the classification are given in Appendixes C through K at the end of the book. I believe that these analyses demonstrate the power of concept analysis to understand concepts and detect potential difficulties in learning them. However, the power is in the *doing* and not in the final product. Your own analyzing of concepts that you have difficulty teaching will benefit your instruction far more than reading about the following analyses conducted by me and my students.

CLASSES OF SCIENCE CONCEPTS

Analysis of science concepts reveals that they differ in many ways. Some concepts are more difficult than others, and the difficulty of the concept is often related to whether it can be learned by accretion, tuning, or some kind of restructuring. In addition, when restructuring through patterned generation or schema induction is required to learn the concept, the difficulty of learning the concept is often determined by whether the pattern of events that define the conceptual category can be derived directly from sensory perception or only indirectly through complex elaboration of perception.

I have developed a classification of concepts that is based on characteristics that influence the difficulty students have learning the concept. These categories are listed in Table 10.1.

CONCRETE CONCEPTS

Concepts that represent classes that have numerous perceptible instances and attributes that are easily perceived (what Gagné (1977) called concrete concepts) present few learning difficulties, even when learning them involves schema induction. Examples of such concepts are bird, plant, beaker, man, and Arrhenius acid. Concrete concepts normally develop naturally as described in Chapter 9. Natural concept development may result in misclassification of instances that differ markedly from our prototype, but these difficulties can be corrected by instruction focusing on the critical and variable attributes of the concept (i.e., by tuning).

Although research is still underway to determine the most efficient procedures for teaching concepts (Dunn, 1983; Finley and Stewart, 1982; Gabel, 1993; Hakerem, Dobrynina, and Shore, 1993; Schonemann, 1983; Tessmer, Wilson, and Driscoll, 1990; Williamson and Abraham, 1993), several effective procedures for

Table 10.1 Classification of Concepts

Concept Class	Characteristics	Examples
Concrete concepts	perceptible examples, perceptible attributes	bird plant beaker man Arrhenius acid
Invisible examples	no perceptible instances	atom molecule universe God angstrom photon
Invisible attributes	critical attributes are not perceptible	element compound
Dual concepts	learned at two levels	element compound
Principle concepts	classification requires knowledge of principles used to test instances	mole mixture
Attributes or properties	name attributes or properties	mass weight charge frequency oxidation number
Symbolic concepts	names form of symbolic expression; symbolic form governed by explicit rules	sentence formula chemical symbol paragraph equation number sentence word
Processes	concept is a process rather than an object	melting oxidation distillation dissociation electrolysis

teaching concepts with perceptible instances and perceptible attributes have been developed and described (Klausmeier, Ghatala, and Frayer, 1974; Markle and Tiemann, 1970; Merrill and Tennyson, 1977). The following section summarizes these recommendations.

RECOMMENDATIONS FOR TEACHING CONCRETE CONCEPTS

In some suitable form, provide information about the critical and variable attributes of the concept. What constitutes *suitable form* varies from one concept to another. In some cases, a definition is a suitable beginning; in other cases, focusing on characteristics of an example or contrasting an example with a nonexample may be more effective. Teachers must use their own judgment because no one procedure will be best for every concept. What *can* be said is that some means must be found to call the learner's attention to the things that matter and the things that do not. For example, in teaching a student what we call a *pipet*, we might show a variety of pipets and describe their common characteristics.

Provide an opportunity for the student to construct the desired concept and to differentiate it from related concepts by successive tasks that require the student to make judgments about examples and nonexamples of the concept. For example, you might say, "Here is another piece of glassware that we use to measure liquids, but we call it a buret. How does it differ from a pipet?"

Presentation of instances should be sequenced so that prototypic examples are presented early and far out examples presented later. For example, transfer pipets might be presented early, followed by measuring pipets, and finally, medicine droppers or other unusual forms of pipets.

The presentation of nonexamples should be just the opposite; that is, the more obvious nonexamples should precede the less obvious ones. For example, beakers and graduated cylinders might be contrasted with pipets before burets are contrasted.

Instructional examples and nonexamples should be selected so that all critical and variable attributes are clarified. For example, both measuring and transfer pipets should be shown and size should be varied so that all are clearly known as pipets. Nonexamples such as burets and separatory funnels might be shown to clarify the differences between pipets and other glassware used to measure or transfer liquids.

Once the concept has been taught, examples and nonexamples *other than those used in instruction* should be used to test understanding. For example, students might be shown photographs of ancient glassware with instructions to name the items shown as they would be classified today. Justification for the classification would strengthen the question.

No teacher is likely to go through such a formal procedure to teach the concept of pipet. We usually are not particularly concerned about how well such concepts are developed. However, we are concerned about other concepts, and this procedure has been demonstrated to be effective.

CONCEPTS WITH INVISIBLE EXAMPLES

Atom is an example of an important class of concepts with no perceptible instances. It is an invented concept, but we imagine that the concept we construct corresponds to something in external reality that we cannot experience directly. Atoms are invisible.

Atom is, in many ways, a concept like *beaker* or *bird*. It names an object. However, an important difference exists between concepts with invisible examples (atom, molecule, nucleus, universe, angstrom, light year, and God) and concrete concepts like *beaker* and *bird*. Beakers and birds I sense directly; atoms and angstroms I do not.

The primary value of concept analysis is to gain insight into difficulties that students have learning concepts and to suggest strategies for teaching difficult concepts. To see how this might help, consider the concept of atom. What is an atom? What are its critical attributes; that is, what makes an atom an atom rather than a molecule, a chemical symbol, or an element?

> Please interrupt your reading at this point and do a concept analysis for *atom*. After you finish your analysis, compare it with the one found in Appendix C.

CONCEPT ANALYSIS FOR *ATOM*

My graduate students and I spent a great deal of time arguing before I was satisfied with the analysis for *atom* shown in Appendix C. We had difficulty separating all that we want students to know about atoms from the basic notion of what an atom is. We eventually asked ourselves, "If we were teaching a child what an atom is, what examples and nonexamples would we use? What would we want them to know?"

Our first analysis of atom listed Na and H_2O as an example and a nonexample, respectively, but we soon decided that these are inappropriate for teaching the *concept* of atom. Na is a *symbol* used to represent an atom (as well as other things), but it is not an *example* of an atom. It in no way reveals the features that distinguish an atom from a molecule, electron, or nucleus, and that is the whole purpose of teaching examples.

An example that reveals none of the attributes of the thing it is supposed to exemplify is useless, similar to using *cat* (as opposed to the animal called a cat) as an example of an animal. The three-letter sequence *cat* is little different from the three letter sequences *mat* and *sat*, neither of which is remotely related to animals.

Examples and nonexamples are used to reveal attributes of the concept and to clarify differences between critical and variable attributes. Actual instances can accomplish this task; labels for instances cannot. If the label refers to something that is very familiar, it may serve the purpose because it calls forth from memory a prototype that has the important features. Thus, in Chapter 9 I suggested that you recall the burning of alcohol rather than actually doing it. This recollection saves time and labor, but it is always dangerous—particularly so for beginning students. Even when the example is familiar (such as tree), the feature that one wants to stress may not be salient in the student's prototype.

Return now to the list of critical attributes for *atom*. I list two: An atom is the smallest unit of an element, and it must retain microscopic properties characteristic of the element.

Our first analysis of *atom* listed such things as "made up of electrons, protons, and neutrons", and "cannot be divided by ordinary chemical means". Later we discarded such attributes because *these statements provide information about atoms*—and such ideas will eventually be a part of any well-developed concept of atom—*but they are not central to the concept itself*. They do not tell us what an atom is. As such, they develop the *interiority* of the concept, but they are not what makes an atom an atom.

The second critical attribute in our analysis is an attempt to distinguish between an atom and a part of an atom. In one sense, electrons, protons, and other subatomic particles are parts of an element, and they are all smaller than the atom. In what sense do we say that an atom is the smallest part of an element?

I answer this question by saying that a part of an element must retain properties of the element. But does an atom have color, hardness, index of refraction, density, electrical conductivity, or a melting point? No, it does not. These bulk properties of elements are lost when the element is divided into atoms.

Since our earlier analysis of atom (Herron, Cantu, Ward, and Srinivasan, 1977, p 194), we have tried to circumvent this problem by saying the atom retains the *microscopic* properties of the element: characteristic mass (only true if we ignore isotopes), characteristic combining power, and characteristic absorption of electromagnetic radiation. This statement makes me feel better, but the argument is circular because microscopic, as used in chemistry, simply refers to events at the atomic level. Even our best efforts to be explicit are imperfect under careful scrutiny.

If our purpose for concept analysis were to finally have before us "*the* definitive concept", concept analysis would not be worthwhile. Only the simplest, concrete concepts succumb to such clear-cut analysis. Because our purpose is to detect difficulties in teaching concepts, analysis *is* worthwhile. It encourages us to sort through our own understanding of the concept and bring to conscious attention attributes that are held subconsciously. As we bring these features to consciousness, we can deal with them explicitly and begin to eliminate students' misconceptions as well as our own.

At some level of abstraction, any statement can be questioned. For example, one may ask whether the hydrogen in water exists as *atoms* of hydrogen. It is an empirical fact that the properties of hydrogen in a molecule are *not* identical to the properties of an isolated atom. Such fine distinctions are best left until the basic concept of atom is established and interiority of the concept is being developed through tuning.

CONCEPTS WITH INVISIBLE ATTRIBUTES

A number of concepts have perceptible instances but have critical attributes that are *not* perceptible. *Element* and *compound* are such concepts. Markle and Tiemann (1970) presented a concept analysis of *element* prepared by one of their students and included it in their book as a flawed example of concept analysis (Appendix D). The analysis can also serve to illustrate difficulties in teaching such concepts.

Some of the examples listed in the Markle and Tiemann example are mercury, iron, and chlorine gas; some of the nonexamples are water and salt. Whether the author meant for the student to be presented with actual samples of the substances or whether the student would be presented with the names of the substances was not clear. If the *names* are presented, then the lesson involves word association in which the student learns to associate mercury, chlorine, and iron with the word *element.* This task may be an important component of instruction, but it is not concept learning. (It is, however, all that most entering college students have learned about the concept *element.*) Students have learned a list of names that they call elements, but they neither know why they are called elements nor how they could distinguish an element from a compound or mixture.

Whether or not the student is to be presented with names of elements or actual examples, a problem exists. *Absolutely nothing in the examples enables students to distinguish examples from nonexamples because the critical attributes of the concept are not perceptible in the examples.* Furthermore, selecting examples for which the critical attributes are perceptible is not possible.

The first critical attribute listed in Appendix D is that an element has only one kind of atom. Look at mercury (the word or the real material). Look at water (the word or the real material). Can you perceive that mercury has one kind of atom and water has two kinds? No. Neither can one look at these instances and determine that one can be subdivided into simpler substances whereas the other cannot.

What, then, can be substituted for examples when they do not reveal the critical attributes of a concept? Whatever is used should serve the same educational function as the instances replaced.

RECOMMENDATIONS FOR TEACHING CONCEPTS
WITH INVISIBLE ATTRIBUTES

Appendix E contains teaching suggestions that illustrate one approach to the problem. For some attributes of the concept, actual examples are possible. Switching on a light or striking a match and asking whether the light and heat represent elements focuses attention on the fact that elements are a form of matter; heat and light are not. Showing a solid such as a sheet of copper or a liquid such as water and asking whether it *could* be an element reinforces the idea that elements are a form of matter, but more information is needed to decide whether the matter is an element.

Next, by presenting substances that appear to be homogeneous as well as those that are obviously mixtures and asking students whether these materials *could* be elements leads to consideration of a second attribute of elements, that they are pure forms of matter. Similar lines of questioning can be used to focus attention on all macroscopic properties of elements.

Models or graphics must be used to focus on the microscopic properties of elements. One could begin by presenting piles of disconnected, identical spheres; piles of connected, identical spheres; piles of disconnected, unidentical spheres; and piles of connected, unidentical spheres. Students could then be asked to identify the piles that could represent an element. Once students make a decision, they should explain their answer. During discussion, the essential microscopic feature of elements (consisting of only one kind of atom) can be clarified. Other activities are found in Appendix E.

As you consider these teaching examples or the others found in the appendix, notice that an attempt is made to suggest activities that will serve the same psychological function as normal examples and nonexamples. The general approach may be summarized as follows:

1. In the suggested activities, students are presented with an appropriate stimulus.
2. Students are expected to respond to the presentation.

Many readers may think that a simpler, more efficient, and just as effective approach would be to present the suggested instances and tell students, "This is an example (or not an example) of an element because . . ." rather than asking, "Can this be an element?" *This approach is not the same, and the difference in the two approaches is critical.* When the instructor says, "This is an example . . .", students are not required to activate the schema that they are developing, compare that schema with the stimulus before them, and infer that a match between schema and stimulus does or does not exist. *This active processing on the part of learners is the key to concept learning.* Strategies that do not require such constructive activity accompanied by feedback concerning the accuracy of the construction are not effective unless the student has already realized the necessity of such active learning and makes the comparisons without being asked to do so.

3. Students are expected to make judgments about whether a substance can or cannot be an example of the concept on the basis of the stimulus before them and the schema they are developing.
4. Stimuli presented and the questions asked (usually during the probing stage) focus on a particular attribute of the concept.
5. The set of activities, taken as a whole, focus on all attributes of the concept.

DUAL CONCEPTS

Appendix E contains both a macroscopic and a microscopic definition for element. A characteristic of chemistry is that we often develop concepts at two or more levels of abstraction, and the understanding sought involves both levels. Both levels do not have to be dealt with simultaneously, but explicit attention must be paid to the two levels. In addition, most chemistry concepts are represented symbolically, and the connection between the symbolic representation, the macroscopic concept, and the microscopic concept must eventually be made. A great deal of confusion is fostered by not doing so.

The critical attribute of *element* at the macroscopic level is that the substance under consideration cannot be broken down into simpler substances with new macroscopic properties. Before students can make a decision based on this attribute, they must take the intervening step of asking, "What evidence is necessary and sufficient to say that an observed change in properties is a result of decomposition into simpler substances as opposed to combination to form a more complex substance?" This question is not easily answered. Any effort to answer the question involves sophisticated reasoning and considerable knowledge of chemistry. Thus, concepts such as *element* and *compound* can be difficult for students, not because the critical attribute itself is opaque, but because the knowledge and reasoning skills required to test for the presence of the attribute are extensive.

For any concept that does not have defining attributes that are directly perceptible, classification decisions involve an intervening step that can be taken only with the aid of additional knowledge that is not directly related to the concept being taught. Thus, we can predict that the concept will be more difficult because of its increased abstractness and complexity.

Experts can easily lose sight of how much specialized knowledge they use in making simple classifications. As a result, they believe that students are having difficulty learning simple concepts when, in fact, the difficulty is that students have not learned the specialized knowledge that the expert uses subconsciously.

RECOMMENDATIONS FOR CLARIFYING CONCEPTS TO PEOPLE WITH LIMITED CHEMICAL KNOWLEDGE

One strategy to get around the limited knowledge of beginning students is to play "20 Questions". The teacher supplies the specialized knowledge by responding "yes" or "no" to student questions. For example, the teacher might enter the classroom carrying a sample of solid and ask, "Is this an element" or "Is this aluminum?" The rules of the game are that the student may ask any question that can be answered "yes" or "no". By carefully designing the rules of the game, the teacher can supply data that students would have no way of knowing while requiring the student to decide what information is needed for the classification.

In the final analysis, *knowing what must be decided to categorize some instance as an example or nonexample is at the heart of concept learning.* Once students grasp the decisions to be made, the teacher may move on, even though students may lack information needed to make particular classifications that an expert could make.

PRINCIPLE CONCEPTS

Many concepts are difficult because judgments that must be made in classifying instances require considerable knowledge and sophisticated reasoning. We call these concepts that require knowledge of principles *principle concepts*. Many chemistry concepts fall into this category.

Mole is not a difficult concept even though teachers claim that it is. The International Union of Pure and Applied Chemistry defines *mole* as "the amount of substance of a system which contains as many elementary entities as there are carbon atoms in 0.012 kg of carbon-12" (Lowe, 1975, p 296). The elementary entity may be anything (such as electron, atom, molecule) and must be specified. Although defined as "amount of substance", the critical attribute of that amount is that it contains a specified number of particles.

In making the decision concerning whether a mole is or is not present, students must ask, "Are there 6.02×10^{23} of the specified entities in the sample being considered?" If the answer is yes, it is a mole; if the answer is no, it is not a mole.[1]

The examples normally given when mole is discussed are things like "16 g of oxygen (atom)" or "18 g of water (molecule)." (*See* Groups II and III of the concept analysis in Appendix F.) Unless students have learned a number of principles such as "the atomic mass of an element in grams contains 6.02×10^{23} atoms" and "the molecular mass of a compound in grams contains 6.02×10^{23} molecules", they cannot discriminate such examples from nonexamples. Thus, even though we ask students to discriminate between examples and nonexamples of a *concept*, the student must know and apply a number of *principles* in making the discrimination. This issue is discussed more fully in Chapter 11.

In doing the concept analysis of *mole*, examples such as "3.01×10^{23} bikes (wheel)" or "1.02×10^{21} reams of paper (sheet)" (*See* Appendix F, Group I) seemed to be of value in teaching the concept of mole (as opposed to teaching various principles that are closely associated with mole) because the critical attributes of the concept are more-or-less transparent in these examples. However, it must be understood that the student will be able to generalize only to examples for which he understands principles and rules that enable him to determine that the example contains 6.02×10^{23} elementary entities.

A situation similar to that encountered with mole exists for the concept of *mixture*. (A concept analysis of *mixture* is shown as Appendix G.) Several examples—dirt, dimes and quarters, granite, and brick—could be used to teach the concept of mixture because the critical attribute of heterogeneity is transparent in those examples. However, examples such as milk, gasoline, and air would be of no value because the critical attribute is not perceptible.

We are inclined toward the conclusion that examples and nonexamples are of value only if the critical and variable attributes of the concept are transparent in those examples. Because the transparency of the attributes may depend on the prior learning of the student, it is beneficial to list examples and nonexamples of varying complexity along with some statement concerning the knowledge that must be presumed for the examples to be of value in teaching. Care must be exercised to prevent confusion of the concept being taught and the principles that are needed to discriminate examples from nonexamples.

---

[1]However, semantic problems exist. If 13 eggs are placed on a table and a student is asked, "Are there a dozen eggs here?" the student knowing the concept may respond, "Yes", because he sees that there are 12 eggs plus one more. Another student who knows the concept may respond, "No", because there are *more* than 12 eggs present.

CONCEPTS THAT NAME ATTRIBUTES AND PROPERTIES

Concepts such as *mass, weight, electric charge, frequency, oxidation number, place value*, and *flammable* name attributes or properties of objects. A concept analysis of one such concept, *weight* (Romberg, Steitz, and Frayer, 1971), is shown as Appendix H.

I do not find this analysis of *weight* very helpful. The problem is seen in the examples (10 pounds, 10 tons, and 10 ounces), and the nonexamples (10 feet, 10 dozen, and 10 square yards). These labels are for weight and coordinate concepts, but they are not examples and nonexamples of the concept. Such examples and nonexamples would be of value in teaching labels used to express weight, but they serve little purpose in teaching the concept. Weight is a *force* and must be felt to be understood.

Examples and nonexamples may have little value for concepts such as *weight* and *mass*. An analysis such as the one for *mass* in Appendix I, may do a better job of revealing the essential features of the concept.

The distinction between concept labels and concepts is not trivial. Students in my courses use labels such as cubic centimeter, square centimeter, gram, and pascal with ease, but when asked to draw a square centimeter, show what 10 cubic centimeters would look like, or describe the difference in what is represented by 5 kg and 5 Pa, they are unable to do so. I interpret this failure as a lack of concept learning.

SYMBOLIC CONCEPTS

Chemical symbol, formula, equation, word, and number sentence name concepts that are classes of symbolic representations. Appendix J shows two analyses of chemical symbol: a first attempt that we did not like, and a second attempt that we liked better. The problem with the first attempt can be illustrated by examples of the concept *word*.

Does a student who understands the concept word identify *kat* or *recieve* as examples or nonexamples? If one insists that these should be classified as nonexamples, the concern is apparently more than concept learning. Students are not only being asked to know what a word is, but they are being asked to apply rules that govern the correctness of words.

Merit can be found in separating the *concept* of such ideas as word, formula, and equation from the set of rules that must be understood to judge the acceptability of various examples. For example, the first analysis in Appendix J lists *he* as a nonexample of chemical symbol, whereas *He* is an example. The discrimination here is based on the *rule* that correct chemical symbols begin with a capital letter. This confusion between the critical attributes of the concept and the rules that govern acceptable style in representing examples of the concept is also seen in the list of attributes.

The second analysis of chemical symbol shown in Appendix J does a better job of separating the critical attributes of the concept from the rules for acceptable style. Such analyses are likely to be more valuable in teaching concepts that involve symbolic representations.

PROCESSES

Distillation, melting, electrolysis, dissociation, and oxidation name concepts that are processes. Such terms refer to classes of behavior, and some students of learning do not call them concepts. We have seen no concept analyses of such concepts in the literature, and we have done few of them ourselves. Still, this class is an important type of concepts in science, and analysis seems to present no special problems. A concept analysis for *melting* is presented as Appendix K.

SUMMARY

The goal of chemistry teaching that is most commonly cited is getting students to solve novel problems, but research on problem solving reveals that the most common impediment to problem solving is a lack of conceptual understanding. This chapter has considered a variety of chemistry concepts to see why they are poorly understood.

Concept analysis was introduced early in this chapter, and that tool was used to uncover sources of difficulty in concept learning. Concept analysis involves identification of the critical and variable attributes of a concept and the selection of examples and nonexamples that can be used to clarify what the critical and variable attributes are.

No right analysis exists for a concept. What is included in an analysis related to instruction at a beginning level may be quite different from an analysis related to instruction at an advanced level. In either event, the value of the analysis is in the insights gained in the process of doing the analysis rather than in the final product.

Concept analyses of concepts introduced in a beginning chemistry course led to eight classes of concepts. The characteristics that distinguish the eight classes are related to the difficulty of teaching each class.

Concrete concepts are easy to teach because they have perceptible examples with perceptible attributes. Examples and nonexamples that clarify the critical attributes of the concept are easy to present to students. Most concrete concepts are so easily learned that we spend little or no time trying to teach them, and the learning is left entirely to informal experience.

Some chemistry concepts (e.g., atom, ion, and molecule) name objects that have *invisible examples*, making clarification of the concept by presenting examples and nonexamples impossible. Various pseudoexamples, such as computer animations and physical models, can sometimes be used in place of examples, but care is required to prevent confusing characteristics of the model (color, for example) with characteristics of the concept itself.

Still other chemistry concepts name entities that have perceptible examples but *invisible attributes*. A piece of copper, for example, is a clearly perceptible example of an element, but looking at the example does not reveal that it is made up of a single kind of atom—the most important attribute of an element. Such concepts cause difficulty because an indirect test must be applied to separate examples from nonexamples. These tests may be difficult to understand and use. One way around the difficulty is for the instructor to provide the information that would result from such a test and allow the student to make the judgment about whether an instance is an example.

Concepts such as *element* are further complicated by the fact that we expect students to understand the concept at both the microscopic and macroscopic level. Such *dual concepts* effectively require that we teach two different concepts and the relationship between them.

Principle concept is the label given to concepts that require knowledge of principles before use can be made of the concept. *Mole*, for example, is a straightforward
concept. It simply names the amount of substance containing a particular number of
particles. However, deciding the number of particles in a sample requires knowledge
of many principles.

Concepts that name *attributes* or *properties* have no unencumbered examples. A
mass or a weight cannot be shown; the only possibility is to show something that
has mass or weight. To understand the concept, the property must be disembedded
from the object possessing the property.

Symbolic concepts such as *word* or *chemical formula* are probably as easy to teach
as concrete concepts. However, care should be taken to keep the concept separate
from the rules associated with proper expression of the symbols.

Many important concepts in chemistry name *processes* such as melting, burning,
and boiling. Analysis of such concepts appears to present no special problems and
can suggest useful examples for teaching the concept and differentiating it from
other processes.

11

Difficulties in Teaching Concepts

As seen in Chapter 10, concept analysis is a powerful tool. However, like any powerful tool, it should be reserved for those chores that require it. Without denying the utility of a drag line or backhoe for moving dirt, we still use a broom on the kitchen floor!

Students learn hundreds of concepts during an introductory chemistry course: *beaker, spatula, heat, acid, indicator, ideal gas, proportional, oxidation, mole, equilibrium*, and *aldehyde*, to name a few. Most of these concepts are taught and learned with relative ease. We have little interest in knowing why they are learned easily, and preparing a concept analysis is a waste of time.

This chapter describes general pitfalls in teaching chemistry concepts, and it suggests ways to avoid them. It also suggests ways to determine whether concepts are understood by students. As was the case in the previous chapter, most of what is said derives from attempts to analyze concepts that students find difficult, but the material in this chapter is more general.

One of the first concepts that I analyzed was *mole*, and one of the first things that I learned is that little of the students' difficulties have anything to do with the concept. Thus, the first considerations in teaching concepts are instructional decisions.

INSTRUCTIONAL DECISIONS IN CONCEPT LEARNING

ARE YOU TEACHING A CONCEPT?

Concepts are not taught in isolation, and we can easily confuse what we are doing. *Mole*, one of the most important concepts in chemistry, is a case in point.

Most chemistry teachers complain that *mole* is difficult to teach. Most chemistry students complain that *mole* is difficult to learn. It is neither. However, the multitude of skills, principles, and concepts that teachers have in mind when they mention mole can be difficult to teach and difficult to learn.

The perception that students have difficulty with the concept of mole arises because students are unable to answer questions such as, "How many moles of water are in a liter of water?" Although this question involves the concept of mole, it does not test for an understanding of the concept. Being able to answer the question requires far more than concept learning.

First, students must *recall* facts such as the formula (or mole mass) of water, the milliliter equivalent of liter, and the density of water (or the mass of a liter of water). Doing so requires knowledge of conventions and facts rather than knowledge of the mole concept.

Second, students must recall rules such as "The molecular mass of any compound expressed in grams is equivalent to one mole of that compound," and, "The mass of a substance is the product of its volume and density." Neither of these rules (or principles) is part of the mole concept. Although the first rule relates mass to moles, the rule can be (and often is) applied without understanding what a mole represents. Conversely, a person may understand the concept of mole without knowing this relationship between mole and mass.

Third, students must apply these rules by using other rules for mathematical calculation. The question can be answered correctly without having an adequate concept of mole. Conversely, it can be missed by students who understand the concept perfectly.

WHICH CONCEPT DO YOU MEAN?

Once concepts are separated from the facts, rules, and principles with which they are entangled, other difficulties arise. One such difficulty is that the same label may be used to refer to many concepts. Generally the context in which the word is used indicates which concept is meant, but not always. The problem can be especially difficult when the student knows one concept and the concept label is then used to refer to another concept. Terms such as *force* and *work* in physics, or *precipitate* and *strong* in chemistry are examples.

Difficulties caused by different concepts under the same label are not confined to technical concepts that share names with everyday words. We teach three or four concepts of *acid* in chemistry, and we use acid in reference to all of them. We talk about an acid solution, about water acting as an acid, about hydrogen acetate being a weak acid, and about BF_3 behaving as an acid. Our exact meaning must be inferred from the context in which the word is used.

HOW FAR SHOULD YOU GO?

Once we are clear about the concept being taught, we must decide how far to go in discussing a concept. A judgment must be made when doing a concept analysis. The decision must be made by each teacher, and it depends on the purpose of the lesson and the students being taught. However, some rules of thumb can guide such decisions:

1. Never make the idea more complicated than necessary.
2. Do not introduce concepts that call for distinctions that the student cannot make.
3. In simplifying, do not teach what must later be unlearned.

Often, conflict occurs between the first two rules and the last. Complicated ideas are taught to beginning students *so that they will not have to unlearn simplifications later*. By the same token, bad ideas are taught to keep things simple. When is teaching less than the absolute truth acceptable? When is it wrong to do so? Consider decisions that must be made when teaching about atoms and molecules.

The concept analysis for *atom* (*see* Appendix C) contains no information about the structure of atoms. Consequently, nothing in the analysis would allow students to distinguish *atom* from coordinate concepts such as *ion*. Furthermore, the critical and variable attributes that are listed could lead to conclusions that are incorrect.

The microscopic property of mass is *not* constant across all atoms of an element, and the relative mass of a given atom is *not* the same as the relative mass of the element. Isotopes exist. If students learn the concept of *atom* as outlined in Appendix C, are they not learning what must be unlearned? I do not think so.

The key question is whether students can develop the schema that they ultimately need through tuning and patterned generation rather than schema induction. If tuning and patterned generation occur, we need not be concerned; if what is taught necessitates schema induction later, we should reconsider.

Only minor tuning is involved in altering "all atoms of an element are identical" to a schema that includes isotopes. Similar tuning is involved when students learn that "atoms are spheres capable of bonding together to form molecules" and later learn about the arrangement of electrons within that sphere and how electron configuration affects bonding. In both cases, the later learning involves the addition of information to what is already present. It does not require discarding what has been taught and starting over with a radically different model. This process is particularly true if the discussion of orbitals begins with the spherical shape of the set rather than the very different shapes of orbitals within the set.

A different situation exists when students learn that electrons are particles that travel in fixed orbits around the nucleus and are later asked to think of electrons as waves or probability distributions. I see no way to construct the new schema by tuning the old one, and the relationships implied by the new model are so different from those suggested by the old one that creation of the new schema by patterned generation seems unlikely. The conflict between the pattern suggested by the original model and the new one creates major difficulties for students. They already have in place a model that seems adequate. Students are understandably reluctant to discard an old, familiar idea for a new and unfamiliar one. This kind of *unlearning* is likely to conflict with the Principle of Least Cognitive Effort, and the student will hold tenaciously to the older concept as long as possible. By the time the student realizes that the old model is inadequate, so much of the evidence for the new model has been presented that it is poorly learned.

WHERE DOES THE CONCEPT FIT?

Concept analyses list supraordinate, coordinate, and subordinate concepts of the concept under analysis. This listing helps focus on where the target concept fits within a larger instructional setting, a part of the pedagogical content knowledge discussed in the following section.

Saying that *elementary particle* is a supraordinate concept for atom (*see* Appendix C) suggests other concepts to which students may relate *atom*. Students whose concept of *atom* is a tiny particle fail to discriminate an *atom* of oxygen from a *molecule* of oxygen. Other coordinate concepts that are potentially confusing are *ion*, *kernel*, and *nucleus*. Subordinate concepts are the various kinds of atoms (sodium atom, oxygen atom, etc.) and *isotope*.

Listing coordinate and subordinate concepts sometimes reveals questions about a concept analysis that have teaching implications. Nothing in the analysis for *atom*, for example, suggests isotopes or a difference between *atom* and *ion*. Whether these omissions represent ideas that should be taught now or later depends on overall curriculum development.

PEDAGOGICAL CONTENT KNOWLEDGE

All of the instructional decisions outlined in the preceding section require what Shulman (1987) calls pedagogical content knowledge. According to Shulman, all teachers need three kinds of knowledge: First, they must know general principles of teaching and learning (pedagogical knowledge). Second, they must understand the subject matter they are teaching (content knowledge). Finally, they must understand how to teach the specific ideas of their subject (pedagogical content knowledge).

APPLICATIONS OF PEDAGOGICAL CONTENT KNOWLEDGE

Knowing how far to develop concepts at a particular grade level is an example of pedagogical content knowledge. Making such decisions requires an understanding of how students learn, but a deep understanding of content is also required. This understanding is also required for decisions about where a particular concept fits among the constellation of concepts comprising the discipline, how concepts should be sequenced, what demonstration or experiment is best suited to clarify a concept, and so forth.

Graduate students in science education and undergraduate students in elementary education frequently complain about science requirements that present content that exceeds what they expect to teach. They often pass required courses without comprehending the content. This lack of comprehension causes difficulties when these teachers make instructional decisions.

Often, alternative ways to present concepts at an appropriate level exist. One procedure may lead to the development of schemas that must be replaced during subsequent instruction, whereas alternatives may lead to schemas that merely need to be altered through tuning or patterned generation. Unless teachers are aware of the content that students are likely to encounter during later instruction, they often make poor instructional decisions.

RESPONSIBILITY FOR PEDAGOGICAL CONTENT KNOWLEDGE

University and college chemists are prone to blame, unfairly, education faculty when concepts are poorly taught by elementary and secondary teachers. Education faculty can teach general pedagogical principles, but they cannot teach the pedagogical content knowledge required to decide how specific concepts should be taught. Only content specialists have the content knowledge required to do that. If chemistry education at lower grade levels is to improve, chemists (and other scientists) must accept their obligation to teach courses for teachers that provide pedagogical content knowledge as well as appropriate content knowledge.

Determining what content knowledge is appropriate for prospective elementary, middle school, and high school teachers is one of the most difficult tasks facing chemists who teach courses for teachers. Asking what concepts these teachers will teach and what they must understand about the concept themselves is not enough. Serious thought must be given to how the concept develops over time, and teachers should be aware of how the concept they teach can affect subsequent learning.

When especially difficult concepts are taught, teachers should be shown how initial lessons can be kept simple without distorting the concept to the extent that later learning requires induction of a new schema rather than enlarging and tuning an existing one. One of the values of concept analysis is that it encourages just this kind of critical thought.

SPECIAL PROBLEMS WITH ABSTRACT CONCEPTS

PSEUDOEXAMPLES

As indicated in Chapter 10, the standard strategy for teaching concepts is to present an example, focus attention on the features that make it an example, present a matched nonexample, focus attention on the feature that prevents the nonexample from being included in the concept class, and continuing until the student can accurately and reliably classify new instances that are presented. This procedure is impossible if no perceptible instances exist, and it is complicated if the concept is an attribute or property rather than an object or event. What does an instructor do, then?

In teaching concepts that have no perceptible instances or imperceptible attributes, the critical question is, "What stimuli can be presented to students to elicit responses that allow assessment of students' understanding of the critical and variable attributes of the concept?" The most straightforward approach is to use what I call pseudoexamples. *Pseudoexamples* are substitutes for actual examples that are designed to reveal the critical and variable attributes of the concept.

The pseudoexamples listed in the analysis for *atom* are best guesses concerning presentations that would clarify the concept. Those instances at the top of the list (animated *pictures* of atoms) should be most effective because the attributes of *atom* are fairly transparent; these pseudoexamples are less abstract than the more commonly used pseudoexamples at the bottom of the list (chemical symbols or formulas). The major disadvantage of animated pictures is the ease with which students confuse the *representation* with the concept it is intended to represent.

All pseudoexamples have features that can be confused with the entity they represent. (Most chemistry teachers who use atomic models have encountered students who learn that oxygen atoms are red and carbon atoms are black!) Care must be taken to vary these features across pseudoexamples so that the student can appreciate that they are not characteristic of the concept being taught.

DISEMBEDDING ATTRIBUTES

In the case of concepts that name attributes, the problem is slightly different. One cannot give an example of a property or attribute. You cannot show *mass* or *weight* in the same sense that you can show *chair* or *bird*. You can show something that *has* mass or weight, but this demonstration is qualitatively different from giving an example of the concept itself. When one shows a desk chair to exemplify the concept *chair*, the example is a member of the class *chair*; when one uses a desk chair to exemplify the concept *mass*, the concept is embedded in the object along with other properties such as color, shape, and texture.

The problem is similar to the problem of developing logicomathematical knowledge. As indicated on pp 42–43, concepts of number and grouping cannot be presented in the absence of other information that has nothing to do with the concept being constructed. A child sees two marbles or two people or two hands. The *twoness* must be disembedded from the physical knowledge about marbles, people, or hands, and it must be incorporated as a separate schema.

Similarly, when pseudoexamples are used to teach physical knowledge about objects that cannot be perceived, the information concerning the abstract concept (*atom*, for example) must somehow be disembedded from such physical knowledge as the color and texture of models, the two dimensions of diagrams, and the inflexibility of macroscopic solids.

SUITABLE SUBSTITUTES FOR EXAMPLES OF PROPERTIES

We must address the question of what can substitute for examples and nonexamples of attributes and properties. Appendix I shows an analysis for the concept *mass*. It does not follow the standard format for concept analysis that was introduced in Chapter 10 because that format is not very helpful for concepts that name attributes.

In place of ordinary examples and nonexamples, Appendix I substitutes information about how mass is detected, how it is measured, and teaching activities that help clarify the concept. The first teaching example suggests that students be shown a horn and be asked if it has mass. The horn could then be blown and the student asked if the sound it produces has mass or if the mass of the horn has changed. Students might then be asked what evidence exists to support their opinion that the thing named does or does not have mass (i.e., to identify the critical test for the attribute named). Similar to the examples and nonexamples for concrete concepts, these activities have the essential characteristics for concept learning:

1. The student is presented with some kind of stimulus.
2. The student is expected to respond to the stimulus.
3. The kind of response expected involves an interpretation or an inference.
4. The interpretation or inference is based on a comparison of the stimulus to the schema that the student has developed for the concept.
5. Each instance (or substituted activity) is selected to focus the student's attention on a particular attribute of the concept.
6. The set of instances (or learning activities), taken as a whole, requires the student to focus on *all* critical attributes and all potentially confusing variable attributes of the concept.

A complete set of learning activities is not given in Appendix I, but the suggestions made should clarify what should be taught. As in the case of all concepts, students understand concepts that name attributes and properties when they can

- define the concept,
- list critical and variable attributes of the concept,
- describe how the attribute is detected or measured,
- list concepts with which the target concept is easily confused, and
- list instances in which the attribute is observed and list instances in which the attribute is absent. (These lists are comparable to examples and nonexamples.)

As in other cases, activities of the kind used to teach the concept should be effective in testing for understanding of the concept. However, these activities may be awkward to use in the normal testing situation. Consequently, I have included in the concept analysis of *mass* a section giving test questions that seem appropriate for standard tests. Presumably, these particular questions were not encountered during instruction.

Although the discussion presented here is in terms of the development of schemas for science concepts such as mass, weight, electric charge, frequency, and place value, the suggested strategy is the best that I can recommend for the development of schemas for logicomathematical knowledge such as proportional reasoning, controlling variables, and correlational reasoning.

DETECTING UNDERSTANDING

An important part of teaching concepts is detecting sound understanding and identifying misconceptions. Occasionally, test questions or informal interviews must be designed to check understanding, but a great deal can be learned by attending to what students say during class discussion.

ATTENDING TO WHAT STUDENTS SAY

Beginning teachers (and perhaps experienced ones) overlook clues that concepts are misunderstood. For example, they allow student statements such as, "Take two atoms of water," or, "The area of the paper is 625 centimeters," to pass without comment, or they restate what the student said and substitute the correct word for the incorrect word used by the student. The assumption is that the student *misspoke*, which is a reasonable assumption, because everyone occasionally says something other than what was intended.

Also, students commonly use the wrong term when they are confused about the concept labeled by that term. A few pointed questions such as, "Did you really mean to say 'two *atoms* of water?' " followed by, "Why does it not make sense to say 'two atoms of water?' " will reveal whether a misconception actually exists. Asking other class members whether they see anything wrong with the way the idea was expressed can reveal whether others are confused.

Carefully reading laboratory reports, homework, and test papers can also provide indications of misunderstood concepts. Making notes and following up in the manner outlined above can reveal whether students misunderstand the concepts or are having trouble expressing their ideas.

LIMITATIONS OF VARIOUS ITEM FORMATS

At the end of Chapter 9, you were asked to take a test covering *concept* (p 119). The items on that test were typical of items commonly used to test conceptual understanding. Some were more useful than others.

PROVIDING DEFINITIONS. The first item on the concept test asked you to write a definition for *concept*. How do you feel about this output as evidence of understanding? We often feel that we understand something that we cannot define. By the same token, we know that a definition that has little meaning can be memorized. Consequently, definitions have limited value in assessing conceptual understanding.

LISTING EXAMPLES. The second and third items on the test asked you to list examples and nonexamples of concept. Such exercises can help determine whether your concept of *concept* is identical to mine, but they have the limitation that everything listed could agree with my definition and yet we might disagree about other instances.

When students select examples and nonexamples, they select those about which they are most confident. Students avoid mentioning ideas with which they are not sure, because they do not wish to be wrong. Consequently, important characteristics of my concept of *concept* may not be revealed by this exercise simply because I avoided difficult cases.

IDENTIFYING EXAMPLES. If *I* select the examples and nonexamples, I can force you to make some of those difficult decisions to reveal differences between your concept and mine. This strategy is defined in item four (p 142). The instances were selected to focus on characteristics that distinguish *concept* from ideas that are not concepts. In the test on p 119, I checked a, b, e, h, i, l, m, and n as concepts. I did not check the others. I doubt that your response was the same. The differences between your list and mine result from two factors: our concepts differ, and we interpret instances differently.

What you called concept at the end of Chapter 9 and what I call concept are not the same. What happened when you checked items in the list is similar to what happens when you ask a beginning science student to classify various situations as examples of work. For example, if you ask a beginning student whether work is done when a person pushes on a brick wall but fails to move it, the response will probably be "yes" because the activity conforms to the common sense definition of work as "an expenditure of effort." You, however, would respond that no work is done because the wall does not move; a force is not applied through a distance. The two concepts of *work* are different.

Your classification of items in question four of the concept test may differ from mine because of differences in interpretation of items in the list, even though we have the same concept of *concept*. For example, some individuals include Na in the list of examples because Na is a chemical symbol, and chemical symbol *is* a concept as defined in this book. I did *not* classify Na as a concept because it is an *example* of the concept *chemical symbol* rather than a *name* for the concept. Similarly, I did not classify Dudley Herron as a concept because it is an *example* of the concept *man* (or *person*, or *professor*, or *author*, etc.). This distinction between a particular *instance* of a concept and the concept itself may seem as picayune to you as the fine distinctions made about concepts such as *work* appear to science students, but there are implications for teaching, as pointed out in Chapter 10.

PROVIDING EXPLANATIONS. Asking students to explain why they made a particular choice on a test, as was done in items five and six on the concept test, sometimes reveals correct thinking that led to an incorrect choice. Although difficult to evaluate, this device provides one of the better indications of conceptual understanding.

As you can see, all common efforts to test conceptual understanding have limitations—including performance testing, portfolios, and other procedures currently advocated as alternatives to traditional test formats. No new test formats are offered here, but limitations of standard test questions are discussed along with suggestions that may increase awareness of how students understand concepts.

IMPORTANT PROPERTIES OF TEST ITEMS

WHAT DOES THE ITEM TEST?

Consider the stoichiometry problem discussed in Chapter 7:

> The chloride of an unknown metal is believed to have the formula MCl_3. A 2.395-g sample of the chloride is dissolved in water and treated with excess silver nitrate solution. The mass of the AgCl precipitate formed is found to be 5.168 g. Find the atomic mass of M, the unknown metal.

This problem is a good one, and students who solve it probably understand a great deal of chemistry. But what about those who do not solve the problem? Where might they go wrong?

The problem statement includes words for many concepts: *chloride, metal, formula, sample, dissolved, solution, mass, precipitate*, and *atomic mass*. The problem can be solved without understanding some of them—*metal*, for example—but excerpts from protocols for students solving this problem indicate that misunderstanding (or at least, misrepresenting) other concepts can be disastrous.

One student represented "the chloride is dissolved in water" by this partial equation:

$$Cl_2 + H_2O \rightarrow$$

One may reasonably infer that this student did not understand "chloride" and might not have understood "dissolve."

Another student represented the same phrase by the following:

$$2MCl_3 + 3H_2O \rightarrow 2M(OH)_3 + 6HCl$$

What does "dissolve" mean to this student? Or perhaps the appropriate question is, "What does an equation mean to the student?" Other examples could be given, but these two examples illustrate how much success at complex intellectual tasks such as problem solving depends on clear understanding of concepts.

How we test understanding of concepts is critical. Far too often, teachers ask students to define a word (e.g., "What is a chloride?") and infer from the result that the student does or does not understand the concept named. As indicated above, this kind of testing is inadequate.

The essential part of understanding concepts is the ability to categorize correctly, and the following question is closer to the mark:

> 1. Circle any of the following that could be described as a *chloride*:
> a) Cl_2 b) KCl c) Cl_2(aq) d) $KClO_3$ e) ClF_3 f) $SnCl_4$ g) HCl

A further indication of understanding is the ability to explain why a particular classification has been made. This kind of information can be obtained by asking students how they decided which formulas should be circled. Another approach is to ask questions such as the following:

> 2. In each of the following pairs, the formula in bold type represents a chloride, but the other does not. Explain why.
>   a) ClF_3; **KCl**
>   b) $Cl_2(aq)$; **$SnCl_2$**
>   c) NaOCl; **NH_4Cl**
> 3. Chemists refer to CH_3Cl as methyl chloride and HCl(g) as hydrogen chloride. Would you categorize these compounds as chlorides? Why or why not?

FOCUS THE ITEM ON THE TASK AT HAND

I have tried to use examples that show how items that test understanding of concepts may vary in difficulty. But what is the appropriate difficulty for test items? I take my cues from the complex tasks that I expect students to perform. For example, if students are expected to solve the unknown metal chloride problem given previously, an appropriate test of the student's understanding of chloride in the context of that problem might be the following item:

> 4. Instructions in a laboratory manual say to "add a small amount of chloride." Which of the following reagents would you add?
>   a) $Cl_2(g)$
>   b) $CCl_4(l)$
>   c) $NH_4Cl(s)$
>   d) NaOCl(aq)
>   e) None of these would do.

Although success on this or any other test item requires knowledge that goes beyond the concept of immediate interest, the understanding tested is closely related to the understanding required for success on the metal chloride problem. This correlation is not the case for the question on the mole concept presented earlier in this chapter: How many moles of water are in a liter of water? This question requires far more than understanding *mole*. A question that comes closer to the *concept* of mole is:

> 5. Which of the following represent one mole of the elementary entity in parentheses?
>   a) 6.02×10^{22} cars (car)
>   b) 3.01×10^{23} bikes (wheel)
>   c) 1.20×10^{21} reams of paper (sheet)
>   d) 1.20 gross of pencils (pencil)
>   e) 16.0 g oxygen gas (O atom)

This question calls for classification behavior, which is central to concept learning. Students must decide whether the instances are examples or nonexamples. To do so, students must know the critical attributes, ignore the variable attributes, and apply this knowledge to the task of classification.

Even in this question, one must know *more* than the concept. In 5a the student must know about exponential numbers, in 5b the student must know that a bike has two wheels, in 5c the student must know the number of sheets of paper in a ream, in 5d the student must know the number in one gross, and in 5e the student must know the mole mass of oxygen atoms.

We hold no knowledge in isolation. All attempts to measure knowledge of one kind require students to use knowledge of a different kind. The best we can do is work to keep our measures as close to the behavior of interest as possible. We must constantly ask whether peripheral knowledge required in a question (the number of wheels on a bike or the number in a gross, for example) is *well known* by the student.

My point is not that the second question about moles is more important than the first; it is probably less important. Rather, my point is that what we first perceive as difficulties in learning *concepts* are often something else altogether. Unless we ask questions that focus on the *concept*, we are unlikely to learn how concepts are understood.

USING PSEUDOEXAMPLES IN TEST ITEMS

Just as teaching concepts with no examples or no perceptible attributes presents special problems, so does testing for understanding. Testing must rely on the use of pseudoinstances such as those used in instruction. Figure 11.1 presents examples of test items that I use in the absence of more authentic visual representations.

The following diagrams represent models for pure elements, pure compounds, mixtures of elements, and mixtures of compounds. *Use them to answer questions 6–9.*

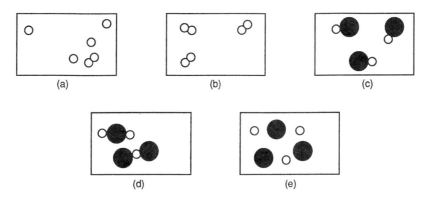

Figure 11.1. Test items on coordinate concepts of *atom* and *molecule*.

6. Which diagram(s) could represent a pure element?
   A. (a)
   B. (b)
   C. (e)
   D. both (a) and (b)
   E. none of these

7. Which diagram(s) could represent a pure compound?
   A. (b)
   B. (c)
   C. (d)
   D. both (b) and (c)
   E. none of these

8. Which diagram(s) could represent a *mixture* of *elements*?
   A. (c)
   B. (d)
   C. (e)
   D. both (d) and (e)
   E. all of these

9. Which diagram(s) could represent a *mixture* of *compounds*?
   A. (c)
   B. (d)
   C. (e)
   D. both (d) and (e)
   E. all of these

Testing examples must differ from teaching examples. If they do not, knowing whether students have only memorized the teaching instances or have actually developed the concept is impossible.

Many authors recommend that testing examples be listed in a concept analysis. I have not done that because there is no real difference in teaching and testing examples. However, it is important to collect a large sample of instances. They will be needed for teaching, testing, and reteaching.

ANALYZING TEST RESULTS

Data from tests can be analyzed in several ways. If carefully constructed multiple-choice items are used and machine-scored, the proportion of students selecting each distracter will provide important clues. If free-response items are used, careful reading of a random or representative sample of papers will provide similar information.

Although a great deal of information about student misconceptions can be obtained from incorrect choices on carefully prepared multiple-choice exams, the distracters must be selected with diagnosis in mind. One of the best ways to identify useful distracters for multiple-choice items is first to ask the question in a free-response format. When the free-response tests are graded, look for common errors or misconceptions and tally them. If what went wrong is not clear from a student's response, ask the student to explain how he or she went about answering the question when the test papers are returned. Then use the common errors to construct distracters for multiple-choice questions.

After several years of this activity—less, if you share items with colleagues—you will have a sizable bank of good multiple-choice questions and understand common misconceptions and errors well enough to construct suitable multiple-choice questions without going through the preliminary step of giving free-response items first.

What I have just described is a general procedure for using routine tests to obtain detailed information about misconceptions. Once that information is in hand, the instructional implications are often obvious. You will find yourself saying things such as, "I'm not surprised that many students would use CCl_4 when asked to 'add a small amount of chloride.' I have taught them to name binary compounds as chlorides, but I have not discussed the difference between ionic compounds that yield chloride ions in solution and covalent compounds that do not."

The reasons why a large number of students hold a particular misconception or why a particular error occurs frequently are not always clear. I find that a formal analysis of the concept often reveals difficulties that are not obvious from the test results.

IMPLICATIONS OF LEARNING THEORY

Little has been said about learning theory in this chapter. In as much as this book is an attempt to relate theory to practice, let us now consider how the suggested practices are related to the theory presented in earlier chapters.

We assume that learning is goal-directed, that we select goals that have survival value, and that choices among goals or paths to the same goal adhere to the Principle of Least Cognitive Effort. If we accept these assumptions, learning will be increased if we can reduce the effort required to learn and increase the benefits.

Students are hampered in learning some chemistry concepts because they see little use for them. For example, the concepts of *atom* and *molecule* have little value unless one needs to interpret events that are more easily rationalized in terms of a microscopic model. As I examine standard introductory textbooks, I find much material that an 18-year-old would never use outside of the chemistry classroom. To the extent that my criticism is valid, modifying the curriculum to stress concepts that students can use *here and now* and *in the "real" world* will enhance learning.[1]

Difficulties described in this and the previous two chapters can be summarized by saying that learning what we cannot experience is difficult. Feuerstein described this problem in terms of level of abstraction. Level of abstraction describes the "cognitive distance" between our mental activity and the object of that activity. Concepts such as *tree* have numerous perceptible examples and are close to experience; they are less abstract than concepts such as *atom* that have no perceptible examples and cannot be experienced directly.

The strategy suggested for teaching abstract concepts such as *atom* is to reduce the abstractness of the concept by using animated sequences and models that are closer to the concept than words or mathematical equations. Unfortunately, this strategy can go only so far. Research suggests that we overcome the difficulty of abstractions by building schemas (what Piaget calls logical operations) that enable us

---

[1]Anecdotal support is found in a recent student comment. In the summer of 1994, we began using *Chemistry in Context* as the text for an introductory course. Students who were retaking the course commented that "this text is more difficult than the one used last semester, but I like it much better. It tells you what chemistry is good for and why it is important to learn."

to reinterpret the signals coming through our senses in terms of more abstract schemas such as number, proportion, probability, reversibility, serial order, and control of variables. In effect, we learn to translate words and other stimuli into images, we learn to abstract general properties from objects and relationships, and we learn to manipulate these abstractions as though they were real things. The more sophisticated the logical operations that students have developed, the easier they learn concepts such as those encountered in chemistry.

This book emphasizes the ideas that learning is a constructive process and that the teacher's role is to mediate learning. This mediation consists of providing opportunities for directly experiencing phenomena, asking students to explain what they see, providing language and describing models capable of explaining what is experienced, ensuring that students have ample feedback concerning the match between the schema they construct and the concept held by others, and providing opportunities for students to apply new concepts in a variety of settings. The teaching strategies suggested in this chapter are consistent with that theoretical position.

SUMMARY AND DISCUSSION

A great deal has been said about concepts in Chapters 9–11, and we need to pull ideas together. When we speak of concepts, we are talking about *classes of objects, events, or other entities that we group together on the basis of shared characteristics.* Each class is referenced by a particular name or symbol. Concepts constitute a large part of knowledge. They are the basic building blocks for principles, rules, and other kinds of knowledge.

Most concepts develop naturally as a result of informal experience. Natural concepts form as a result of detecting features that are common to objects and events that we encounter. However, we are not conscious of these features, and concept schemas do not list concept features. Rather, schemas store in memory a prototype for the concept. This prototype incorporates the idealized features of the concept and allows us to identify instances as examples when we encounter them.

Most information can be stored at various levels; we can develop concept schemas for elements and incorporate information about sodium, iron, oxygen, and so forth within that schema. Conversely, we can develop very particular schemas for sodium, iron, oxygen, and so forth. We hold concepts in memory at a basic level. We then operate on these basic-level concepts to generate supraordinate and subordinate categories as needed.

Our basic-level concepts are compromises that provide the greatest amount of information for the least cognitive effort. What level this is depends on the intellectual tasks we perform. The basic level for an expert differs from the basic level for a lay person because the two groups perform different intellectual tasks with the concepts.

Insisting that lay people develop the same abstract concepts that are basic-level concepts for experts is counterproductive. The effort required to develop these concepts is great, and the savings in cognitive effort from using them is nil.

Although most concepts develop naturally, the process has limitations. Concepts such as *atom, molecule, mass, acceleration, mathematical limit, square root,* and *probability* have features that are undetected through experience. Some concepts are invented and have meaning only through social convention. Such concepts cannot develop naturally. They must be invented.

The normal procedure for teaching concepts is to describe the distinguishing features (critical attributes) of the concept and present examples to illustrate those features. When this procedure is done without consciously analyzing the concept, forgetting certain features or selecting examples that lead students to believe that attributes are critical when they are not is easy. For example, students may believe that elements must be solid at room temperature because only elements with this characteristic were used as examples.

Concept analysis is a formal procedure to identify the critical features of a concept, consider variable attributes that might be confused with these critical features, and then select examples and nonexamples that could be used to clarify the concept. By carefully selecting the instances, presenting them to students, and asking students to decide whether the instance is an example, students can be led to construct a schema for the concept that corresponds to the one held by others.

Teaching based on careful concept analysis has been demonstrated to be effective, but difficulties exist. First, concept analyses require a great deal of work—more than teachers could do if the procedure were followed for every concept taught. A reasonable approach is to use the procedure for concepts that cause difficulty and to let other concepts develop naturally.

Other difficulties with the recommended procedure are that it is ineffective when the concept has no perceptible examples, when the attributes that we want to show are not perceptible in the examples, when the examples require students to apply principles that they do not know, or when the examples lead students to confuse the concept with labels for the concept. These difficulties were discussed in this and the previous chapter, they were illustrated with examples from chemistry, and suggestions were made for overcoming them.

12

Relationships Among Concepts: Propositions, Principles, and Rules

I have given a great deal of attention to concepts—classes of objects and events that are the basis of thought. I have argued that we should distinguish concepts from relationships among them to clarify learning difficulties. I do not wish to suggest that relationships among concepts are unimportant. To the contrary, such relationships constitute a large part of knowledge. This chapter focuses on relationships among concepts and how we go about learning them.

Vast literature exists on concept learning, but I am aware of little research pertaining to learning propositions and principles. Consequently, this chapter is far more limited than the previous three. In research based on an information-processing model, most relationships among concepts fall into the category of declarative knowledge. Declarative knowledge connects concepts in the form of propositions and adds interiority to concept schemas. Most research on learning relationships among concepts has been done with concept maps.

CONCEPT MAPS

"*Propositions* are elemental linear strings of information units that are connected grammatically" (Lambriotte et al., 1989, p 349); or as Novak and Gowin put it, "*Propositions* are two or more concept labels linked by words in a semantic unit" (1984, p 15). Novak and Gowin see "concepts and propositions composed of concepts [as] the central elements in the structure of knowledge and the construction of meaning" (p 7). They represent propositions in semantic networks called *concept maps*. Many others have found concept maps (also called semantic networks, semantic maps, and knowledge maps) useful in teaching propositions.

A review of semantic maps by Lambriotte et al. (1989) indicated that the largest body of work on knowledge maps has been in reading comprehension. A meta-analysis of 23 research studies in this area revealed that their value has been insignificant to moderate, but recent work based on mapping strategies developed at Texas Christian University (TCU) and Cornell University suggests that concept maps can be useful in a variety of learning situations.

EXAMPLES OF CONCEPT MAPS

The maps developed at TCU (Figure 12.1) and Cornell (Figure 12.2) are similar, but they differ in details. Novak and Gowin suggest that concept maps should always be hierarchical, but the TCU team argue that "it [is] convenient and even necessary to sometimes make non-hierarchical maps" (Lambriotte et al., 1989, p 352). The lines joining concepts in the TCU maps are arrows showing the direction of the link; the Cornell group omits arrows except in cases where direction cannot be inferred from the hierarchical arrangement of the map. Letters standing for a limited set of words showing the type of link are used at TCU: a for analogous to, p for part of, t for type of, c for characteristic, l for leads to, e for evidence for, and so forth (*see* Figure 12.1). Letters on directed links stand for the kind of linkage; dotted links indicate sequential events (Dansereau, 1985, p 223). The Cornell group writes on the lines linking concepts any words needed to clarify the proposition (*see* Figure 12.2). These and other differences in the mapping strategies employed by the two groups are spelled out in Lambriotte et al. (1989) and Novak and Gowin (1984).

LIMITATIONS OF CONCEPT MAPS

Concept mapping, like other heuristics, requires time and effort to learn. Unless students are convinced that a payoff will result in terms of less effort later or greater comprehension, they will not invest the time and effort. Furthermore, the quality of student-produced maps varies greatly, even after extensive training (Lambriotte et al., 1989, p 362).

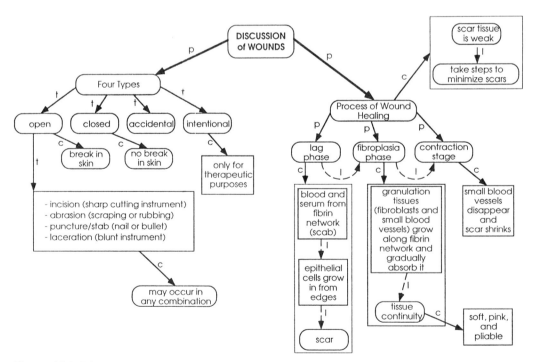

Figure 12.1. TCU concept map of a nursing textbook chapter. Key concepts are shown in ovals; descriptive information is in rectangles.

To find out how students use concept maps, Briscoe and LaMaster (1991) introduced mapping early in a biology course and interviewed students at the end to see how they used maps. Many students did not. The students cited the effort required to construct maps and pointed out that they could pass most exams by memorizing, which they found easier. Students who found the maps useful typically constructed one map when a topic was taught; a second map two or three days before an exam; and a final, organized map after filling in gaps revealed by the second map.

EXPERT-PRODUCED MAPS

Because of the difficulties with student-prepared maps, the TCU group's recent research has focused on expert-produced maps. They have used them in a variety of ways to improve recall of factual information. Cliburn (1990), whose concept maps derive from the work at Cornell, also reports positive results from using expert-produced maps. Cliburn begins with an overview map showing an entire unit. This map is followed by smaller, more detailed maps for each topic as the topic is presented. When new units are introduced, he uses *comparative maps* showing relation-

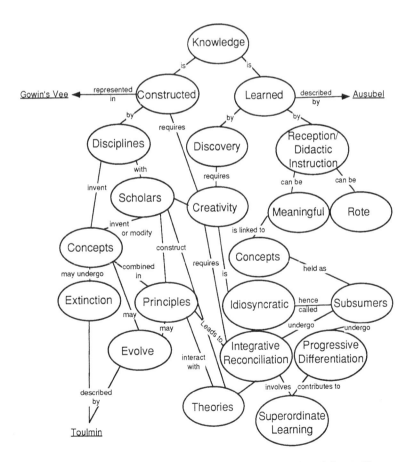

Figure 12.2. Concept map showing major ideas presented in Novak and Gowin. Key concepts are shown in ovals; appropriate linking words form key propositions (Novak and Gowin, 1984, p 2).

ships between the unit just finished and the unit he is introducing.[1] He uses color-coded maps that join individual concept maps to trace concepts throughout a unit and facilitate what Ausubel (1963) and Novak and Gowin (1984) described as *integrative reconciliation*. Cliburn studied the effectiveness of his procedures and found that those who were presented concept maps learned more and retained more than those who only heard lectures and used text.

RESEARCH ON CONCEPT MAPPING

Among the conclusions drawn from TCU-based research are the following:

1. Students are able to acquire more effective metaknowledge about the domain to be studied from a brief exposure to a map (4 min) than from a brief exposure to text. That is, they appear to have a better idea of how to budget their study time and are able to predict their subsequent free-recall performance more accurately.
2. Procedural/process maps (e.g., the flow through the digestive system) appear to be more effective than conceptually descriptive maps (e.g., a hierarchical description of the renal system). The predominant processing route in the procedure/process maps may reduce confusion and nonproductive processing.
3. The procedural/process maps are generally more effective than their text counterparts, especially for the main ideas and structure of a knowledge domain. The parts of the maps that are visually more salient appear to be recalled better.
4. There is some evidence that maps may inhibit the processing of content-related pictures. The spatial processing of the maps may interfere with the visual processing of the pictures.
5. Students with low prior knowledge about a subject area benefit greatly from the use of maps during lecture.
6. Maps facilitate cooperative interactions among students with high verbal ability.

. . . [W]e are . . . not sure about when it is more effective to engage a learner in map generation as opposed to providing an "expert" map to be interpreted. (Lambriotte et al., 1989, pp 362–363)

Other research has used concept maps to compare biology textbooks (Lloyd, 1990), to examine the effect on student anxiety (Olugbemiro et al., 1990), to compare the value of concept maps and text as aids in cooperative teaching (Patterson et al., 1992; Rewey et al., 1989), to trace students' constructions of knowledge over time (Beyerbach and Smith, 1990), to assess changes in understanding that result from teaching experiments (Wallace and Mintzes, 1990), and as aids for science curriculum development (Starr and Krajcik, 1990; Novak and Musonda, 1991).

MAPPING TOOLS

The effort required to construct concept maps is a practical constraint that limits their usefulness in classroom settings. Several researchers have used computer soft-

---

[1]Although Cliburn does not describe the theoretical basis for comparative maps, the parallelism in the maps should facilitate *patterned generation* of new schemas as discussed in Chapter 9.

ware as an aid in developing various kinds of maps (Beyerbach and Smith, 1990; Fisher, 1990; Fisher et al., 1990). Various tools do different tasks. In describing SemNet, one of the better developed software packages, Fisher (1990) said:

> Computer-based semantic networks differ from paper-and-pencil maps in that they are *n*-dimensional; each concept can be linked to many other concepts; relations are bidirectional; representations can include images, text, and sound; and nets can be very large. Disadvantages of SemNet networks include (a) the difficulties in obtaining a clear overview and (b) the homogeneous nature of the representations, in which all links look alike. Advantages include the ability to integrate ideas across a large knowledge base, the ease and rapidity of net creation, the ease with which elements (concepts, relations, or propositions) can be found within nets, and the utility of nets as self-study tools. Concept mapping and semantic networking are complementary strategies that can be used effectively in tandem to help students learn, to help teachers teach, and to support cognitive research (p 1001).

Fisher et al. (1990) drew these conclusions from their research with SemNet:

> The test scores of students using SemNet as a study tool in an introductory biology class increased steadily, relative to the scores of students not using Sem-Net. Further research is necessary to determine if these increases are attributable to depth of processing, time on task, or other factors. . . . As students create or review their nets . . . they see just one frame or screenful of information at a time. This frame serves as a "chunk" of information that theoretically can be encoded by the student in long-term memory. . . . Students tend to have difficulty interrelating concepts widely separated in a text or series of lectures. . . . When students build semantic networks as they learn, concepts encountered later in the course are often brought into close juxtaposition with concepts encountered earlier, thus reducing the physical proximity (and therefore the conceptual distance) between them (p 351).

Concept maps can aid learning declarative knowledge of relationships among concepts. But, like behavioral objectives, cooperative learning, mastery learning, learning cycles, and numerous other teaching tools that have been shown to enhance learning, the value of concept maps depends on variables that change from one classroom to another. Teachers must experiment with such tools to learn how to use them in their own classrooms.

PRINCIPLES AND RULES

The declarative knowledge shown on most concept maps is not difficult to learn if the concepts connected on the maps are understood. There are, however, principles (Gagné (1977) calls them rules) that cause special difficulties. The following rules are representative:

- A body at rest will remain at rest and a body in uniform, straight-line motion will continue its uniform motion unless it is acted on by an unbalanced force.
- The total pressure of a gas is equal to the sum of the partial pressures of its constituents.
- At constant temperature, the volume of a gas is inversely proportional to its pressure.

- The kinetic energy of an object is directly proportional to its mass and to the square of its velocity.
- Equal volumes of gases at the same temperature and pressure contain the same number of molecules.
- When a system at equilibrium is subjected to a stress, the system responds to reduce the stress.
- When the equations for two or more chemical reactions can be added to produce the equation for another reaction, the enthalpy change for the reaction represented by the final equation is the sum of the enthalpy changes for the reactions described by the equations that are added.

> As an exercise, work with two or three other people to identify other chemistry principles that students find difficult.

WHY PRINCIPLES ARE DIFFICULT TO LEARN

Students have difficulty with these principles for a variety of reasons:

1. The concepts inherent in the principle are poorly understood. (*See* Chapter 11 for a discussion.)
2. Many chemistry principles are expressed as mathematical relationships, and students inadequately understand the mathematical statements of the principles.
3. Many principles in chemistry involve proportional relationships, and students lack proportional reasoning. (Problems associated with proportional reasoning are discussed in Chapter 15.)
4. A variety of language problems can impede understanding. (These problems are discussed in Chapter 13.)

Principles such as those stated may or may not be clearly represented by concept maps. My attempt to represent important principles pertaining to ideal gases is shown as Figure 12.3. Although that map could undoubtedly be improved, the map alone would probably not be instrumental in clarifying the various principles alluded to in the map. However, small-group discussions aimed at elaborating and clarifying the relationships represented by the map *might* be effective. To my knowledge, no research has examined this question.[2]

> As an exercise, suggest hypotheses that could account for student difficulties with the principles that were identified in the previous exercise. Think of ways to test your hypotheses when you encounter students who experience difficulty with the concept.

Research by Starr and Krajcik (1990) suggested that small-group discussion such as that suggested previously can be a powerful tool in clarifying difficult relationships. In their research, teachers used concept maps as a tool for science curriculum development. The primary benefit seemed to result from discussions among teachers as they produced successive revisions of their curriculum maps.

---

[2]Patterson et al. (1992) and Rewey et al. (1989) reported research in which concept maps were used in a cooperative learning environment, but the focus of their research was recall of factual information rather than understanding of difficult principles.

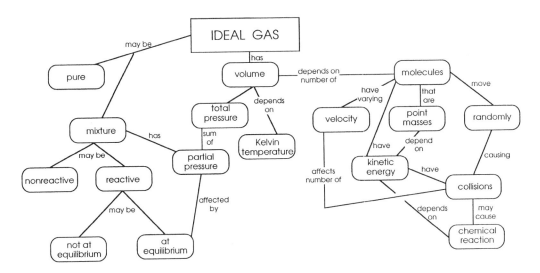

Figure 12.3. Concept map showing principles related to ideal gases.

> The revision process . . . [creates] a learning experience in which the teachers increase their knowledge of their own individual thinking processes and structures, the understandings of their co-workers, and also, what they can anticipate for student learning. In addition, the revision process is an opportunity for teachers to consider and discuss the importance of individual concepts, the placement of the concepts on the map (including the relationships between concepts), and the propositions which are used to connect concepts (p 997).

It may be possible to analyze chemistry principles and separate them into classes that present similar learning difficulties, much as classes of concepts were identified in Chapter 10. I am aware of no attempt to do this. Until such research is carried out, all that I can recommend is that each principle that you have difficulty teaching be analyzed to identify and eliminate potential sources of difficulty as outlined in Appendix I. Zoller (1990) made similar recommendations. Although we have not developed special procedures for analyzing principles, the kind of analysis suggested for attributes such as mass (Appendix I) should be helpful. Gagné (1977) described mass as a "defined concept" and argued that defined concepts are learned in the same manner as principles and rules.

EVALUATING PRINCIPLE LEARNING

Chemists seem to have little trouble writing test items that require students to demonstrate understanding of principles and rules. Several such items were given in Chapter 11 as examples of items that test for *more* than concept learning. Nothing is wrong with those items, so long as we understand that they provide information about student understanding of rules and principles rather than individual concepts.

Tessmer, Wilson, and Driscoll (1990) argued that concept learning should be expanded to include relationships among concepts—what I prefer to call propositions, principles, or rules. Their article includes several suggestions for assessing student understanding of principles:

Performance measures that reflect "higher" kinds of cognitive activity are using a concept (e.g., writing stories or sentences, role playing) and generating inferences about a concept (e.g., theorizing, drawing implications, understanding context). These uses and inferences reflect a conceptual-tools view of . . . learning, where the inferences drawn from concepts and uses of the concept are part of the purposes and determinants of . . . learning (p 48).

As an exercise, decide how you would determine that students understand each of the difficult principles that you identified in the section of this chapter called **Principles and Rules**.

SUMMARY

This brief chapter calls attention to knowledge such as propositions, principles, and rules that state relationships among concepts. The body of research pertaining to learning such relationships is limited. Most recent research pertaining to relationships among concepts focuses on concept maps.

Concept maps are two-dimensional diagrams showing concept labels connected by lines with words written on each line to describe a relationship between the concepts. Concept maps can be constructed in several ways. The techniques developed at Texas Christian and Cornell Universities are among the most popular, and differences between them appear to be unimportant.

Evidence suggests that students who use concept maps to organize learning in knowledge-rich domains such as science are able to recall more factual information than those who only listen to lectures and read texts. Both student-prepared and expert-prepared maps have been used successfully, and it is still not clear when one may be better than the other. The time and effort required to prepare concept maps is so great that many students refuse to complete the task.

The principles and rules that cause the most difficulty in chemistry have yet to be adequately represented by concept maps, but small-group discussion of even crude maps of such principles might be an effective way for students to clarify their understanding of those principles. Research is needed to find out if this method works consistently.

13

The Role of Language in Teaching Chemistry

Throughout this book, I have emphasized the idea that learning is a constructive process. I may have left the impression that the role of the teacher is insignificant. This, of course, is not true. However, the effectiveness of the teacher is enhanced when the teacher is aware of how much the internal operations of the learner modify the external presentations of the teacher. At no point does this awareness seem more important than in the case of language.

This chapter describes psychological research related to language comprehension, reviews research in chemistry education on misunderstanding language, and suggests teaching practices that enhance understanding.

HOW WE UNDERSTAND LANGUAGE

Although people have used language for centuries, we are a long way from understanding how people comprehend language or how they learn to use it. However, research in linguistics and cognitive science is beginning to provide information about what is important.

LANGUAGE PROCESSING

Presumably you understand what you are reading now, but what enables you to understand? Several things are involved:

1. You decipher symbols. You recognize one shape as the letter *a*, and another shape as the letter *e*. You distinguish a *d* from a *p* and so forth.
2. You recognize words that denote ideas: *cat* stands for an animal with defined characteristics; *sat* stands for past action from a standing to sitting position.
3. You apply rules to decipher sequences that denote more involved ideas: "The cat sat" makes perfectly good sense, but "Sat cat the" does not; it violates the syntactic rules you have learned.

BOTTOM-UP PROCESSING. The section called **Learning as Adaptation** in Chapter 5 (p 43) called attention to the fact that all learning is influenced by information coming from the external environment as well as schemas that we have previously built in our heads. In reading, the external environment is the printed symbols that we interpret. *Bottom-up processing* refers to the way these symbols and other external stimuli evoke knowledge in memory to suggest hypotheses concerning meaning (Brown, Collins, and Harris, 1978, p 12). It is the process of interpreting sensory information coming from the external world.

TOP-DOWN PROCESSING. Bottom-up processing obviously takes place. What is less obvious is top-down processing, which accompanies it. *Top-down processing* refers to the way existing schemas compete with one another to provide the best account for the words we hear or see (Brown, Collins, and Harris, 1978, p 12). Top-down processing is governed by existing knowledge and by contextual cues that lead us to relate particular segments of knowledge to the message under consideration. The way that context affects meaning is easily illustrated.

"I want to buy an apple" spoken in a fruit market elicits a different meaning from the same words spoken in a computer store. Various meanings of *apple* coupled with the context in which it is used leads us to infer whether fruit or a computer is desired. This example is as trivial as it is obvious. However, research on top-down processing provides less obvious and more significant examples. The richness of what we understand about such simple words as *buy* is not immediately obvious. What does *buy* suggest?

DEVELOPING MEANING

IMPLICATIONS OF BUY. First, *buy* implies a transaction between two or more agents. Some agent called a *buyer* and another agent called a *seller* must exist. The transaction involves some kind of transfer. The buyer is expected to exchange some kind of token (money) for something (goods) currently in the possession of the seller. At the conclusion of the *buy* transaction, the money is in the possession of the seller, and the goods are in the possession of the buyer.

None of these facts about *buy* is explicit in the word, but these facts are understood and used as we interpret sentences that incorporate the word. In the sentence "I want to buy an apple" the "I" in the sentence is easily identified as the buyer, and apple is seen as the goods presently in the possession of the seller. We are likely to infer, at least for the present, that "I" has money to exchange for the apple. Such inferences are made subconsciously and derive naturally from previously learned meaning about *buy*, other words in the sentence, and transactions that take place in the real world. All communication takes place in this kind of semantically rich environment, and meaning is determined as much by inferences based on previous knowledge as by the words we hear or see. This semantic richness is not acquired overnight.

DEVELOPING SEMANTIC MEANING. In discussing research on children developing understanding of possession verbs (give, take, trade, pay, buy, sell, and spend money), Norman, Gentner, and Stevens (1976) concluded thus:

> It takes children approximately five years to progress from the state in which they use words like 'give' and 'take' properly to the point where they use the

entire set of possession verbs properly. . . . Children must learn a number of different concepts—some dealing with social conventions, some dealing with language, and others dealing with the physical properties of objects and locations. All these concepts must be understood, and then, for proper linguistic labels to be assigned, they must be interrelated (pp 182–183).

Once we acquire information about the linguistic rules that apply to utterances and information about word meaning, we apply these notions with considerable fluency. We do so subconsciously and with little effort. For example, we readily read "T/–\E C/–\T," interpreting the first /–\ as an "H" and the second one as an "A," even though the actual symbol is neither. We know what *belongs*, and we put it there. Similarly, enough redundancy usually exists in communication for us to infer what belongs, even when something else is there. If I substitute an x fox evxry xhixd lxttxr in xhix sextexce, xou xtixl mxy bx abxe tx rexd ix (Anderson, J., 1980, pp 43–45). You would be less successful if the words and message were less familiar.

None of us deduces meaning from written materials very efficiently until processes such as those mentioned previously are carried out automatically. The following quotation from Adams (1980) sums this phenomenon up nicely:

> In G. Stanley Hall's words (1911), true reading only occurs "when the art has become so secondarily automatic that it can be forgotten and attention be given solely to the subject matter. Its assimilation is true reading, and all else is only the whir of the machinery and not the work it does (p 134)" (p 13).

Typically, students who study chemistry are good readers; they process English automatically. They decode well, and they have large vocabularies. Still, chemistry students have difficulty reading their chemistry textbooks. New ideas are introduced so rapidly that students overload working memory. In processing a sentence, students encounter many unfamiliar words (or familiar ones in an unfamiliar context). These unfamiliar words require students to quit processing the sentence to process the word. Many such pauses result in students reaching the end of the sentence, only to realize that they have forgotten the beginning. If the content of each sentence is complex, the additional demand on working memory may make the thought of the sentence difficult to place into larger segments of meaning.

"True reading"—G. Stanley Hall's kind—can only take place if we have the right concepts in memory and if those concepts are activated. How these factors operate is illustrated by experiments conducted by John Bransford and his associates.

IMPORTANCE OF CONTEXT FOR MEANING. In one of Bransford's studies, students were asked to read the following passage. They were told that they would be asked to recall the passage later.

> The procedure is actually quite simple. First you arrange things into different groups. Of course one pile may be sufficient depending on how much there is to do. If you have to go somewhere else due to lack of facilities that is the next step; otherwise you are pretty well set. It is important not to overdo things. That is, it is better to do too few things at once than too many. In the short run this may not seem important but complications can easily arise. A mistake can be expensive as well. At first the whole procedure will seem complicated. Soon, however, it will become just another facet of life. It is difficult to foresee

any end to the necessity for this task in the immediate future, but then one never can tell. After the procedure is completed one arranges the materials into different groups again. Then they can be put into their appropriate places. Eventually they will be used once more and the whole cycle will then have to be repeated. However, that is a part of life (Bransford and McCarrell, 1974, p 206).

You will not be surprised that students who read this passage had poor scores on both comprehension of the passage and recall of what it said. The passage does not make much sense. However, once you know the passage is about washing clothes, it is quite sensible. Students who read the passage after being told the subject of the passage scored well on both comprehension and recall. *Being able to make sense of what is read is important for remembering as well as understanding.*

It is tempting to dismiss Bransford's study as a cute trick, but one that has no relevance to chemistry. We would not think of including a passage such as this in a book or in a lecture without identifying the subject as washing clothes. Quite true, but we frequently include passages in chemistry books and chemistry lectures about things that we name without ensuring that the word calls forth the concept that provides meaning to the passage. I repeat here, for those who may have missed the point in Chapter 10: I have many students who use units such as square centimeter or cubic centimeter and terms such as area, volume, pressure, density, and chemical change without any apparent awareness that they do not know the meaning of the words or that constructing meaning for the words is essential for continued learning. *In large measure, teaching reading is teaching concepts, and that part of the teaching of reading is the responsibility of the chemistry teacher, not the reading teacher.* Furthermore, not understanding the concepts labeled by words is the reading difficulty that is most prevalent in chemistry classes.

Understanding at the time we process information is important. Students who were told the subject of the clothes washing passage *after* reading it did no better on tests of comprehension and recall than students who were never told the subject.

Bransford and McCarrell (1974) reported the results of similar experiments that supported their conclusions "(a) that [people] do make cognitive contributions while comprehending; (b) that certain contributions are prerequisites for achieving a click of comprehension; (c) that knowledge of abstract constraints on entities and relations plays an important role in determining [people's] contributions; and (d) that meaning is the result of such contributions and is best viewed as something that is 'created' rather than stored and retrieved" (p 201).

Many researchers obtained results similar to Bransford's (Barclay, 1973; Craik and Lockhart, 1972; Dooling and Lachman, 1971; Woodward, Bjork, and Jorgeward, 1973). The implication of the studies is that we store information in terms of *meaning* rather than literal messages. We can, of course, process words at a surface level. We are forced to do this with nonsense syllables or even *potentially* meaningful messages such as the clothes washing passage without the title. However, *the evidence is strong that when we process information at a surface level, it is difficult to retain.* We are equipped to process information at a deeper level of meaning. This ability sometimes causes trouble.

Distortions of information read have been found in numerous studies. Bartlett's well-known study (1932) was described earlier. As Mary diSibio (1982) notes, "Bartlett noted that [students] were especially prone to rationalize, reconcile, omit, or distort just those features of the story that would likely be most unusual for British readers" (p 151). These errors in recall increased over time.

Clearly, we try to make sense of what we read, and the only way we make sense of anything is in terms of what we already know. *If what we read is inconsistent with what we know, we either distort the message to make it conform to our knowledge and expectations or we fail to comprehend the message at all.* The implication of the previous sentence is obvious in the research on misconceptions in science.

LANGUAGE PROBLEMS IN CHEMISTRY

Several studies disclosed language-related difficulties of chemistry students. These difficulties are as follows:

- lack of understanding of familiar words used to convey meaning in chemistry,
- lack of understanding of technical terms introduced in the study of chemistry,
- ascribing a familiar meaning to a common word used in a technical sense,
- using everyday meaning to draw incorrect inferences about chemical events, and
- failing to learn the conventions applied to specialized chemical language to the level of automatization required to "read chemistry" fluently.

In the early 1970s, Gardner (1972) surveyed Australian students to assess their understanding of ordinary words used in science. The 599 words in Gardner's study were among the 20,000 most frequently encountered words of English (Thorndike and Lorge, 1944), and the items on his test were meant to test understanding in an everyday context rather than a specialized science context.

Gardner's work was extended in England (Cassels, 1976; Cassels, 1980; Cassels and Johnstone, 1980). The essential outcome of these studies is that a substantial number of students do not understand common words used to teach and test students in science.

Just what one should make of these studies depends on one's point of view. At what point should we become alarmed? Few words in either study were completely understood by all students. (That is, virtually none of the multiple-choice questions used in the studies was answered correctly by 100% of the students.) On the other hand, by the time students were in the fourth form (16–17 years old), more than 70% of the students surveyed seemed to understand the vast majority of the words included in the survey (Cassels and Johnstone, 1980, p 9). Is this outcome good or bad?

Normative data such as the proportion of a national sample who respond correctly to particular items do not seem important to me. None of us teaches a national sample of students, and what words are misunderstood is influenced by local conditions. How much fluctuation may exist is suggested by data from one of Cassels' studies (1976, pp 13–15).

CULTURAL EFFECTS ON WORD MEANING

Data for a school drawing from a low socioeconomic population were compared with data from a school drawing from a high socioeconomic population. The proportion of students in the respective schools who understood some of the words varied considerably: 18% vs. 68% for *converse*, 33% vs. 68% for *spontaneous*, and 48% vs. 86% for *correspond*.

American readers will appreciate how common usage can affect results of such studies from these items taken from Cassels and Johnstone (1980, p 42):

---

1. The pupils were told to *revise* their work. They had to
    a) complete it.
    b) put it into a folder.
    c) put it into paragraphs.
    d) punctuate.
    e) read it again to learn it.
2. If you *revise* something you
    a) write a summary of it.
    b) change or alter it.
    c) read it quickly.
    d) copy it out.

Between 94 and 100% of the British students gave the correct answer (e) for item 1, but only 5–9% gave the correct answer (b) to item 2. The results would almost certainly be reversed for an American sample. Americans use *revise* only in the sense of item 2 and use *review* in the sense of item 1.

LINGUISTIC CUES TO WORD MEANING

The meaning of a word is influenced by contextual cues and knowledge of what fits as much as by the meaning of the word in isolation. Once again, the point is illustrated by examples from Cassels and Johnstone (1980, pp 43–45).

Item 3 was answered correctly by 100% of students in year 1 through year 6 of secondary school (12–18-year-olds), but item 4 was answered correctly by only 51% of students in years 1–2 and by only 87% of students in years 5–6.

---

3. To start one of those egg timers with sand in them, you have to *invert* it. You have to
    a) shake it.
    b) turn it upside down.
    c) lay it on its side.
    d) put more sand in it.
4. To *invert* an object means
    a) to turn the object upside down.
    b) to make an object that no one else has made.
    c) to turn the object on its side.
    d) to decorate the object.

The correct answer to item 3 is easily inferred from real-world knowledge of egg timers. In fact, the item would almost certainly produce a majority of correct responses if the stem for the item ended after "with sand in them" and leaving out the word that is supposedly tested. In contrast, the stem for item 4 contains no information that permits the student to relate *invert* to any real-world knowledge. In essence, the item tests the student's understanding in isolation.

Items 5 and 6 show how knowledge about speech patterns enables us to select the correct word for a sentence even when understanding of the word is incomplete.

> 5. Mrs. Smith applied to the firm to become an _____ to sell cosmetics.
>   a) agent
>   b) agile
>   c) active
>   d) avenue
> 6. An *agent* of change is something which
>   a) is changed itself.
>   b) stops a change.
>   c) is not changed at all.
>   d) is an enemy of change.
>   e) brings about the change.

Subconscious knowledge of syntax is enough to eliminate *agile* and *active* from consideration in item 5, and few students will have heard *avenue* used in reference to a person. Linguistically, *agent* is the plausible choice, and linguistic cues were undoubtedly used by many of the 96–98% of students who selected *agent* as the correct response to item 5.

In contrast, nothing in item 6 gives away the meaning of *agent* when used in the particular sense "agent of change," and the proportion of students responding correctly to this item ranged from 48% of students in years 1–2 to 100% of students in years 5–6.

IMPORTANCE OF SCIENCE CUES TO WORD MEANING

The point of the preceding discussion is to emphasize the role of real-world and linguistic knowledge in giving meaning to words in context. Equally important is the knowledge that words that *are* understood in isolation will not be seen to apply in a science context unless the science context is also understood. Again, here are some examples from Cassels and Johnstone:

> 7. If I *detect* something I
>   a) find it out.
>   b) write it down.
>   c) throw it away.
>   d) leave it behind.
>   e) cancel it out.
> 8. The use of a flame allows hydrogen gas to be
>   a) displaced.
>   b) distorted.
>   c) decomposed.
>   d) detected.
>   e) dehydrated.

Between 90 and 100% of the students surveyed responded correctly to item 7 by indicating that detect is a word that is understood. However, only 40% of the years 1–2 students and 79% of the years 5–6 students responded correctly to item 8. Have they suddenly forgotten what it means to detect? This is unlikely. It is more likely that students

- have never seen a lighted splint stuck into a bottle of hydrogen gas as a test for hydrogen,
- do not associate this startling experience with the fact that it is a way of detecting hydrogen, and
- recall flames being used to heat gases so that they expand and become displaced or recall flames being used to decompose and dehydrate various chemicals.

A chemistry teacher with years of experience will have no difficulty associating the stem of item 8 with its intended answer, but the students' basis for such an inference is far more tenuous.

These examples make clear the idea that our common complaint that English (or math) teachers have not done their jobs is unfair. Many—perhaps most—of the reading problems that students have in chemistry are problems that the English teacher could not possibly anticipate or correct. *Every teacher must be a reading teacher because the most important part of reading is the top-down process by which existing schemas impose* particular *meaning to words in a given context.* The context in which chemistry students impose meaning to words is a chemistry context, and only chemistry teachers can teach what inferences are appropriate in that context.

ORDINARY WORDS WITH TECHNICAL MEANING

The preceding examples illustrate problems of using an ordinary word in a scientific context without changing the meaning of the word. Additional problems occur when the ordinary word takes on new meaning. The following item, taken from a college chemistry exam, illustrates the point.

Which of the following contains the *strongest* acid?
  a) 2 M NaOH(aq)
  b) 1 M NH$_3$(aq)
  c) 1 M HCl(aq)
  d) 2 M HC$_2$H$_3$O$_2$(aq)

As used in everyday speech, strong often means "concentrated" (as in strong coffee) or "powerful" (as in strong detergent), but in chemistry it means "highly dissociated" (as in strong electrolyte). Here, students' understanding of a word from real-world experience hinders rather than helps in the understanding of the word in a new and unusual sense. Even when the new meaning has been learned, the more familiar and well-practiced meaning may win out as top-down processing imposes meaning in the chemistry context.

In everyday speech, the greater familiarity with the content of communications and the larger number of contextual cues from which to infer meaning allow students to get along without totally understanding familiar words. But when the same words are used in chemistry, less redundancy, more unfamiliarity, and new meanings for the words exist. In addition, tasks that we expect students to perform in science require a *precision* in meaning that goes beyond what is required in other settings. In particular, any message conveyed in a context that provides redundancy (i.e., where meaning can be obtained from words, facial expressions, intonation, pictures, or other sources), a person may perform very well without a precise understanding of the words used. In a context that is devoid of redundancy, precision in word meaning is important. As a result, the student may fail to comprehend without realizing why.

OTHER IMPEDIMENTS TO COMPREHENSION

NOT KNOWING BUT NOT KNOWING YOU DON'T KNOW

Ann Brown (1980) cited the following example taken from Holt's classic book, *How Children Fail.*

> Holt's lucid description of children's mystification over school problems includes many examples of metacomprehension failures. For example, faced with the task of listing verbs that end with a *p*, one fifth-grader became obviously upset, repeating "I do not get it," but was totally unable to say why she failed to understand. Holt then asked the child if she knew what a verb was and gave her some examples; immediately the child went to work. Holt suggests that the child did not ask what a verb was simply because she did not know herself that she did not know (p 458).

Although I know of no controlled studies of such behavior in chemistry, I have already mentioned the fact that students in my remedial college chemistry course use area, volume, pressure, density, and similar terms without knowing what they mean *and without knowing that they do not know*. For example, students multiply the length and width of a sheet of paper to find its area, but they cannot say what the resulting product is telling them about the paper. Similarly, when asked how to find the size of a surface such as a table top, or what units would be used to describe the size of the surface, or what they are telling me about the table (distance, area, volume, mass, etc.), these students cannot respond. When told that the room they are standing in has a volume of 2000 cubic feet and asked what that says about the room, I have had college students respond, "It tells me how much there is on all the walls and the floor and the ceiling." When students who have correctly calculated the density of several objects are handed two objects of the same size but different masses and are asked which has the greater density, they are often unable to say. Similar problems are reported by Mughol (1979), Osborne and Cosgrove (1983), and others.

The number of students who make a *particular* error of the kind just described is not large (I estimate fewer than 5% of the students that I teach), but the number of students who make the *kind* of error just described is large.

The idiosyncratic occurrence of such errors makes them difficult to detect and remediate, but the problem is real and needs attention. As Goetz and Armbruster (1980) stated: "Teachers should realize that different students interpret the same text differently and tests must be designed to reveal not only whether or not a student has understood, but *what* that student has understood" (p 217). The examples of test questions cited from the Cassels and Johnstone (1980) study and the discussion of testing in the previous chapter show how difficult these requirements of teachers can be.

Chemistry teachers are aware of the need to teach technical terms, and most textbooks include word lists, definitions, and other aids for teaching such terms. However, I am not convinced that we are adequately aware of how long it takes to understand relationships involving a word, the ways a word may and may not be

used, and the subtle changes in word meaning when words are used in various contexts. I am reminded of Norman, Gentner, and Stevens's (1976) estimate of five
years for children to progress from correct use of *give* and *take* to correct use of
other possession verbs such as *trade, pay, buy, sell,* and *spend money.* Although I am
aware of no research pertaining to the issue, inferring that months or even years
could be required for students to progress from the point at which they first
encounter *equilibrium* in chemistry until they understand the various ways that the
term may be used to describe chemical processes does not seem unreasonable.

ONE WORD, MANY MEANINGS

The meaning of words evolves over time. This fact is no less true of technical terms
than any others. However, technical terms are frequently taught as though they have
a fixed, commonly accepted meaning. When a word (concept) is introduced, students are frequently taught *the* definition of the word (concept). When the word is
later used in a new and different context, the altered meaning of the word is frequently ignored and students are left to sort out the meaning for themselves with
few contextual cues to aid them. The problem is further compounded because
chemistry teachers may be unaware that their use of a term is not accepted by all
chemists.

I am aware of few studies of chemists' use of technical terms, but the following
results from an informal investigation suggest the nature of the problem. Discussion
of a concept analysis of *ionization* revealed considerable disagreement about the
meaning of the term. As a result, a quiz consisting of the following items was constructed, given to graduate students and faculty at Purdue, and subsequently published in the *Journal of Chemical Education* with the suggestion that responses be
sent to the author for analysis (Herron, 1977a):

> The equations written below represent various chemical processes. Some of
> these would be described as ionization, some would be described as dissocia-
> tion, and others would be described by other terms.
>
> Place an 'I' in the blank if you would describe the process as ionization. Place a
> 'D' in the blank if you would describe the process as dissociation. If the process
> is neither ionization nor dissociation, write in the term (if there is one) which
> describes the process (p 758).

__ 1. $Na(g) \rightarrow Na^+ + 1\ e^-$

__ 2. $Cl(g) + 1\ e^- \rightarrow Cl^-(g)$

__ 3. $Fe^{2+}(g) \rightarrow Fe^{3+}(g) + 1\ e^-$

__ 4. $F^-(g) \rightarrow F(g) + 1\ e^-$

__ 5. $HCl(g) \overset{(H_2O)}{\rightarrow} H^+(aq) + Cl^-(aq)$

__ 6. $NaCl(s) \overset{(H_2O)}{\rightarrow} Na^+(aq) + Cl^-(aq)$

__ 7. $HAc(aq) \rightarrow H^+(aq) + Ac^-(aq)$

__ 8. $SO_3(g) + H_2O(l) \rightarrow 2H^+(aq) + SO_4^{2-}(aq)$

__ 9. $H_2(g) \rightarrow 2H(g)$

The responses to the quiz (Herron, 1978a; Driscoll, 1978) left little doubt that
chemists differ in their opinions about what should be described as ionization. If
you are in doubt, personal interviews of your colleagues will be more instructive

than a repetition of the statistics obtained in our survey. Personal interviews provide ample opportunity for insertion of "yes, buts" and other qualifiers that reveal the *nature* of the disagreements, and those qualifiers are more important than the actual numbers of people who agree or disagree. Such discussions are also likely to help students learn the various meanings given to technical terms, a point developed in the section on overcoming word problems beginning on p 173.

CHEMICAL LANGUAGE. Special problems occur when we consider chemical language—formulas, equations, and other symbolic representations used in chemistry. Here the problem is compounded because both the events and their symbolic descriptions are unfamiliar to students. Teaching chemical language is like teaching a foreign language and limiting discussion to the culture in which that language is spoken. Using a new language to talk about *familiar* events is difficult enough; when the language and events are both unfamiliar, little opportunity exists to do more than process symbols at a surface level. The connection between the symbols and the previously learned ideas, which would facilitate processing at a deeper level of meaning, does not occur.

CONVENTIONS IN CHEMICAL LANGUAGE. Chemists easily forget how much about their language is not obvious, just as native speakers of any language are unaware of the linguistic rules applied automatically.

A chemist would surely recognize OH_2 as water, even though the formula does not follow normal convention. However, the transformation to produce $2HO$ would be interpreted as something quite different by the chemist. How does the student learn the linguistic or semantic rules that allow OH_2 but not $2HO$?

Consider the following equations:

1. $Sb + I_2 \rightarrow SbI_3$
2. $2Sb + 3I_2 \rightarrow 2SbI_3$
3. $2Sb(s) + 3I_2(sol) \rightarrow 2SbI_3(sol)$
4. $I_2 + Sb \rightarrow SbI_3$
5. $Sb \rightarrow I_2 + SbI_3$
6. $Sb + SbI_3 \rightarrow I_2$
7. $Sb_2 + 6I \rightarrow 2SbI_3$
8. $Sb + I_2 \rightarrow SbI_2$

Chemists would accept any of the first four equations as suitable (though incomplete) descriptions of the reaction that takes place when solid antimony and solid iodine are placed in toluene and refluxed for several minutes. However, they would not accept the last four. How do students learn why this is so?

How, for example, do students learn that the transformation in equation 1 to form equation 4 is just fine, but the transformation in equation 1 to form equations 5 or 6 does not make chemical sense? So long as students see equations as equations—that is, as long as the processing is at the *surface level* of syntax—they will have difficulty understanding why one is acceptable, whereas the other is not. However, once they are clear about *semantic meaning*—that is, they clearly connect the *symbols in the equation* with the *things we do with chemicals*—the transformations of equations that are allowed and not allowed are self-evident. *Connections between equations (symbolic representations of chemical processes) and the real-world knowledge of chemical processes enable us to read equations meaningfully.* Many teachers (and most textbooks) give little attention to these connections.

What about equation 7? Both the antimony and iodine placed in the flask are in chunks containing thousands of thousands of atoms. Why is the antimony represented by Sb and the iodine by I_2? Such conventions in symbolic representation are not obvious, and even when students associate what we write with what we do, inferring why a particular convention is followed is not always possible. Instructors know of evidence that iodine exists as diatomic molecules in both gas and liquid phases. They also know that metals exist in aggregates of various sizes but none of constant composition. Given this knowledge, representing the metal by its symbol with no subscript is sensible (What subscript would you use?), whereas the iodine is represented as a diatomic molecule. These representations are as close to the reality in the flask as we can get, and that reality is what we aim for in equations. Until students are told such things, the symbolism used in chemistry appears very mysterious.

The error in equation 8 is all too common and looks very sensible to anyone processing information at a surface level. Why do chemists say it is wrong? Again, students must be told where chemists get their information if chemistry language is to be sensible. When students see us write formulas for the products of reactions, they frequently infer that the information is coming from the formulas for the reactants. Until they realize that it is *independent* information obtained from empirical data, painstakingly acquired from analysis of reaction products, they are likely to infer incorrect formulas from information in the "chemical sentence" rather than consult a reference or ask someone what the formula is.

LEARNING CHEMICAL LANGUAGE

Throughout this discussion, I have stressed the connection between the language of chemistry, the macroscopic events it is used to describe, and the microscopic models on which the language is structured. I have done so because without that kind of connection, no language exists.

THE PRESS OF NEW SYMBOLS

Now consider beginning students trying to comprehend chemistry equations. First, the symbols and the rules for organizing them into sentences are new. Consequently, students have difficulty deciphering the symbols. Second, even though students have seen many chemical reactions, they have had little experience thinking about them in terms of one kind of matter being transformed into another kind, the way an equation describes chemical change. Neither are they accustomed to explaining chemical change in terms of a reordering of atoms, which is the explanatory notion underlying the *spelling* of chemical words. Thus, the schemas needed to assign meaning to chemical language are poorly developed if they exist at all. Third, clues such as the title in Bransford's clothes washing story are absent and cannot assist students in making the connection between the symbols and the appropriate schemas in memory. So what happens? Like the students in Bransford's study who read the story without a title, chemistry students are forced to process chemical language at a superficial level and to focus on the symbols rather than the ideas the symbols convey. The material is poorly understood, poorly retained, and frequently confused.

There is much to learn about chemical language, as I have suggested. We must learn what symbols mean, what they do not mean, and how they are related to macroscopic and microscopic events. Also, rules of inference must be learned.

RULES OF INFERENCE

"I want to buy an apple" provides no explicit identification of buyer, seller, money, and goods, but we all have sufficient experience with the English language and the concept of *buy* to make reasonable inferences about their existence and identification when we hear this sentence in context.

Chemical equations make no mention of containers, solvents, reaction time, temperature, amounts of reagents used, possible side reactions, and other information important in carrying out a chemical reaction. An experienced chemist can infer much of this information from an equation, within limits, just as in ordinary English. *What inferences are reasonable and how they are made is important knowledge about chemical language. They must be taught.*

I am again reminded of G. Stanley Hall's suggestion (cited in Adams, 1980) that true reading occurs only when the art has become so automatic that we can devote our entire attention to the message and leave the machinery of reading to whir along unnoticed. I am also reminded of how difficult it is to teach someone to read, even after a spoken language is well developed, or to teach an adult to converse in a foreign tongue. These tasks are not trivial.

Given the time and effort required to reach Hall's level of true reading, how important is teaching every student who wishes to know chemistry the language of chemistry? Given our present knowledge in chemistry education, we do not know what difficulties would be encountered by teaching chemistry for general education without formulas and equations. Still, when we consider the difficulties of developing *proficiency* with the language, this approach is worth exploring.

OVERCOMING WORD PROBLEMS

In the preceding discussion, failure to comprehend words was attributed to the following:

- failure to decipher symbols,
- lack of schemas that are capable of giving meaning to deciphered symbols, and
- failure to link the symbols with appropriate schemas.

Therefore, efforts to overcome word problems must attack these sources of difficulty.

DECIPHERING SYMBOLS. Most chemistry students can read in the sense that they can decipher simple English. However, certain deciphering skills are unique to chemical language.

I use the following questions to detect difficulties in deciphering chemical language. Similar questions are incorporated in the textbook and in classroom activities. Few students have difficulty with such questions by the time they appear on exams, but they are included to identify students who need additional help and to emphasize that such mundane skills are important.

1. All of the following represent hydrogen. Using a circle to represent a single hydrogen atom, draw diagrams to illustrate what is meant by each representation.
    a) 2H b) H_2 c) $2H_2$
2. Describe by words and diagram the difference in what is represented by CO and Co.
3. A student balanced the equation, $Sb + I_2 \rightarrow SbI_3$, as $Sb + I_2 \rightarrow SbI_2$. The equation is balanced. Explain why it is wrong by describing the difference in what is represented by the student's answer and the correct equation, $2Sb + 3I_2 \rightarrow 2SbI_3$
4. Write ordinary English sentences for two possible translations of the following equation: $2H_2 + O_2 \rightarrow 2H_2O$
5. The equation in 4 can be balanced as $H_2 + 1/2\ O_2 \rightarrow H_2O$
    When balanced in this way, one of the translations possible for the equation in 4 does not make chemical sense, but the other one does. Which one does not make sense? Why?
6. Solid copper is normally represented as Cu and solid sulfur is normally represented as S_8. Neither representation is a totally accurate representation of what is present in the solids. What is different about the two representations? Why are they different?
7. OH_2 is a "wrong" representation for water, even though this formula would suggest the same molecule as H_2O. Why is the first formula considered "wrong"?

OTHER SYMBOL SYSTEMS USED IN CHEMISTRY. The reader will undoubtedly think of other conventions concerning chemical language that must be clarified, but the ones just listed are sufficient to illustrate the kind of clarification needed. Similar clarification is required for other kinds of symbolic representations. The following questions suggest several of these areas:

8. The concentration of a salt solution was described as $8\ gL^{-1}$. Show an equivalent way of representing the units and describe the concentration of the solution in ordinary English.
9. The graph given in Figure 13.1 summarizes measurements made on a sample of nitrogen gas. What measurements on the sample were used to construct the graph?
10. What conclusion(s) about the behavior of nitrogen gas can you draw from the graph?
11. The line in the graph does not pass through all points. Why is the line drawn in this manner? What assumption would you make about the measurement represented by the point that is farthest from the line?

TRANSLATING LANGUAGES

Closely related to the skill of deciphering symbols is the skill of translating larger units of thought from one language to another. In chemistry we use at least three languages: ordinary English, chemical equations, and mathematical equations. The following questions illustrate some of the translating skills that students need to be successful in chemistry. These examples are meant to be suggestive rather than exhaustive.

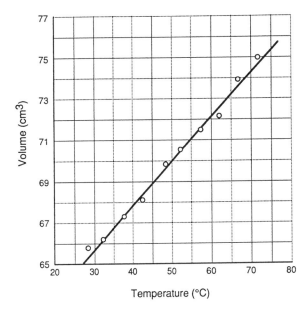

Figure 13.1. Measurements made on a sample of nitrogen gas.

1. If P represents pressure, V represents volume, and k represents a constant, which of the following equations indicates that the pressure of a gas is *directly proportional* to its volume?
   a) $P/V = k$ b) $P/k = V$ c) $V/k = P$ d) $k/P = V$ e) $k/V = P$ f) $PV = k$ g) $Pk = V$
2. Write a mathematical expression that says volume, V, is directly proportional to mass, M, and inversely proportional to temperature, T.
3. Write a chemical equation that describes the chemical reaction that you just observed. (This question may follow any demonstration that is sufficiently simple that students should be able to describe it.)
4. List some of the information that was observed but not included in the chemical equation that you wrote. (The list might contain such things as the kind of container in which the reaction took place, other chemical species—air or a solvent, for example—that were present but not included in the equation, temperature, light, mass of reactants actually present, and so forth. The point of the exercise is to make students aware of the fact that much information is systematically omitted from chemical equations.)
5. Use a chemical equation to say, "The ice melted."
6. Use a chemical equation to say, "The sodium chloride dissolved in water."
7. Use a chemical equation to say, "Hydrogen chloride gas reacted with water to form hydronium ions and chloride ions."
8. Translate the following equation into ordinary English:
   $2Sb(s) + 3I_2(sol) = 2SbI_3(sol)$

I cannot claim that correct answers to these questions depend only on "deciphering skills." Concepts are also involved. Still, as in most reading tasks, meaning comes only when the reader can simultaneously apply knowledge about how the symbols are used (deciphering) and what the symbols mean (concepts).

SEMANTICALLY POOR READING

Several deciphering skills related to ordinary language should be emphasized in chemistry. Many words are easily confused because they are lexically similar. *House* and *horse* look very much alike. If contained in a list of nonsense words to be read rapidly, they are easily confused. As they are encountered in language, semantic and syntactic cues allow us to read rapidly with little chance of error: "The boy rode the horse" and "The boy entered the house."

Unfortunately, these cues that students use so effectively in ordinary speech may be insufficient in chemistry. We may say, "Dissolve the chlorine in water," or "Dissolve the chloride in water." We can send for a bottle of potassium nitrate or a bottle of potassium nitrite. We can describe concentration in terms of molarity or molality. Experienced chemists have little difficulty with such words because they have learned to use other cues (usually a large body of chemical facts) that suggest which word may be coming next, and they have learned to look carefully when no cues are available. Beginning students must be taught that technical language, unlike ordinary English, is deliberately concise, precise, and devoid of redundancy. They must develop new reading skills, particularly the habit of carefully attending to detail and frequently checking to verify that the correct meaning has been obtained.

Chemists expect to see molality in a problem dealing with colligative properties and molarity in a problem dealing with titration. When they get potassium nitrate from the shelf, they read the label twice and check the formula before using the contents. They know that when the laboratory manual says "titrate with permanganate," it means that potassium permanganate or some other source of permanganate *ion* should be used. Beginning students need instruction on such matters, and the instructor must be alert to confusion that arises because students lack the wealth of knowledge that guides experienced chemists.

WORD-ATTACK SKILLS

Good readers have learned to recognize new words formed by adding prefixes and suffixes to root words, but they still find these skills difficult to use in science, where the roots as well as the derived words are unfamiliar. Time must be spent discussing common roots and the meaning of words derived from them: *act* [L. *agere*, to do], react, reactive, unreactive; *organ* [Gr. *organon*, an instrument], organism, organic, inorganic; *equal* [L. *aequus*, even], equi-, equilibrium, disequilibrium; *thermo-* [Gr. *therme*, heat], thermal, thermometer [+ Gr. *metron*, measure], thermonuclear [+ L. *nucleus*, a kernel], thermoplastic [+ Gr. *plassein*, to form], thermos, thermostat [+ Gr. *-states*, stationary], thermodynamics; *dynamic* [Gr. *dynasthai*, be able], dynamo, dynamite; and *certain* [L. *cernere*, decide], certify [+ *facere*, to make], uncertain, uncertainty.

Discussion of word histories such as those given in *Picturesque Word Origins* (1933) can add a human touch to the teaching of science as well as improve the student's understanding of words and help students develop word-attack skills.

I have never seen a book comparable to *Picturesque Word Origins* that specifically deals with science words. If such a book exists, I would like to know about it so that I can add it to my kit of teaching tools. However, much valuable information can be gained from a standard dictionary, as suggested by the words I have listed above, all of which came from my desk dictionary.

Teachers should not have all the fun of building word histories. Students should be asked to help in the historical search and share their results with the class. Some will find it far more enjoyable than balancing equations or naming compounds, and the skills they develop in the process will serve them well after they have forgotten all the chemical facts we drum into their heads.

WORD GAMES

Word games are useful tools for learning. By using common prefixes, suffixes, and roots, students can make up words and ask classmates to decide what they mean. Class discussion of whether the new word is a good one or not will increase students' understanding of roots, how they can be used to form new words, and how one can use such knowledge to make reasonable inferences about the meaning of totally unfamiliar words.

Some readers will undoubtedly react to the preceding suggestions with a string of expletives followed by the exclamation: "I'm here to teach chemistry! How can I do that if I spend all of my time teaching English?"

This point is well taken. All teachers must make judgments about how to spend their limited time, and only the classroom teacher has the information required to make the best decision about when to teach vocabulary and when to develop concepts and skills. Still, no teacher can safely ignore the fact that success in teaching chemical facts, concepts, and skills depends on students understanding the language we use. If teachers do not have time to teach the technical vocabulary they are accustomed to using, they should reduce the number of technical terms they use. This practice is not as difficult as it may seem. "In water" can substitute for "aqueous"; "break apart" can be used for "dissociate"; "positive and negative electrode" can substitute for "anode" and "cathode"; "make" can substitute for "prepare"; "melt" for "fuse." Our language is often pompous. It need not be.

GENERAL LANGUAGE DEVELOPMENT

SCHOOLWIDE APPROACHES TO LANGUAGE DEVELOPMENT

Although reading skills such as those discussed here cannot be left in the hands of the English teacher, time can be saved if these skills are addressed by the faculty as a whole rather than by each teacher acting independently. More progress can be made with less effort if the faculty adopts uniform policies concerning reading and writing skills. A school can prepare a handbook of reading and writing style to be used by all teachers as they correct written papers or prepare reading assignments. The handbook can include specific suggestions concerning reading and writing issues to be addressed by English teachers, chemistry teachers, and others. If these suggestions are conscientiously followed, students benefit, and the payback is greater than from comparable effort exerted without collective action.

Students can do much of the work needed to learn the history of words on their own. Too much cannot be made of this point. We are too ready to demonstrate our superior knowledge by answering students' inquiries when students actually need to develop skills that enable them to find answers themselves. We can easily tell students (and in many circumstances, we should), but telling them the meaning of a word increases their information without improving the intellectual skills that allow them to derive meaning on their own.

READING AS THINKING

Reading specialists emphasize that reading is a thinking process and that readers must learn to be good thinkers. Readers must learn to use contextual cues, syntactic cues, semantic cues, and word-attack skills to derive meaning. Such skills do not develop automatically, but teachers can encourage their development by the way they respond to students who ask for help with words. The following suggestions are adapted from Stauffer (1975).

1. Ask the students what *they* think the word means. Students often make good hypotheses about word meaning but need confirmation. If we are able to confirm their hypotheses, we reinforce their faith in the processes they are using. If their analyses have led to incorrect hypotheses, we can often show them where their thinking went astray, thus providing corrective feedback concerning the *logical process* they are using.
2. If students have not done so, ask them to read to the end of the sentence or the end of the paragraph to see if any contextual cues can help. If the word is in some chemical context, such as the description of some chemical change, ask the student to describe the process to see if that description provides useful cues. This practice not only encourages students to relate language to the chemical events that are described, it also provides you with useful information about how well the student understands those events.
3. If the word is derived from some root (chemical or other), ask students if they recognize the root and know its meaning. Then see if they can guess the meaning of the derived word.
4. Encourage the student to use a dictionary.
5. You need not wait for students to ask the meaning of a word. When you use a new word or one has been used in a reading assignment, initiate the process described above by asking students what they think the word means. If this questioning is done in a group setting, discussion among students can reveal the tools that students are using to derive meaning and can provide excellent opportunities for them to practice rational thought. (Other procedures that help students develop rational thought are described in Chapter 16.)

The suggestions given here are incorporated in what Stauffer described as a directed reading–thinking activity. The closeness with which the procedures mimic those associated with scientific thought is interesting.

1. Students are encouraged to *predict* what a story will be about by looking at the title, headings, or illustrations—actually, anything that might provide a cue. As they read, they are encouraged to process the ideas to see if their predictions were correct and to make new predictions as the information they obtain forces them to give up their initial predictions.
2. Students are encouraged to *look for proof* that their ideas are right.
3. Teachers encourage this thinking process by asking students *what they think* (both before and after the reading).
4. Furthermore, teachers ask students *why* they think so.
5. Teachers then insist that students *prove their point* by referring to facts in the reading and by making rational inferences.

Stauffer emphasized the idea that this thinking process rather than the calling of words is the essence of reading, and he associates poor reading with failure to develop such skills.

An appealing suggestion that I have never tried was reportedly advanced by Richard Binns (reference unknown). The suggestion is that books be published with the left-hand page blank for students to write on. Students are encouraged to write on the left-hand page their understanding of what is said on the right-hand page.

Stauffer's suggestions concern the processes used to derive meaning. As such, they are just as applicable to other exercises from which we expect meaning to accrue—listening to a lecture, watching a demonstration, or performing a laboratory investigation.

LISTENING AS THINKING

Research on listening in the late 1950s (Nichols and Stevens, 1957) led to recommendations that are almost identical to those made by Stauffer. After extensive interviews of good and poor listeners, Nichols and Stevens reported several characteristics that separate the two. First, they pointed out that poor listeners allow the medium to interfere with the message. Poor listeners "turn off" the speaker because of the speaker's looks, delivery, or poorly prepared illustrations. Poor listeners are so busy picking apart the speaker that they have no time to attend to what the speaker says. By contrast, good listeners notice the same faults, but they quickly set these aside to see if anything worthwhile can be found in what the speaker is saying.

Once listeners overcome their own prejudices and attend to the message, another problem derives from the difference between the speed at which we think and the speed at which we speak. The average person processes information at approximately five times the normal rate of speech. Unless listeners are interested in what the speaker is saying, this differential allows the mind to wander and eventually not return to the lecture. Typically, according to Nichols and Stevens, we all tune in for a few seconds and then tune out to think about something else. Poor listeners repeat the process for a while, tuning in and out until some reverie becomes more interesting than the speech. Only at the end of the lecture do poor listeners realize that they have missed the last half!

Good listeners use the differential between thought speed and speech speed to organize and reinforce the message delivered. First, they habitually predict what the speaker will say next. Second, they listen intently to see what is actually said. Third, they compare what is predicted with what is said, thereby reinforcing the idea if the prediction is correct and asking why the speaker developed the speech in the different way when the prediction is wrong.

Nichols and Stevens's suggestions concerning good listening are similar to Stauffer's suggestions concerning good reading, and they are almost identical to what happens in an effective demonstration.

CHEMISTRY AS THINKING

Chemists who use demonstrations effectively almost always present some situation and ask students to predict what will happen. Students are invited to see if their prediction is true, and they are asked to explain why their predictions were right or wrong. Effective laboratory investigations have similar properties.

ENCOURAGING READING IN CHEMISTRY. Stauffer's suggestions for developing reading skills are useless unless students read. Many chemistry teachers complain that students will not read assignments and depend entirely on lectures for information.

Again, this point is well taken. Students do not read when they should, and teachers must encourage them. An important step will be writing chemistry textbooks in more understandable language. Another step will be our own insistence that students get information from reading when they are perfectly able to do so.

Students frequently ask questions about correct laboratory procedure when the answer is given in understandable language in their manuals. The problem is not that they are incapable of understanding what is written. Asking is easier. The students' goal is to finish the lab, and the principle of least cognitive effort dictates that they ask rather than read. The teacher's goal is that students become independent learners. Consequently, students should be encouraged to see if they can find the answer by reading. If they cannot, the teacher can then answer the question.

Students often fail to read assignments because they know that the instructor will tell them everything they need to know in the lecture. Teachers should make clear to students the fact that certain information for which students are responsible will not be covered in lecture. This may be the first time that students encounter this phenomenon. Such insistence, coupled with periodic tutorial or recitation lessons designed to help students learn to read the language in textbooks, can lead to less dependence on oral instruction and more independent reading.

LABELS VERSUS CONCEPTS. The preceding discussion describes techniques that help students attach meaning to the labels (words) for ideas (concepts). The techniques will be effective so long as the ideas to which the labels belong are already in students' heads or other ideas can be put together to form the appropriate idea. These requirements are often met, but not always. Teachers must differentiate between words and concepts. In some cases, teachers need to teach words; in other cases, they need to teach concepts.

To illustrate, let me tell you about durians. A *durian* is a fruit found in Southeast Asia. It is about the size of a pineapple, but it is covered with sharp spines. It has the odor of rotten eggs mingled with decaying rats. Inside the hard rind are several cavities containing a large seed surrounded by gelatinous flesh. The flesh is the part that is eaten. It is almost as sweet as saccharin. According to those who enjoy the fruit, the first time you eat it, it is awful. The second time, it is tolerable. The third time, it is delightful. I would not know. I only ate it once!

I have used concepts that you know to tell you about durian. You would certainly be able to distinguish a durian from a peach, apple, grape, or other familiar fruits. Still, you do not have a concept of durian in the same sense that you have a concept of fruits you have seen, smelled, and tasted. Certain aspects of the concept *durian* will only become clear through experience.

NECESSITY OF EXPERIENCE

We trust words too much and rely on them to convey more meaning than they can convey. Words do more after relevant experience than before. Readers who are familiar with durians will attach more and different meaning to the preceding paragraph than those who have never heard of the fruit. Depending on their bias, they may even feel that the paragraph is an adequate *explanation* of the concept. They will sense the sight, the smell, and the taste. Not so for those who have never encountered the fruit.

Those who are familiar with chemical phenomena are easily deluded into believing that our *description* of chemical events is equivalent to the event itself. For the sake of efficiency, words are used when experience is more instructive.

I have frequently sat in the back of a classroom watching teachers *teach* a concept such as acid by using nothing but words and symbols written on the board. They tell students that acids taste sour (which is probably acceptable because students have tasted sour things), that they react with metals to produce hydrogen gas (which is not acceptable if the students have never seen the reaction), that acids react with bases to form water (which is unlikely to be sensible even if students have seen the process), and so forth. Teachers may write symbols on the board to represent the acid by hydrogen ions accompanied by some other ion and then write an equation representing neutralization.

I do not wish to suggest that the words and the symbolic descriptions are not important. They are. They represent the language used to convey meaning about acids. But they are not the same as the *concept* of acid. The concept is closely tied to the experiences described, and just as in the case for durians, one will not substitute for the other. To be meaningful, words must be *associated* with experience, and this association cannot happen in the *absence* of experience.

Acids can easily be placed in front of students as acids are discussed or placed in contact with metals (those that *do not* react as well as those that do) as students are encouraged to describe what they see (they will not all see what the words describe). Students can then build a language to summarize the observations. Only through processes such as these are students able to attach adequate meaning to *acid* and H^+. Without such experiences coupled with words and other symbols used to describe them, students learn to use words when given appropriate cues; however, the words do not convey the meaning that chemists attach to them.

Once words are associated with the meaning derived from experience, these words can be used to develop meaning of other words; *sour*, learned through experience outside of chemistry, can be used to attach meaning to acid.

INSUFFICIENCY OF EXPERIENCE

Just as experience is a necessary part of acquiring meaning, it must never be seen as sufficient. Few of us see differences among monkeys in a pack or cows in a herd, but the experienced zoologist or farmer will. Most of us recognize snow when we see it, but not the eight varieties named by the Eskimo. Similarly, beginning chemistry students do not see many features of chemical events that are the basis for meaning that chemists derive from those events. Students can be taught to see those features, just as an Eskimo could teach us to recognize all kinds of snow, but seeing chemical reactions is no more sufficient than seeing snow.

The features that require attention are not always clear to one who is familiar with a concept. Eskimos may point to varieties of snow and name them with ease, yet be unable to call attention to the subtle differences that lead to the classification. They have almost always known, and they apply their classification schemas unconsciously. So do experienced chemists.

Subconscious application of knowledge is desirable—even necessary—in our area of expertise, but this application can stand in the way when teachers want to make experts of others. To accomplish this, teachers must be conscious of what they normally do subconsciously. They must teach explicitly or students must apply generalized learning skills (general intelligence) to infer what is going on. The students' process is important, but teachers can help by teaching explicitly.

The best procedure I have found for consciously attending to how I deal with familiar concepts is concept analysis. By deliberately listing the critical and variable attributes of a difficult concept and then testing this list against those things that I know as examples or nonexamples, I can see more clearly the classification rules applied subconsciously. I am then able to teach the concept to students.

Still, concept analysis must be used with intelligence and judgment. Concept analyses tend to be rigid formalizations, whereas the concepts they describe tend to be fluid. Most words are used in many contexts, and the precise meaning is context-dependent. Consider the various meanings associated with *acid*:

- When I say the wine is a little acid, I say something about its sour taste, but I do not expect it to react with zinc to produce hydrogen.
- When I say that water can act as an acid, I mean that it can donate protons to other species, but I do not expect it to neutralize strong bases.
- When I hear about acid rain, I infer that the pH is below 7 but not much else.
- When I describe BF_3 as an acid, I mean that it can act as an electron pair acceptor, but I do not expect it to furnish hydrogen ions.

Whether one chooses to describe the various meanings associated with *acid* as varying concepts of *acid* or varying usages for the same concept is a matter of choice. Making sure that students understand these meanings and are able to use contextual cues to attach the correct meaning to the word in each case is not. It is what teaching language is all about.

SUMMARY

This chapter on language is closely associated with Chapters 9–11 on concept learning. Language is what we use to convey meaning of concepts, and this chapter has emphasized that "many a slip twixt the cup and the lip" may occur.

The meaning of all language is context-dependent; and the meaning we attach to language, written or spoken, results from top-down processing under the control of existing schemas as much as from bottom-up processing of symbols and sounds. Reading is an inferential process. What we read "between the lines" is as important as the words on the line.

Spoken language and ordinary writing are characterized by redundancy that allows readers and listeners to infer meaning even when a particular symbol or context is unclear. This redundancy is not true of technical language. Technical language is sparse, and minor changes in symbolic representation produce major changes in meaning. If students are not aware of this potential and do not know which deviations in symbolic representations are acceptable and which ones are not, they make errors and become confused.

Words and other symbols are used to convey meaning derived from experience but do not substitute for experience. Chemists must help students connect the symbolic language of chemistry with the macroscopic world of experience and the microscopic world in which much chemical thought takes place. The symbolic language was invented, after all, as a bridge between the macroscopic and microscopic worlds of chemistry.

This chapter contains several suggestions for teaching word meaning and other reading skills. Some chemistry teachers argue that this instruction is not their job, but much of teaching reading is teaching concepts that only chemists know. Besides, the intellectual skills used to make sense out of words are the same as those used to make sense out of the physical world, and is that not what chemistry is about?

14

Generalized Intellectual Skills

As demonstrated in the previous chapters of this book, we use many general skills in learning. We make inferences, relate one idea to another, distinguish relevant from irrelevant information, consider various possibilities, order information and events, compare current tasks with previous ones, decide what to attend to and what to ignore, decide the order in which we will carry out activities, and decide how long to persist before giving up or seeking help.

The kind of knowledge that we are talking about here goes by many names that depend on the author and the particular tasks under discussion. In one frame of reference, knowledge is divided into *declarative knowledge* and *procedural knowledge*. Declarative knowledge, introduced in Chapter 2 and alluded to several times throughout the first half of this book, is what we normally refer to as *information*. It includes such statements as "The symbol for sodium is Na", "My name is Dudley Herron", and "Most elements are solids at ordinary temperatures."

Declarative knowledge is sometimes subdivided in order to focus on particular features; for example, the symbols for sodium and my name are sometimes referred to as "social knowledge" because names are arbitrary, social conventions. By contrast, the physical state of elements at any given temperature is governed by nature and is not social knowledge.

Procedural knowledge describes procedures that are applied to existing knowledge to produce new knowledge. Another name for procedural knowledge is intellectual skills. Reading and writing represent procedural knowledge. So do designing an experiment, representing a physical system in terms of a differential equation, and deciding how to go about tackling a novel problem.

Procedural knowledge is a large category, and people normally focus on only a few of the generalized intellectual skills that fall within that category. People who study problem solving often speak of the *strategic knowledge* used in attacking the problem. Others talk about *executive strategies,* comparing metacognition, which governs how we perform intellectually, to the way a corporate executive governs a company. I am most interested in what Piaget (and others) describe as *operational knowledge*. Operational knowledge operates on other knowledge to transform it in some way. Our operational knowledge governs how we make sense out of the world.

Although I focus on the operational knowledge described by Piaget in this and the next two chapters, most of what I have to say pertains to all generalized intellectual skills. Whatever label is used, this knowledge is qualitatively different from the declarative knowledge that I focused on in the chapters on concept learning, language, and principle learning. We must be concerned about generalized intellectual skills, and we must help students develop them. In this chapter I hope to convince you of the following:

1. Generalized intellectual skills, such as those described by Piaget as logical operations and by others as procedural or metacognitive knowledge, develop as a result of genetic and environmental factors.
2. Whatever the influence of genetic factors, individuals of normal intelligence can develop the logical thought characteristic of science; that is, these individuals are capable of what Piaget described as formal operations.
3. Generalized intellectual skills (formal operations and other higher order skills) can and should be taught.
4. A dynamic relationship exists between content and intellectual process that limits the generalizability of any knowledge, including formal operations.

SCIENTIFIC REASONING

Piaget described the highest stage of intellectual development that he observed among children and adults as formal operations. Formal operations are such processes as thinking in terms of possibilities rather than actual events; systematically considering all possibilities; accepting the logical necessity of all other things being equal before one can establish cause and effect relationships; developing hypothetical, logical connections in the form of "if . . . then . . . therefore"; and using proportional, correlational, and combinatorial reasoning. Formal operations are so clearly related to success in science that *formal operations* and *scientific reasoning* are commonly used interchangeably.

How common is scientific reasoning? In 1971, McKinnon and Renner published findings that only 50% of students entering college had developed formal operations. The results stirred the science education community. The study was repeated with various tests, with students of various ages and interests, and across cultures (*see* Chiappetta, 1976, for a review). Although the proportion of students who demonstrated formal operations reasoning varied from one study to another, the general conclusion was the same: A substantial portion of the adult and school-age population does not use those generalized intellectual skills that are required to solve novel problems in technical fields; these people are not "formal operational."

Many learning psychologists and science educators were uncomfortable with these results. Several conducted studies that seemed to cast doubt, not only on the validity of studies such as McKinnon and Renner's, but on Piaget's theory as a whole. Novak (1990) summed things up like this:

> In the community of cognitive psychologists during the past few years, there has been a growing recognition that Piaget underestimated the cognitive capabilities of children. . . . [N]ewer theoretical views are beginning to guide our

interpretation of cognitive development in children. Donaldson (1978), Carey (1985), Driver (1983), Bruner and Haste (1987), Flavell (1985), and Matthews (1980, 1984) are a few of the leading scholars who now confirm that children have substantial cognitive abilities—far beyond those suggested by narrow interpretations of Piaget's work (p 938).

Some chemists may interpret the research summarized in this quote as an indication that they can forget about how adolescents think, that the only real difference between children and adults is their store of accumulated information. These chemists may believe that they can teach by telling. Such an inference is ill-advised. Piaget's theory of intellectual development and, in particular, his description of formal operations reasoning continue to have important implications for chemistry teaching.

No amount of reading about Piaget's work will substitute for seeing children respond in the way he describes. If you have never interviewed children using Piagetian tasks, please do. Several tasks are described in Appendix L. If you accept the challenge, arrange to present the tasks to children who differ widely in age (e.g., 5, 8, 11, 14, and 18 years, respectively). The dramatic change in response over time is of most interest to teachers.

The question of *why* responses to Piagetian tasks change over time is still an open question. Those who conduct research in chemistry education should be aware of the controversy and the literature pertaining to this question, and I have included a sample in Appendix M. Appendix M outlines Piaget's stage theory, describes factors that affect responses to Piagetian tasks, and reviews some recent research aimed at explaining response patterns.

CURRENT UNDERSTANDING OF INTELLECTUAL DEVELOPMENT

The discussion of research in Appendix M leaves no doubt about the fact that responses to Piagetian tasks are affected by modifications in the task, the way it is presented, and the language used in the questions. Neither is there doubt that, to the extent that Piaget's stage theory implies sharp boundaries distinguished by qualitatively different mental structures rather than the gradual acquisition of increasingly more powerful logical operations (*see* Herron, 1978c, for a different interpretation), *stages* do not exist.

Task modifications that are required for young children to solve class inclusion tasks beg the question of *why* such modifications are necessary, and the answer is far from clear. Children understand the words used in all of the questions, but when some question forms that are sensible to adults are asked, children answer inappropriately without being aware of their error. (As we will see in the next chapter, so do adults.) What we know can be summarized in this fashion:

1. Nobody questions the results of Piagetian tests. Children do respond differently than adults. Many questions and situations that are trivial for adults are incomprehensible to children and adolescents (or at least incomprehensible when presented in the normal adult manner).
2. By modifying Piagetian tasks so that the materials and language are within the range of the child's everyday experience, children often respond correctly, but they still cannot handle some logical forms that are trivial for adults.

3. The explanation for the observed differences in performance of children and adults is still uncertain. Piaget's explanation has been challenged, and several alternative hypotheses have been advanced.

RESEARCH ON MISCONCEPTIONS IN CHEMISTRY[1]

The research discussed in Appendix M is about things that children appear not to understand. As such, it can be viewed as a subset of a larger body of research on misconceptions. The difference is that children's failure in the previous research appears to depend on reasoning; on the surface at least, success is unrelated to domain-specific knowledge. The misconceptions discussed here clearly depend on domain-specific knowledge.

A substantial body of research documents misconceptions about fundamental chemistry concepts. For example, balancing chemical equations has been a standard part of the chemistry curriculum for years, but evidence suggests that students do not understand what equations should be telling them. Yarroch (1985) gave 14 students in the top third of two classes four simple equations to balance. As expected, all of the students were successful at balancing the equations, but only five could draw diagrams that accurately depicted the molecules and groups of molecules represented.

Similarly, in a series of studies, Nurrenbern and Pickering (1987) found that students could solve routine problems concerning gas laws and stoichiometry reasonably well, but when asked to identify diagrams that would represent the distribution of gas molecules in a container or draw diagrams to represent the changes in molecules taking place in the stoichiometry problems, student performance dropped drastically. Many others have documented common difficulties that students have "thinking atoms and molecules" (Ben-Zvi, Eylon, and Silberstein, 1982, 1986; Bleichroth, 1965; Cros and Maurin, 1986; Gabel, 1993; Glassman, 1967; Mitchell and Kellington, 1982; Novick and Nussbaum, 1978, 1981; Pfundt, 1981, 1982; Pickering, 1990; Sawrey, 1990).

Gorodetsky and Gussarsky (1986, 1987), Wheeler and Kass (1978), and others documented common misconceptions concerning equilibrium; still others studied common misconceptions about mole, volume, heat, temperature, energy, and a multitude of other concepts commonly taught in chemistry (Al-Kunifed, Good, and Wandersee, 1993; Albert, 1978; Anderson, 1965; Andersson, 1980, 1986; Barke, 1982; Boyd, 1966; Buell and Bradley, 1972; Cachapuz and Martins, 1987; Caramaza, McCloskey, and Green, 1981; Dennis, 1957; Doran, 1972; Driscoll, 1978; Driver, 1981; Driver and Easley, 1978; Duncan and Johnstone, 1973; Engel, 1981; Erickson, 1979, 1980; Gabel and Enochs, 1987; Geddis and Jaipal, 1993; Gennaro, 1981; Gilbert, Watts, and Osborne, 1982; Gorodetsky and Hoz, 1985; Hackling and Garnett, 1985; Hall, 1973; Harlen, 1968; Haupt, 1952; Helm, 1980; Herron, 1978a; Hewson, 1984, 1986; Johnstone, MacDonald, and Webb, 1977; Mali and Howe, 1979; McClelland, 1975; Milkent, 1977; Novick and Mannis, 1976; Osborne and Cosgrove, 1983; Robertson and Richardson, 1975; Rowell and Dawson, 1977; Saltiel, 1981; Schmidt, 1987, 1991, 1992; Selley, 1978; Shayer and Wylan, 1981; Solomon, 1982, 1983; Stavy and Berkovitz, 1980; Swan, 1980; Vien-

---

[1]The first half of this section is adapted with permission from Herron, 1990, pp 40–42.

not, 1979; Voelker, 1975; Watts, 1982). If we are to improve learning in chemistry, we must understand the sources of such misconceptions and learn how to overcome them.

I find it useful to divide misconceptions into two categories. One category deals with what happens in the physical world. Students believe that heavy objects always sink in water, that the bubbles seen in boiling water are hydrogen or oxygen, that rapidly boiling water is at a higher temperature than gently boiling water, and that mass changes when matter melts or boils. These ideas are simply contrary to empirical facts.

Other misconceptions deal with students' explanations of what happens in the natural world. For example, several studies were done concerning students' conceptions of heat (Albert, 1978; Erickson, 1979, 1980; Hewson, 1984; Shayer and Wylan, 1981). In general, students understand heat in terms of a caloric theory of heat, and as Fuchs (1987) pointed out, an internally consistent theory of thermodynamics can be developed around the caloric idea. The students' conception works; it explains what happens in the natural world. However, the explanation differs from the one accepted in science. The point is that many ideas labeled as misconceptions are not misconceptions about what happens but alternative *explanations* of what happens. In most cases those explanations are logical from the students' point of view, are consistent with their understanding of the world, and are resistant to change.

The studies on misconceptions are far too numerous to review in detail. However, I call attention to characteristics that are common to many concepts that cause difficulty.

1. Many misconceptions are related to concepts that involve *proportional relationships*: density, equilibrium, mole, acceleration, and rates of various kinds.
2. Many misconceptions are related to *theoretical models that require the student to interpret observations in terms of something that cannot be experienced directly*: explanations in terms of genetics and evolution, explanations in terms of an atomic model, and explanations in terms of probabilistic models.
3. Many misconceptions are related to *difficulty in following chains of logical inference* (if . . . then . . . therefore reasoning).

Other generalizations can be drawn from the research on misconceptions. These three generalizations were cited because of their relationship to Piaget's description of formal operations. They represent the kind of observation that has led many to argue that successful science teaching requires attention to the development of such skills (Arons, 1984; Abraham and Renner, 1985; Herron, 1975, 1976, 1978d; Renner and Stafford, 1972; Shayer and Adey, 1981; Kamii, 1979; Karplus, 1977; Lawson, 1985; Lawson, Abraham, and Renner, 1989; Lawson and Wollman, 1976; Wollman and Lawson, 1978).

EDUCATIONAL RESPONSE TO PIAGETIAN RESEARCH

Two kinds of responses have addressed the research on misconceptions in science and the observation that many misconceptions are related to formal operations. One response is an effort to match the level of the task to the cognitive ability of the student. The other is to speed up intellectual development.

COGNITIVE LEVEL MATCHING

The best example of cognitive level matching is Shayer and Adey's (1981) work in England. Shayer and Adey administered a battery of tests aimed at identifying the level of intellectual development used by schoolchildren. They coupled this testing with an analysis of existing curriculum materials to determine the logical operations that would be required to understand the material. They then compared the match between the demands of the curriculum and the ability of the students for whom the curriculum was designed. They found that the match was not good and counseled that the curriculum be changed.

The research on Piagetian tasks makes clear that student responses are sensitive to changes in language and task presentation. At times these effects are substantial, but they are not always predictable. Consequently, using a student's response to one task (or set of tasks) to predict his or her response to another task (or set of tasks) that has the same logical structure is difficult.

The proportion of a particular age group that will respond correctly to Piagetian tasks varies tremendously. Shayer and Adey's data show that as many as 85% of 14-year-olds in some schools demonstrate formal operations reasoning on Piagetian tasks, whereas no more than 20% in an average school demonstrate that level of response. Similar results were obtained in research at Purdue University (Lehman, Kahle, and Nordland, 1981). Such variation suggests that any matching that takes place must be done by the classroom teacher.

Analysis of curriculum materials to judge their logical requirements is no trivial task. Little evidence suggests that this task can be done reliably or that the procedure is cost effective. Because of these problems, cognitive level matching is difficult to do.

RESEARCH ON TRAINING

A great deal of research has been done on how intellectual development takes place and what might promote such development. Most of the research is based on the constructivist view of learning outlined in Chapter 5.

The constructivist picture of intellectual development suggests the following:

> Intellectual development is a process of restructuring, not just an accumulation of knowledge and skills. Each new type of structure, each new way of thinking that develops, is better than the last in the sense that it makes the child more capable of dealing intelligently with his environment (Rohwer, Ammon, and Cramer, 1974, p 108).

What factors affect this restructuring? How can we promote it?

THEORIES IN ACTION. Several suggestions can be derived from research reported by Karmiloff-Smith and Inhelder (1977). They investigated how children solve the problem of balancing blocks on a thin rod.

Blocks of various kinds were used. Some were uniform and balanced at their center. Other blocks were made by gluing large blocks onto thin strips so that they were obviously nonuniform in weight. These blocks, of course, did not balance at their centers. Still other blocks looked uniform, but were weighted at one end by concealed metal plugs so that they did not balance at their centers.

Children first attempted to balance blocks with no apparent theory to guide them. So long as they were successful, they continued in this manner, but they learned little about what affected balancing. *Not until children experienced failure and asked why they failed did learning take place.*

Children began to learn when they adopted a working hypothesis (what the authors call a theory in action) in response to early failure and used this theory to guide further exploration. This theory in action constitutes operational knowledge. It directs the child to operate in a particular way to make sense of his or her experience. When the experience resulting from that manipulation confirms the theory in action, confidence is gained and the theory becomes consolidated as useful operational knowledge.

One of the more interesting observations is what happens after a theory in action becomes consolidated and *contradictory* evidence is encountered. *Rather than discarding the theory, the child ignores the contradictory evidence.* (Compare this behavior with adult responses to the Water and Wine problem described on p 205 and the behavior of Pat described on pp 189–190.)

If subsequent experience leads to occasional failure of the theory, less effort is expended by ignoring the failures or by treating them as exceptions rather than by discarding the existing theory and developing a new one. However, if the number of exceptions becomes large, less effort is required to develop a more powerful schema than to deal with exceptions.

If we ignore the evidence that contradicts our inadequate operational knowledge, how do we ever develop more powerful operational schemas? Once again, the Karmiloff-Smith and Inhelder study provides some promising suggestions.

Once students adopted the theory that "blocks balance at the middle", they were unable to balance nonuniform blocks, *even though they succeeded in balancing these during the trial-and-error stage of their work.* They seemed unable to attend to the perceptions that guided them earlier—the feel of which side was heavier. Rather, they looked at the block, tried to balance it at the center, and discarded the block as *impossible.* However, when they were asked to balance such blocks *blindfolded,* the visual perception required by their theory was blocked, and they reverted to the perception of "feel" and successfully balanced the blocks. Thus, *the strategy that helped the child overcome the inadequate operational strategy was to block the information on which it depended while providing information on which the adequate operation is based.*

Another factor that seemed to promote the replacement of the inadequate theory with a more adequate one was to make the information on which the more adequate schema is based very obvious. Children learned to balance the blocks that were obviously nonuniform first, and finally generalized their new operational schema to the more difficult task of balancing those blocks that appeared uniform but were weighted at one end.

IMPLICATIONS. Two points seem relevant here. First, progress is first made in those situations where information contrary to theory is so obvious that it is difficult to ignore. Second, the older, consolidated theory is not immediately discarded in the face of contradictory evidence, but is first adjusted and eventually replaced over a protracted period of time.

The difficulty of assimilating information that contradicts a consolidated theory is seen in a case study of "Pat", reported by Scott (1987).[2] Pat, a 14-year-old of average intelligence, was studying solids, liquids, and gases in Scott's class. Pat

---

[2]In keeping with the policy adopted in this book, the name used by Scott has been replaced by one commonly used for either males or females.

started with a continuous model of matter but adopted a particle model as a result of instruction. However, Pat took several weeks to accept the idea that particles could be far apart with nothing between. A story that Scott told about a goldfish that died in distilled water because the water contained no air seemed to prove to Pat that air could not be between particles of water, as Pat had supposed. However, even after accepting the proof, Pat did not give up the idea until much later. Scott used the incident to call into question the utility of critical experiments in casting doubt on a student's current understanding (Herron, 1990, pp 43–44).

One final point needs to be made in connection with the Karmaloff-Smith and Inhelder research. The authors speak of children testing their theories in action, but they make clear that *this procedure is not explicit and conscious* among students at the concrete operational stage. The students vary things and observe what happens, but this process is not part of a deliberate plan. They cannot say explicitly what they are doing and why. By contrast, *students who use formal operations intentionally seek counterexamples and consciously test hypotheses. They know what they are doing and why.*

THE ROLE OF COGNITIVE CONFLICT

In the Karmiloff-Smith and Inhelder research, the cognitive conflict resulting from an experiment not turning out as expected appears to be an important factor in the development of better theories. A great deal of other research supports this idea (Doise and Mugny, 1984; Kubli, 1983; Perret-Clermoat, 1980).

Inducing cognitive conflict is frequently mentioned as a strategy for promoting intellectual development. Science teachers have used discrepant events, counterexamples, and experiments with unexpected outcomes for years. Now they are adding social interaction to their bag of tricks.

As students consider a task, they frequently disagree about how it should be attacked and how data should be interpreted. Such interaction can stimulate individuals to reconsider their own position in light of what others have to say. However, social interaction does not always lead to such reconsideration.

Doise and Mugny (1984) discussed a series of experiments that explored conditions of social interaction that lead to intellectual development. The primary conditions are that *true cognitive conflict must exist* and *students must remain engaged*. If group participants are too different in their thinking, less able students may capitulate or more able ones may take over. Without real engagement, no progress is made. But when students in a group are at the same intellectual level, they often proceed without conflict, again without progress. If conditions are arranged so that students have different perspectives so that conflict exists (e.g., each group member has results from different experiments that appear to be contradictory), progress can be made even though they use the same reasoning.

When conflict does exist, progress depends on how students respond to it. One response focuses on the task, and progress is made through social coordination of different viewpoints concerning the task. Another response focuses on individuals and relations among them. Faced with disagreement, one may give in to the other "to be nice" or simply accept contradictory information without any effort to resolve it.

> In this case, sociocognitive conflict is displaced by a social adjustment. The conflict is resolved by focusing not on the task and the social coordination of viewpoints, but on the functioning of the group itself. In this case, cognitive

progress has little chance of emerging (Doise and Mugny, 1984, p 123; *see also* discussion of task and ego involvement in Chapter 18).

Several experiments led to the conclusion that group interactions are most beneficial when students are just beginning to develop a concept. Once some progress has been made, individuals may progress about as much as groups. Also, disadvantaged students can progress rapidly through interactions in a group. This progression does not mean that they catch up with their more advantaged age-mates because the same conditions that produce progress for the disadvantaged help the advantaged. It does suggest, however, that the retarded growth is due to environment rather than to genetics.

In still other experiments, adults participated and performed in different ways. When the adult essentially said, "Don't you think you better reconsider what you did?" the student capitulated and did not progress intellectually. However, students who argued with the adults did progress. Doise and Mugny made the point that an adult who avoids imposing himself or herself and allows the child to be actively involved may be more effective in precipitating progress than an advanced student who has no patience and just bowls the other child over.

> It was regularly found that subjects who deferred to others did not progress. However, it is possible to counteract this effect by involving the child in a sociocognitive conflict situation which makes . . . compliance difficult. This is the case when the child is systematically challenged without also placing a response model before him, or when two adults advocate two mutually opposing viewpoints which are also incorrect (p 161).

An example of sociocognitive conflict without providing the student with a response model is described by Kuhn (1981). Kuhn met with 4th- and 5th-graders for 20 to 30 minutes each week for 11 weeks. Students were shown variations of the combinatorial reasoning task (*see* Appendix L) and were then asked the following questions:

---

1. What do you think makes a difference in whether it turned red or not?
2. How do you know?
3. Can you be sure what makes a difference? Why?
4. Do the others have anything to do with it? Which ones? How do you know?

Experimental subjects were then asked: "Are there any other ways of doing it you'd like us to try to find out for sure? Let's set up the ones you'd like to try." Students then planned experiments they wanted to try. Before doing the experiments, they were asked these questions:

---

5. What do you think you will find out by doing these?
6. How do you think it will come out? Why?

The experiments were then performed before the following questions were asked:

7. What do you think about how it turned out?
8. What did we find out?
9–12. Questions 1–4 were repeated.

Other students did the same thing, but when the experiments had to be planned, the experimenter said: "Watch. I'm going to set up some other ways . . ." The experimenter then did the same experiments suggested by the experimental subjects. Both immediate and delayed posttests indicated intellectual development on the part of the first group of students.

A study reported by Perret-Clermoat (1980) led to the conclusion that social interaction was effective only when the less able member of the pair could follow what the more able member was doing. The point is not that a difference in interpretation existed—that was to be expected. Rather, the point is that what the more able member of the pair *did* to arrive at the different interpretation (counting in this case) was not understood by less able members who did not profit from the interaction. Because the less able members could not follow what was done, constructive cognitive conflict could not take place. Perret-Clermoat concluded, "It would appear that in order to be able to participate in a given social interaction—and therefore to benefit from it—subjects must already have acquired abilities which are called upon during the interaction" (p 140).

Research such as that just cited is beginning to provide some powerful suggestions for promoting intellectual development. The Learning Cycle and the Generative Learning Model outlined in Chapter 6 incorporate many of them. Case (1985) described a slightly different model that is similar to my suggestions for teaching proportional reasoning in Chapter 15. All of these models are helpful, but we still have much to learn about the best way to promote intellectual development.

SUMMARY

Near the beginning of this chapter I indicated that I hoped to convince you of these conclusions:

1. Generalized intellectual skills, such as those described by Piaget as logical operations and by others as procedural or metacognitive knowledge, develop as a result of genetic and environmental factors.
2. Whatever the influence of genetic factors, individuals of normal intelligence can develop the logical thought characteristic of science; these individuals are capable of what Piaget described as formal operations.
3. Generalized intellectual skills (formal operations and other higher order skills) can and should be taught.
4. A dynamic relationship exists between content and intellectual process, and this relationship limits the generalizability of any knowledge, including formal operations.

Now that I have described some research related to these conclusions, it is appropriate to consider whether they are justified.

Strong evidence suggests that operational knowledge develops as we grow older. The evidence is also strong that at a given age, considerable variation exists in the operational knowledge held by various individuals. Although this evidence is consistent with the statement that such knowledge develops as a result of genetic and environmental factors, proving that this variation is the case beyond reasonable doubt is difficult. Like most of the nature versus nurture arguments, this one is still being debated.

A great deal of research on training was reported in this chapter, and that research is consistent with the idea that individuals of normal intelligence develop operational knowledge. Whether *all* normal individuals can become facile with formal operations is still debated. However, sporadic successes at teaching various operational knowledge have led most cognitive psychologists to focus on *how* rather that whether such learning can be facilitated. Chapter 15 provides recommendations based on my understanding of the process.

Perhaps the clearest conclusion emerging from research is that content and process are so closely related that we are unlikely to deal with the learning problems in one of these areas without considering problems in the other. As much as we would like to separate the two for simplicity, total separation appears to be impossible.

Another important theme of this chapter is that the exact reason that children give *silly* responses to questions that are simple for adults is elusive. The wording of questions, amount of practical experience, and other variables clearly have an effect. Unless teachers are prepared to ask probing questions to learn the cause of a student's silly response, teachers are likely to spend a great deal of time *teaching* with little learning taking place.

Research demonstrates that we all have difficulty believing that others are unable to follow some logic, once that logic is clear to us. Consequently, teachers must go through some kind of exercise to reveal the inappropriate (from the teacher's point of view) logic used by students at various ages. Conducting the clinical interviews suggested in Appendix L is one way that this might be done. The research literature is filled with similar tasks that can be used for the same purpose. If you have never given tasks of this kind to students over a wide age range, doing so now will be far more informative than reading the rest of this book!

15

Scientific Reasoning

In Chapter 14 I introduced the idea of generalized intellectual skills and distinguished them from the declarative knowledge discussed earlier. Particular attention was given to Piaget's conception of operational knowledge and research challenging his notion that operational knowledge develops in stages. The formal operations that characterize Piaget's last stage of intellectual development are used so often in science that they are often called scientific reasoning. In this chapter we take a closer look at formal operations, and in particular, proportional reasoning.

In view of the outstanding questions about Piaget's stage theory, I begin by clarifying my own position: I find Piaget's distinction between physical knowledge and logicomathematical knowledge extremely helpful. I find his description of the trends observed in logicomathematical knowledge over time convenient. However, *stages* has never suggested to me the abrupt, totally distinct states of mind that some people infer from Piaget's stage theory (Herron, 1978c). Clearly, changes in operational knowledge that guide our thinking are gradual and, normally, undramatic. How people at different stages of intellectual development respond to typical Piagetian tasks and, I might add, formal science instruction, *is* dramatic.

I use Piaget's language without apology. However, you should keep in mind that this language says nothing about the brain or how it functions. The same can be said about other terms in common use. The classification of knowledge as declarative, procedural, operative, or executive focuses on the way knowledge is used. As Case suggested, "At different points in time, and depending on the problem in question, the same structural unit might serve any one of these functions" (Case, 1978b, p 186). In other words, our description of ways that knowledge is used implies nothing about the way it is stored in the brain or how it got there.

FORMAL OPERATIONS

CHARACTERISTICS OF FORMAL REASONING

At the heart of formal operations is the ability to think in terms of possibilities and abstractions rather than concrete entities, but there is more. Thought is more carefully planned; possibilities are considered systematically, and the thinker is conscious of those possibilities that have been considered and discarded. The consideration of possibilities itself is conscious; it is characterized by an "if . . . then . . . therefore" chain of inference.

Young children use "if . . . then . . . therefore" reasoning. They know that "If I drop the glass, then it may break; therefore, I must be careful", or "If I am naughty, then I will be punished; therefore I shall be good." However, before formal operations, such logical chains are closely tied to knowledge derived from direct experience, and the thought process is not formalized and is seldom conscious.

The reasoning of formal operations goes beyond that of young children to consider possibilities that are farther removed from direct experience. It considers the consequences of consequences or, as Piaget suggests, operations on operations. It is often deliberate, planned, and formal.

To take an example from chemistry, the ability to "think atoms and molecules" seems to require reasoning at a formal level. By "thinking atoms and molecules", I mean the ability to observe some chemical phenomenon, imagine what might be taking place at the microscopic level, and then consciously consider the likely effects on atoms and molecules of some hypothesized change in the system.

Concrete operations are sufficient to consider what will happen at the macroscopic level when the temperature in a familiar chemical system is increased, but formal operations are required to analyze an unfamiliar system of unknown composition. Here the nature of the system must be postulated, and the consequences deduced. Then an alternative postulation and its consequences must be produced. The procedure must be iterated until all plausible postulates have been considered and the most likely one accepted. Both the number and kind of possibilities considered will depend on a person's store of domain-specific knowledge. This *content dependence* of logical operations is taken up later.

Another of the formal operations is proportional reasoning. It is particularly important in chemistry because so many chemistry facts are described in terms of proportions. Formulas and equations are statements of proportional relationships, and all stoichiometry is based on proportions. Rate equations, equilibrium constants, gas laws, and most other mathematical laws are statements of proportional relationships. Defined concepts such as density, pressure, concentration, and reaction rate all describe proportional relationships; and proportional reasoning seems necessary to understand these concepts as a chemist does.

NATURE OF PROPORTIONAL REASONING

Because of the pervasiveness of proportional reasoning in chemistry and evidence that proportional reasoning is a major source of difficulty for students, understanding exactly what is meant by the term is important. Toward that end, I ask that you complete the following exercise taken from Karplus et al. (1977).

THE RATIO PUZZLE. Figure 15.1 is called Mr. Short. We used large round buttons laid side-by-side to measure Mr. Short's height, starting from the floor between his feet and going to the top of his head. His height was four buttons. Then we took a similar figure called Mr. Tall, and measured it in the same way with the same buttons. Mr. Tall was six buttons high.

> Now please do these things:
> 1. Measure the height of Mr. Short using standard paper clips in a chain. The height is ____.
> 2. Predict the height of Mr. Tall if he were measured with the same paper clips. ____
> 3. Explain how you figured out your prediction. (You may use diagrams, words, or calculations. Please explain your steps carefully.)

The following are some typical student responses to the ratio puzzle. Read them and compare them with your own. Look for similarities and differences between type A and type B responses.

Student A_1 (Henry, Age 14)

Prediction for Mr. Tall: 8.5 clips

Explanation: "If he is 2 buttons taller I guess he is 2 clips bigger, which would make it 8.5."

Student A_2 (Norma, Age 12)

Prediction for Mr. Tall: 8.5 paper clips

Explanation: "Mr. Tall is 8.5 paper clips tall because when using buttons as a unit of measure he is 2 units taller. When Mr. Short is measured with paper clips as a unit of measurement he is 6.5 paper clips. Therefore, Mr. Tall is 2 units taller in comparison, which totals 8.5."

Student A_3 (Delores, Age 17)

Prediction for Mr. Tall: 8 paper clips tall

Explanation: "If Mr. Short measures 4 buttons or 6 paper clips (2 pieces more than buttons), then Mr. Tall should be 2 paper clips more than buttons."

Student A_4 (John, Age 16)

Prediction for Mr. Tall: 9 clips (pencil marks along Mr. Short)

Figure 15.1. Mr. Short. (Reprinted with permission from Karplus et al., 1977.)

Explanation: "I estimate the middle and then one-fourth of Mr. Short. That's about the size of one button. I measured the button with my clips and found 1.5. So then I counted out six times 1.5 buttons and got nine."

Student A₅ (Jim, Age 14)
Prediction for Mr. Tall: 12 clips
Explanation: "Mr. Tall was 2 buttons taller than Mr. Short. The buttons must be larger than the paper clips. So I doubled Mr. Short's height in paper clips for Mr. Tall's height."

Student B₁ (Harold, Age 18)
Prediction for Mr. Tall: 9.5
Explanation: "Figured it out by seeing that Mr. Tall is half again as tall as Mr. Short, so I took half of Mr. Short's height in clips and added it on to his present height in clips and came up with my prediction."

Student B₂ (Betty, Age 16)
Prediction for Mr. Tall: 9.5 paper clips
Explanation: "I figured that the ratio of paper clips to buttons to be approximately 1.5:1, so two more buttons would make approximately 3 more clips. Since it's a little more than 1.5:1, he is approximately 9.5 clips tall."

Student B₃ (Inez, Age 16)
Prediction for Mr. Tall: 9.5 clips
Explanation: "Mr. Tall is 1.5 times the height of Mr. Short, as measured with buttons, and if the measurement techniques were identical would be 1.5 times Mr. Short's height with any measurement medium. Assuming that the measurement techniques are identical, Mr. Tall's height in clips is 1.5×6.33, which is 9.5 (I think)."

Student B₄ (Jean, Age 13)
Prediction for Mr. Tall: 9.2 paper clips
Explanation: "The ratio using buttons of height of Mr. Short and Mr. Tall is 2:3. Figuring out algebraically and solving for x: $2/3 = 6.5/x$ gives you 9.2 as the height in paper clips."

Student B₅ (David, Age 14)
Prediction for Mr. Tall: 9 paper clips tall
Explanation: "I figured out by figuring that Mr. Small is 2/3 as tall as Mr. Tall."

ERRORS ON THE RATIO PUZZLE. The ratio puzzle is a problem in the sense described in Chapter 7. As we have learned, there are many ways to go wrong on any problem, and in the absence of information that might be obtained from an individual interview, we cannot be certain about how any of the answers shown above were obtained. Still, let me do the best that I can to interpret the information shown.

Apparently, all of the students dealt with the problem successfully; that is, the responses suggest that they understood what the problem said, identified the goal of the task, represented the problem in some manner, and *solved* the problem. Students in both A and B groups got a wrong answer, but none of the wrong answers appear to be due to pitfalls in problem solving such as misreading the problem, failing to understand the goal, or overloading working memory. Apparently, some computational errors occurred in both groups, so computational errors is not what distinguishes the two groups. Rather, the difference seems to be the way students were thinking about the problem. The A students compared Mr. Short and Mr. Tall by *taking a difference*. The B students compared the two by *forming a ratio*.

SUMS AND RATIOS

Why do some students give an additive response rather than the *sensible* proportional one? As Carey's (1985) analysis discussed in Appendix M makes clear, the answer to this question is not clear. Many people argue that failure on this task is due to a lack of domain-specific knowledge—students do not understand proportions. Depending on what one means by "understand proportions", this hypothesis is reasonable. However, if "understand proportions" means what students normally learn in math class when they study ratio and proportion, that hypothesis is not supported by empirical evidence.

Piaget explained the difference between the A and B responses to the ratio puzzle in terms of *logicomathematical* or *operational* knowledge. As discussed in Chapter 5, Piaget distinguished knowledge that may be derived more-or-less directly from the environment (physical knowledge) from knowledge that must be derived indirectly over an extended period of time by abstracting from events some regularity that is not inherent in the physical objects themselves (logicomathematical knowledge).

Students in both the A and B groups are using logicomathematical knowledge to make sense out of the ratio puzzle, but they are not using the *same* logicomathematical knowledge. Only what Piaget called *proportional reasoning* produces a sensible answer to the ratio puzzle, and, for whatever reasons, only students in the B group used it.

ADDITIVE REASONING. Rather early in life we are confronted with "more than" and "less than" comparisons. Tom is taller than Bill. Alice is older than Flo. India is farther from the United States than England. Sally Mae got more candy than Pedro.

As our experience grows, we may compare comparisons. The difference in Tom and Bill's age is the same as the difference in Alice and Flo's age. The difference in these lines, —— and —, is the same as the difference in these, ——— and ——. If we were to express the equality between such pairs mathematically, it would take the form of $A - B = C - D$. This form of comparison is the one made by the A students in responding to the ratio puzzle. Unfortunately, it does not work there, but the A students either do not recognize that this comparison leads to incorrect predictions, or they have not developed the logical operation needed to make the comparison that does work.

PROPORTIONAL REASONING. Additive relationships inherent in "more than" or "less than" comparisons are not the only ones possible. Multiplicative relationships also pertain, but these are less frequent (or perhaps less salient) in everyday experience. The equality between the following pairs of lines, —— and —— or ——— and ——, is just as real as the equality between the pairs shown in the previous section. If we were to express the equality between these pairs mathematically, it would take the form of $A/B = C/D$. This expression is, of course, a proportional comparison.

The equality between ratios inherent in the proportional relationship is evidently less apparent than the equality between differences because the schema for using the proportional relationship develops later. It may be less apparent because it is farther removed from direct observation (i.e., the sensory data require additional transformation before the equality is revealed). Also, the proportional relationship may develop later because it depends on understanding the operations of multiplication and division, which are built on an understanding of addition and subtraction.

This later development may simply be because everyday experience provides fewer opportunities to use this equality in making sense of things. We know it develops later; we do not know why.

ISSUES CONCERNING THE TEACHING OF REASONING

FOLLOWING ALGORITHMS OR REASONING

SENSIBLE PREDICTIONS. Teaching adolescents to *manipulate* equations such as $A/B = C/D$ and to solve them in terms of one of the variables is not difficult. However, such manipulation does not ensure that the student will recognize when an equality of ratios leads to a sensible prediction, whereas an equality of differences does not. It does not ensure that the student recognizes that equality is preserved even though the values of A and B change, so long as their ratio remains constant. Neither does it ensure that the student associates the equality of ratios with such instances as the taste of foods as we add spices to varying volumes or visual proportion in geometric figures.

EVIDENCE OF PROPORTIONAL REASONING.[1] A great deal of controversy surrounds proportional reasoning (as it surrounds much of Piaget's work). The controversy hinges on what one should accept as evidence. Consider the following questions:

---

1. If apples cost 10 cents each, how much do 5 apples cost?
2. If 2 apples cost 20 cents, how much do 5 apples cost?
3. If 1.7 apples cost 27 cents, how much do 5.4 apples cost?
4. If a bag containing 15 apples costs $1.65 and a bag of 21 apples costs $2.39, which is the better buy?
5. If 5 moles of an unknown metal weigh 115 g, how much do 1.7 moles of the metal weigh?

All five questions involve proportions, but they vary in difficulty. Young children can answer the first question correctly. Slightly older children can answer the second one. The third question will stump several adults. The fourth question is answered correctly by about half of adults, and many people labeled "formal operational" miss number five. By the same token, some chemistry students who *do not* understand proportions answer question 5 correctly. In which case should we say "proportional reasoning" is taking place?

The important consideration is not whether the question is answered correctly but how the person thinks about the question.[2] (Suggestions for doing this exercise are given in Chapter 16.)

Probably no one uses proportional reasoning to answer the first question. The proportional relationship does not need to be recognized to find the answer, and

---

[1]This section is based on observations and informal interviews with students at various ages over a period of approximately 20 years. Although not the result of a single, formal research study, it is based on real data; and readers can easily replicate the findings reported here.

[2]For research on this point, *see* Luchins and Luchins' (1977) discussion of Wertheimer's work (p 32).

the question is easier to solve *without* proportional reasoning: by counting by tens, multiplying by five, or mentally lining up apples and coins to get the proper sum.

The second and third questions are logically the same and seem to require proportional reasoning, but many students who can solve the second question in their heads are stumped by the third. Furthermore, when they are asked how they got the answer to the second question, they are unable to say. Careful probing suggests that the answer to question 2 is derived by mentally forming a series of correspondences: Two apples correspond to 20 cents; one apple corresponds to 10 cents; 2 to 20, 3 to 30, 4 to 40, 5 to 50, and so on. Such correspondences are much more difficult to construct for the third question, and the strategy breaks down.[3]

When students who are able to solve question 2 but not question 3 are shown one of the strategies used by adults to deal with proportional relationships, they do not find the procedure sensible. They may accept the procedure on faith, but they are not confident that the procedure will reliably produce correct answers to similar problems. They do not accept the logical operation as a valid procedure for making sense of experience.

The research literature contains many instances in which correct solutions of questions such as 1 and 2 are taken as evidence of proportional reasoning, but I do not accept this evidence.[4] Until a person is able to see the form of the logic and apply that logic, regardless of the numbers used in the problem, the operation cannot be applied generally. *The ability to generalize the logical operation to a wide range of problems must be developed before proportional relationships used in science can become sensible.* In other words, the logicomathematical knowledge must be disembedded from the context in which it is seen.

To say that proportional reasoning can be generalized does not imply that a person can solve *any* problem involving proportions. For example, a person may still be unable to solve questions 4 and 5. In question 4 one must know what is meant by "best buy", that application of proportional reasoning will lead to the cost of each apple in the two bags, and that the smaller of the unit costs represents the best buy. Question 5 may cause difficulty because certain terms are unfamiliar. Domain-specific knowledge is clearly required *in addition to* proportional reasoning. In other words, proportional reasoning is necessary, but it is not sufficient.

Even *after* logicomathematical knowledge has been disembedded from the context in which it is first constructed, additional learning must take place before that disembedded schema will be spontaneously applied to a wide range of situations in which it *can* be used to "make sense" of experience. Furthermore, even when the schema for proportional reasoning is fully developed and richly elaborated—that is, when a person is "fully formal operational"—situations will occur to which proportional reasoning could be applied without being recognized as such.

---

[3]*Realizing* that a strategy used to make sense of experience is failing us precipitates the development of new logical operations, but the realization alone is insufficient. If the failures are infrequent, we can ignore them. If the situations in which failures occur seem unimportant, we may not make the effort to develop new logical operations. If efforts to develop new operational knowledge are too demanding, we may invoke defense mechanisms rather than endure the frustration of failure. For all of these reasons, intellectual development is difficult, time-consuming, and problematic.

[4]Much of the controversy surrounding Piaget's stage theory hinges on this issue. Many (but not all) of the simplified tasks used to show that young children use logical operations that Piaget attributed to older children can be solved by several strategies. Thus, whether the research results show that young children are capable of more sophisticated reasoning than was once thought possible or whether they reflect solutions that do not require such sophisticated reasoning is difficult to know. A specific case of such confusion is discussed in the next section.

Newell and Simon (1972) gave an interesting example of procedural knowledge not being applied even though it is available. They presented a number scrabble problem that is isomorphic to tic-tac-toe; that is, both problems could be represented in the same way (p 71). Even though I understand tic-tac-toe, I could not solve the number scrabble game until I was told that it is isomorphic to tic-tac-toe. I had the logical operations needed, but I did not recognize that they applied. Newell and Simon argued that number scrabble is much more difficult because the winning triads are not readily available. In tic-tac-toe they are seen directly in the diagram for the game. This example is only one of a vast body of research supporting the conclusion that no knowledge ever generalizes to all possible applications.

I have tried to illustrate what I have in mind by proportional reasoning. At the same time, I have tried to illustrate the idea that performance is influenced by many factors and that inferring intellectual capability on the basis of task performance is difficult. As I have shown, questions 1 and 2 may be answered without proportional reasoning, and questions 4 and 5 may be missed even by those who are capable of proportional reasoning. Also, such questions may be answered correctly *without* proportional reasoning.

The research on problem solving reported in Chapter 7 indicates that the most common approach to problems such as 4 and 5 is the application of a memorized algorithm without any appreciation of why the algorithm produces a sensible result.[5] Application of an algorithm requires nothing more than recall of the algorithm and a cuing procedure that can be used to identify the numbers and where they belong in the algorithm. As indicated in Chapter 16, the entire purpose of an algorithm is to enable us to solve routine problems *with a minimum of thought.*

CAN FORMAL OPERATIONS BE TAUGHT?

The kind of thinking that I have in mind when I talk about formal operations such as proportional reasoning seems to develop over a long period of time as the result of a variety of experiences in many contexts. Although I believe that this kind of reasoning can and must be taught, it cannot be taught quickly in the way that most declarative knowledge is taught. Not everyone agrees.

Some researchers have reported excellent results in their efforts to promote intellectual development. Siegler cites five studies to support his contention that "[a] number of previous studies have demonstrated that even 9- and 10-year-olds can master formal operations problems if provided directive instruction" (Siegler, 1976). Although Siegler's claim may be correct, it and the research on which it is based must be treated with caution. Apparently, disagreement exists about what constitutes formal operational thought. In at least one case the performance accepted as evidence of formal operational thinking is closer to blind application of a rule or algorithm than what most scientists would consider scientific thought.

In a study frequently quoted as evidence against Piaget's theory of intellectual development, Siegler, Liebert, and Liebert (1973) taught fifth-graders to solve Piaget's pendulum task (*see* Appendix L) in a 30-minute training session. Given the difficulty that high school and college students have in recognizing the logical necessity of all other things being equal when they perform an experiment, my graduate students and I saw this result as astounding. We decided to replicate the experiment (Greenbowe et al., 1981).

---

[5]I will argue later that a major reason for this problem-solving behavior is that students who have not developed proportional reasoning have little choice. They cannot make sense out of the problem, so they must resort to algorithms learned by rote or give up.

Siegler kindly assisted by providing samples of the materials he had used in the original study. As soon as we saw the data sheet on which students recorded the results of the training experiments and criterion task, we began to suspect that the accomplishment was not as profound as it had first seemed.

As can be seen in the data sheet shown as Figure 15.2, Siegler and his associates provided a table for their data. The form of the table was exactly the same for the two training tasks as for the modified pendulum task used to test the effectiveness of the training procedure. The way tasks were presented, a definite pattern was established, and it was a simple matter to arrive at a correct answer by examining that pattern. Students probably solved the problem not by examining the effect of one variable while holding all others constant as in the original Piagetian task, but by *fol-*

PROBLEM SOLVING SHEET

Name_____ Age_____ Male_____ Female_____

Problem I: Scale Problem

The 2 dimensions involved are _____ and _____

Balls	Level of Dimension I	Level of Dimension 2	Result
Ball 1			
Ball 2			
Ball 3			
Ball 4			

The important dimension in this problem is _____

Why?_____

Problem II: Thermometer Problem

The 2 dimensions involved are _____ and _____

Glasses	Level of Dimension I	Level of Dimension 2	Result
Glass 1			
Glass 2			
Glass 3			
Glass 4			

The important dimension in this problem is _____

Why?_____

Problem III: Pendulum Problem

Strings	Level of Dimension I	Level of Dimension 2	Result
String 1			
String 2			
String 3			
String 4			

The important dimension in this problem is _____

Why?_____

Figure 15.2. Data sheet used by Siegler, Liebert, and Liebert (1973).

lowing the pattern in the same way that elementary schoolchildren frequently complete the exercises in their math books without considering the logical operations involved in the problems.

When we replicated Siegler, Liebert, and Liebert's study, we observed the same *training* effect found in the original study, but we found little evidence of transfer to another task requiring controlling variables that was given immediately after training. Trained students were no more successful on Piaget's version of the pendulum task than untrained students of the same age when that task was administered two weeks after the training took place (Greenbowe et al., 1981).

Unfortunately, a great deal of learning similar to that produced by the Siegler, Liebert, and Liebert training procedure goes on in classrooms. Many teachers even consider it to be worthwhile, even though overwhelming evidence suggests that little or no transfer occurs beyond the specific tasks used in instruction. We focus far too much on right answers and pay far too little attention to the thought processes behind those answers. Until we are able to get beyond an examination of answers and consider how students arrive at those answers, we will do little to promote intellectual development (*see* Landa, 1975). Neither will we solve the educational problems that prompted the current interest in school reform.

DO INTELLECTUAL SKILLS GENERALIZE?

Just as confusion exists about what is meant by intellectual development, confusion also exists about what is meant by a "generalized intellectual skill". Evidence that individuals who can apply proportional reasoning to solve question 3 on p 200 are unable to apply proportional reasoning to solve question 4 is often taken as proof that operational knowledge is content-specific. This argument has some truth, but the assertion that all knowledge is content-specific goes too far. An analogy may set things straight.

Let us compare a generalized intellectual skill, proportional reasoning, to a generalized psychomotor skill, driving a car. Proportional reasoning is "content-free" in the same sense that driving a car is "vehicle-free".

To begin, we cannot demonstrate proportional reasoning in the absence of some specific content any more than we can demonstrate driving a car in the absence of a motor vehicle. Both skills involve an *operation on something*, and the exact manifestation of the skill is influenced by what is acted upon. Still, to say that the skill is specific to each task would be unreasonable.

We do not say that a person has "learned to drive a 1993 Ford Taurus 5-speed" even when the person's experience has been limited to such a vehicle. Rather, we say that the person has "learned to drive". We expect that the skill will transfer to other vehicles: a 1992 Chevrolet Lumina, a 1978 Cadillac, a 1985 Datsun, and perhaps even to a 1980 Mack truck or a 1975 Dodge bus.

Still, we do not expect the skill to transfer exactly. We would not be surprised to find the transfer accompanied by some jerky starts, uncertain stops, and poor parallel parking. We would not even be shocked to find that our driver is completely stymied by a 1925 Model T, never getting it started or perhaps never finding the starter! Still, we would not interpret such failure as evidence that the person has not learned to drive. We would expect that a minimum of instruction and practice would lead to the same smooth performance in an unfamiliar vehicle that was witnessed in the Ford Taurus in which the person first developed the skill. In other words, we view car driving as a general skill applicable over a wide range of vehicles, but we accept the fact that some additional learning is required each time the skill is applied to a new and unfamiliar vehicle.

Generalized intellectual skills such as proportional reasoning seem to operate in much the same way. In the beginning they are likely to be tied to particular content. The learner may not be able to disembed the logical operation from the content of the problem, just as persons learning to drive find it difficult to focus on what they actually do when shifting gears. As application of the operation to familiar content becomes routine, the operation itself can be focused on to begin to see how it may apply to new content. With practice and appropriate instruction, it can be applied over a wide range of experience. The wider the range of experience, the smoother the transfer to a new situation; however, at no point will the initial application of the generalized skill in an unfamiliar setting be as smooth as its application in a familiar setting. Factors peculiar to each setting (content-specific knowledge, if you wish) must be learned and taken into account before the generalized intellectual skill can be applied.

If the new setting has many elements in common with older settings, a person who has developed a generalized intellectual skill will adapt to the new setting rapidly and without special instruction, but if the new setting is very different (as different as a Model T Ford and a modern car, for example), special instruction may be required before transfer takes place. Our task in chemistry education is to design instruction to maximize such transfer, and review of previously learned operations when new content is introduced is required for such transfer to take place.

THE WATER AND WINE PROBLEM

The ambivalent results from recent research using Piagetian tasks have led educators who work with young children to reconsider what can be taught at an early age. If, under the right circumstances, children and adults think alike, perhaps children can understand more than we have assumed. But for those working with adolescents and adults, the more important implication of "children and adults thinking alike" is in the responses of adults!

Consider this problem adapted from Case (1975): You have a glass of water and a glass of wine (Figure 15.3). Assume that both are pure, homogeneous substances. (If it helps, consider the wine to be pure ethanol.)

1. Transfer one teaspoon of water to the glass of wine and mix thoroughly.
2. Transfer one teaspoon of this contaminated wine to the water. Now both the water and the wine are contaminated.

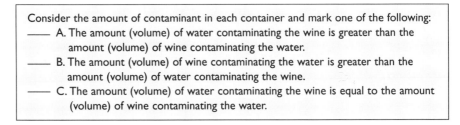

Consider the amount of contaminant in each container and mark one of the following:
_____ A. The amount (volume) of water contaminating the wine is greater than the amount (volume) of wine contaminating the water.
_____ B. The amount (volume) of wine contaminating the water is greater than the amount (volume) of water contaminating the wine.
_____ C. The amount (volume) of water contaminating the wine is equal to the amount (volume) of wine contaminating the water.

If you are not already familiar with this problem, please think about it and mark the answer before you continue reading. Doing so will make points presented in the following sections much more meaningful.

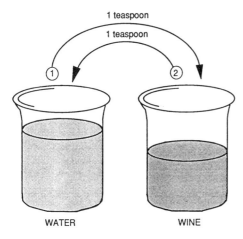

Figure 15.3. A glass of water and a glass of wine.

After you have arrived at your answer, check the footnote to see if it is correct.[6] If you did not get the correct answer, here is a proof:

Let P (for purity) = volume of liquid originally in the glass.
Let I (for impurity) = volume of liquid added to the glass.
Let P_t = volume of purity transferred from the glass.
Let V_1 = total volume of liquid originally in the glass.
Let V_2 = total volume of liquid in the glass at the end.
(Subscripts 1 and 2 index initial and final conditions for other variables as well.)
Proof:
$$V_2 = P_2 + I_2$$
but $P_2 = P_1 - P_t$
Then, by substitution
$$V_2 = P_1 - P_t + I_2$$
but $P_1 = V_1$
So, $V_2 = V_1 - P_t + I_2$
By rearranging, we see
$$V_2 - V_1 = I_2 - P_t$$
But $V_2 = V_1$
so $0 = I_2 - P_t$
and $P_t = I_2$

This proof says that for either glass, the *purity* transferred out of the glass is the same as the *impurity* transferred in. Therefore, the volume of water contaminating the wine is equal to the volume of wine contaminating the water.

Were you able to follow the proof? Did it convince you that C is the correct answer? If not, take some time now to work out your own proof—or prove that both my answer and the proof are wrong!

Once again, please stop reading long enough to work out the correct answer, if you got the incorrect answer before. Subsequent discussion will be more meaningful if you have done so.

---

[6] The majority of adults select A as the correct response, but it is actually C.

I have used the Water and Wine problem in speeches given to audiences ranging from high school students to research scientists. The results are always similar:

1. The vast majority select A as the correct answer.
2. When told that C is the correct answer, few believe it.
3. When shown the proof given above (or one of several alternative proofs that I have used), those who answered incorrectly are not convinced.
4. When given an opportunity to work out the answer for themselves, a large portion of the audience is able to do so.
5. Few people work out the answer in a formal manner.
6. The most common strategy used to solve the problem is to assign a specific volume (say 100 mL and 50 mL) to the water and wine glasses, assign a specific volume to the teaspoon (say 10 mL), and then work out the volume of each component in each glass at the end of each transfer.
7. After people arrive at their own solution to the problem and are convinced that C is the correct answer, they find it much easier to follow the proof given above.
8. Once the correct answer is accepted, many people rationalize their initial, incorrect answer by claiming that they misunderstood the question in the beginning. (The feeling that this was the actual cause of the error was so strong during my first few presentations that I began passing out the written statement of the problem presented above, and I asked members of the audience to describe the task in their own words before answering the question. This modification eliminated the claim that the question was misunderstood, but it had no effect on the error rate.)

My experience with the water and wine problem is reported in the hope that you might experience the phenomenon of not believing what is obvious and simple once the situation is familiar. The insidious difference in our perception of the inherent difficulty of tasks that we *understand* and the perception of the same task by those who do not understand has important implications for teaching.

> I believe we are "at risk" (almost in the medical sense) for egocentric thinking all of our lives, just as we are for certain logical errors. . . . [W]e are usually unable to turn our own viewpoint off completely when trying to infer the others'. Our own perspective produces a clear signal that is much louder. . . . For example, the fact that you thoroughly understand calculus constitutes an obstacle to your continuously keeping in mind my ignorance of it while trying to explain it to me; you may momentarily realize how hard it is for me, but that realization may quietly slip away once you get immersed in your explanation. (Flavell, 1977, pp 124–125. *See also* Johnson et al., 1981.)

The educational implications of the research pertaining to difficulties children and adolescents have with *logical* tasks commonly encountered in science, and the water and wine illustration of how such "misconceptions" interfere with eventual understanding, should be obvious. When we arrive at an interpretation of a problem that is intuitively satisfying, we abandon search for another interpretation. Furthermore, we resist other explanations until something compels us to work through the problem in our own way to arrive at a more satisfying result.

Whether such misconceptions are due to deficiencies in our operational knowledge or to some other deficiency is of little consequence so far as subsequent learning is concerned. In either event, the misconception must be attended to so that the logic inherent in a satisfactory solution is evident to the person who does not understand.

SUMMARY

Because Piaget's formal operations describe reasoning that is used habitually in science, this chapter has focused on those logical operations, particularly proportional reasoning. Many chemistry concepts—formulas and equations, rate laws and equilibrium constants, concentration terms and gas laws, to name a few—involve proportional relationships. They make little sense without proportional reasoning. Other formal operations play an important role in learning abstract concepts and principles such as those that make up atomic theory.

We need to understand that problems involving proportional relationships may be solved without using proportional reasoning; indeed, research on problem solving in chemistry reveals that a majority of students solve all problems by applying a memorized algorithm without understanding the concepts and principles involved. On the other hand, problems may be missed by people who understand proportions but lack some other knowledge required by the problem.

General agreement is emerging among cognitive scientists that generalized intellectual skills (e.g., formal operations) can be taught, but several unresolved issues make it difficult to say exactly how. One issue is what the teacher should take as evidence that a student is actually using a generalized skill such as proportional reasoning. Considerable research in psychology and education has accepted rote, algorithmic performance as evidence of scientific reasoning. Scientists would not. Still, other studies appear to demonstrate real, if modest, gains in intellectual development.

Another difficult issue involves what we mean by a generalized intellectual skill. All knowledge appears to be situated in or tied to the context in which it is acquired to some extent. On the other hand, no knowledge is *totally* context-bound. When we learn to drive a car, that skill is not limited to the vehicle in which we trained; when we learn to read, we have access to books we have never seen. Similarly, when we develop scientific reasoning, we can expect it to transfer to many, but not all, situations other than those within which our learning took place.

As illustrated by the water and wine problem, we never outgrow our susceptibility to *stupid* mistakes. Novel tasks often appear quite simple, and we sometimes assimilate the task to schemas that seem to fit when they do not. We arrive at intuitively satisfying answers that are dead wrong. Until something intervenes to challenge our result, we are unlikely to apply the appropriate schema to the task to arrive at a correct solution. Interestingly, when we come to intuitively satisfying but wrong results, formal arguments seldom sway us. Rather, what seems to be required is something more concrete that we can get our mental meathooks into so that we can think through the problem ourselves.

Chapter 16 will explore this issue further as I discuss how we might go about teaching scientific reasoning in chemistry.

16

Teaching Generalized Intellectual Skills

In Chapter 14, I reviewed research pertaining to intellectual development and ended with guidelines for teaching generalized intellectual skills. There is still much that we do not know about promoting intellectual development. The recommendations outlined in this chapter are based on my experience in teaching chemistry to under-prepared college students. Informal examination of test results before and after these procedures were adopted and personal interviews of students in my classes lead me to believe that the procedures are effective, but these procedures have not been rigorously researched.

In an effort to make the discussion as specific as possible, I have selected three examples, each dealing with different content and a different instructional setting. The first example deals with developing proportional reasoning in an instructional setting that tends to be teacher-centered. As you will see as the discussion proceeds, there are plenty of opportunities to involve students in this approach, but it easily fits a traditional classroom organization. The second example is at the other extreme. It describes an impromptu individual interview aimed at understanding how a student is thinking about metric units and the variables described by those units. As in the first example, the discussion shows how the same interview techniques can be adapted to other instructional settings. The third example is in a laboratory setting. It is, in many ways, similar to the individual interview, but the ideas being probed are much more abstract. Whereas the interview in Example II simply probed to see how the student was thinking, the interview here goes beyond that and includes efforts to *teach* the student the canonical view of the phenomena under investigation.

GENERAL CONSIDERATIONS

All aspects of learning are interrelated. Although I have separated a discussion of concept learning from problem solving, attitudes, and generalized intellectual skills, all of these aspects are involved in the procedures outlined here. Consideration of some issues must precede the actual development of intellectual skills.

ATTITUDINAL PROBLEMS

I have had no success with under-prepared college students without attacking the affective problems that stand in the way of progress. These problems take several forms. One of the most serious problems is that students play "the game of school".

THE GAME OF SCHOOL. In the game of school, people go to a place called school where they perform tasks to earn tokens called grades, which are traded in for certificates needed to get pleasant jobs. Understanding concepts or developing intellectual skills may occur during the game of school, but they are not the goal of the game; grades and degrees are. Thus, when presented with a task such as a stoichiometry problem, waiting for the instructor to explain how to do it and then memorizing the pattern makes far more sense than wasting time struggling.

The focus in the game of school is on getting right answers, and little or no concern is given to why things are right or wrong. This attitude is damning because, as seen in Chapter 14, not until students begin to ask why and test their theories in action does intellectual development take place.

LACK OF SELF-CONFIDENCE. A second attitudinal problem is that students believe that they are stupid and can never do what is demanded (for example, solve stoichiometry problems). Like many teachers, they believe that people are born bright or dull and that's that.

LACK OF RELEVANCE. The third attitudinal problem is that many tasks that we ask students to do are so far removed from their own day-to-day world that they have difficulty understanding why they should learn to do them.

Attitudinal problems and what can be done to overcome them are taken up in the section of this book called **Things in the Affective Domain** (Chapters 17–19).

ASSESSING STUDENT THINKING

Once attitudinal problems are overcome and students are making a conscientious effort to *understand* what I am talking about, I try to present material so that both the students and I are aware of how they are thinking about problems and why they have failed to understand.

I do not administer a formal test to assess students' levels of intellectual development. I am not convinced that this knowledge is useful. Content and process are so confounded that I feel much more confident trying to obtain information about how students are thinking about *particular problems*.

The *rule* rather than the exception is that my students apply an operation (proportional reasoning, for example) in one context, but do not apply the same operation in an unfamiliar context. In such an event, students can be moved rather quickly from the (often inadequate) thought processes they are using to the more adequate thought processes normally applied by experts. However, *I must know the thought processes that students use spontaneously.* Thus, I usually do not need some general classification of students as concrete or formal, but rather I need information about the logical operations activated *in specific instructional settings.*

In the three examples that follow, the amount of attention given to learning how students are thinking varies. In the first example, little is done along these lines. On the basis of previous experience, I was quite confident that I already understood the range of thinking patterns among my students. In the second example, virtually everything I did was aimed at learning how the student was thinking. In the beginning I had no idea what the problem might be, so I probed until I sensed a useful direction to take in instruction. The third example lies somewhat between the first two. It is more representative of the proportion of time I normally spend on assessing understanding and promoting further learning.

EXAMPLE I: PROPORTIONAL REASONING

Proportional reasoning is at the heart of stoichiometry, and my students are bad at it. They cannot understand stoichiometry unless they understand proportions. Developing such understanding is a long and tedious process begun long before the mole is introduced. Indeed, a major consideration in selecting content for my course is my aim to foster proportional reasoning. This consolidation can be seen best by examining the textbook (Herron, 1981a), the instructor's guide (Herron, 1981b), and the laboratory manual (Copes, 1981) written for the course.

INTRODUCING PROPORTIONAL REASONING

One example of the way my commitment to proportional reasoning influences presentations is the way density is introduced. Rather than defining density as the ratio of mass to volume, students are told that "density is the mass of an object of a given size: for example, a cube measuring one centimeter on a side. In other words, you may think of density as the mass of a unit cube" (The context of this introduction is discussed on pp 215–218.) Why is this instruction done?

If you assume, as I do, that knowledge is constructed by the learner and that construction is limited by the tools the student has at hand, you must develop broad, generalizable schemas such as proportional reasoning in terms of schemas that already exist.

INITIAL UNDERSTANDINGS. From previous experience, my students have an intuitive understanding of ratios with a denominator of one. We use this understanding when we say that apples cost 16 cents each, gasoline is $1.25 per gallon, a car is traveling 60 miles an hour, or a person is paid $200 per week. The introduction of density takes advantage of this logical structure, and it takes on the more conventional meaning as proportional reasoning develops throughout the course.

Even the poorest of my students can solve simple problems involving a unary rate: for example, "If apples cost 16 cents each, how much do 5 apples cost?"

For this problem students quickly respond, "80 cents." If asked how they know this answer is right, most respond something like the following: "Each one costs 16 cents, and you have 5 of them. You just multiply." As the discussion beginning on p 200 makes clear, the problem does not have to be changed much before a few students start dropping out.

RELATION TO THEORY. Before proceeding, let me characterize the situation I am describing as it pertains to intellectual development.

1. All of the discussion is in terms of content that is familiar.
2. Students have been exposed to questions having the logical structure of interest.

3. Students can solve some problems having the logical structure of interest, but probably not by using the logical operations I use and want them to develop.
4. Because students can solve some problems having the logical structure of interest, they have a procedure that can be used to check the validity of solutions obtained by using a more formal procedure.

What I attempt to do is help students use their existing intellectual equipment to disembed the logical structure from familiar problems, handle that structure in a more formal manner, and develop confidence that the formal approach can be trusted when applied to more and more complex examples.

I begin this process by presenting the familiar example, calling attention to the logical relationship that is opaque to students, and introducing a language to express the relationship. I then show students how to apply the new language to problems while they use their existing logical operations to verify that the new procedure produces correct results. Let me go through this process in slow motion.

INTRODUCING AN UNFAMILIAR STRATEGY

"You have told me how you know that 5 apples cost 80 cents if each apple costs 16 cents. Let me show you how *I* would describe what you told me:

"You said each apple costs 16 cents. I write it like this:

$$\frac{16c}{1 \text{ apple}}$$

"What is above the line is the cost of the thing below. Next you said that you multiplied the cost of 1 apple by the number bought, 5. I say it like this:

$$16c/1 \text{ apple} \times 5 \text{ apples} =$$

"Now here is the part that may be new to you. I am going to treat the letter *c* and the word *apple* just like unknowns are treated in math. Furthermore, the line dividing the 16c and 1 apple will signify division as it does in math. If there is no line (5 apples, in this example), the quantity is assumed to be above the line or part of the numerator:

$$16c/1 \text{ apple} \times 5 \text{ apples} = 80c$$

"As you can see, my procedure gives the right answer. Now consider the next problem: '5 apples cost 50c. How much do 2 apples cost?' I already know how to find the cost of apples if I know the price of one.

$$\Box c/1 \text{ apple} \times \bigcirc \text{ apples} = ?c$$

"In this case, I want the cost of 2 apples or:

$$\Box c/1 \text{ apple} \times 2 \text{ apples} = 2\Box c$$

"Can you use the information given to find the cost of 1 apple?" (The student *can* do this and will say something such as "divide 50 by 5".) "Right, but let's do it keeping the same language:

$$\frac{\overset{10}{\cancel{50}}\ c}{\underset{1}{\cancel{5}\ \text{apples}}} = \frac{10\ c}{1\ \text{apple}}$$

Now:

$$10c/1\ \cancel{\text{apple}} \times 2\ \cancel{\text{apples}} = 20c"$$

In this early development, working with apples is not important, but working with problems that students understand well enough that they can check the validity of results is important. My students can solve both problems in this illustration intuitively, and they know that the answers obtained by using the new notation are correct. What students see is that *application of an unfamiliar strategy to a familiar problem produces the correct answer, and this outcome suggests the possibility that the new strategy will work for other problems as well.* Practice with additional problems having the same properties increases confidence in the new strategy.

FOCUS ON THE LOGIC

Once confidence in the strategy is established, the strategy is applied to more complex cases: "If 5 apples cost 80c, how much do 1.5 apples cost?" Attention is again drawn to the *form* of the logic previously employed:

"Goal: We want to know the cost of some number of apples:
?c = 1.5 apples
"The strategy is to multiply the number of apples by the unit cost:
?c = 1.5 apples × __c/1 apple
"If we do not know the cost of 1 apple, we use other information to get it:
80c/5 apples = 16c/1 apple
?c = 1.5 apples × 16c/1 apple = 24c"

EXAMINE THE LOGIC OF STUDENT SUGGESTIONS

When this development is done with a group of students, someone usually asks why we need to do all this nonsense when a simpler procedure exists. I ask what the simpler procedure is and then translate the suggestion into the language we are using. One suggestion that comes quickly is to just multiply the 1.5 apples by the ratio, 80c/5 apples:

?c = 1.5 apples × 80c/5 apples = 24c
"The answer is the same, so it seems okay. Let me do a logical check. . . . Well, sure. If I do some arithmetic on the ratio first, the expression is just the one we have been using. Obviously, 80c/5 apples is just another way of writing 16c/1 apple. It appears that your suggestion is a good one and *could* save time and effort."

DO NOT PUSH STUDENTS TO ACCEPT A NEW STRATEGY

Some students who are quite comfortable with the procedure when it is done in steps are not comfortable with the shorter procedure, even after the logical analysis. My experience is that students should never be encouraged to use a procedure that does not make sense to them.[1] Consequently, I encourage students to continue using the procedure with which they are comfortable while others use the newer and more convenient one. Both groups should get the same result, and I encourage them to check to see that they do.

If knowledge is constructed, we can only make sensible constructions by using sensible procedures. Forcing students to use nonsensical procedures when they have a sensible one that works is counterproductive. At the same time, instructors know that as matters become complex, procedures that are efficient and automated (i.e., algorithms) are needed to avoid mistakes.

INFORM STUDENTS OF COGNITIVE LIMITATIONS

I tell students that they will soon encounter complicated problems for which they will want *convenient* procedures as well as sensible ones. I encourage them to keep thinking about the procedures suggested by their classmates to see if those procedures begin to make sense as they gain experience. Students do not need to be encouraged to switch to the more efficient procedure when it begins to make sense. The Principle of Least Cognitive Effort dictates that students will adopt more convenient procedures and use them. To ensure that students continue to consider more efficient algorithms, the logic of each problem done in class is reviewed. The following problem would be encountered much later, but it illustrates how this review is done.

How many grams of carbon dioxide gas can be produced when 10 mL of 1.5 M HCl are dropped on a marble slab?

$$2HCl + CaCO_3 \rightarrow CO_2 + H_2O + CaCl_2$$

?g CO_2 = 10 mL HCl solution × 1.5 mol HCl/1000 mL HCl solution × 1 mol CO_2/2 mol HCl × 44 g CO_2/1 mol CO_2 = 0.33 g CO_2

By pointing to each statement in the solution, the procedure is discussed. "The goal is to find the mass of carbon dioxide produced (*point to ?g CO_2*) when 10 mL of HCl solution react (*point to 10 mL HCl solution*). Because the HCl reacts and not the water, we need to know how much HCl we have in the solution. We learn this by multiplying the volume of the solution by the moles of HCl in 1000 mL. At this point (*point to 1.5 mol HCl/1000 mL HCl solution*), we have calculated the moles of HCl in the 10 mL of solution.

"But we want to know something about the carbon dioxide produced. The only connection between the acid and the carbon dioxide is the one given by the equation (*point to 2HCl and CO_2 in the original equation*). The equation tells us that one mole of carbon dioxide should be produced for every two moles of HCl that react (*point to values in the original equation*). If we assume that all of the HCl in solution reacts, we can find the moles of carbon dioxide produced. That's what we are doing here (*point to 1 mol CO_2/2 mol HCl*). We are multiplying the moles of HCl contained in the solution by the number of moles of carbon dioxide that can be produced by each mole of HCl.

"Fair enough. Now we know the number of moles of carbon dioxide that can be produced from the 10 mL of solution. Are we finished? (*wait for response*) No,

---

[1]This approach seems to be supported by the research reviewed in Chapters 6 and 13 that shows that social interaction has no beneficial effect when students at the lower level of development do not understand what students at the higher level are doing.

this (*point to ? g CO$_2$*) tell us that we started out to find mass. All that remains is finding out what it weighs. If we recall the relationship between moles and mass, we can calculate the mass of each mole of carbon dioxide and then find the mass of any number of moles. That is what we are doing in the final step (*point to 44 g CO$_2$/1 mol CO$_2$*), leading to a value of 0.33 g of carbon dioxide that can be obtained when 10 mL of 1.5 M HCl solution react with an excess of marble."

INVOLVE STUDENTS

This kind of analysis takes on several variations. At times a student is asked to go through the entire procedure. At other times one student starts and another finishes. Any time a student reports using a different procedure, he or she is asked to share it. Other students are asked to check the logic to see if the procedure makes sense. I sometimes challenge a procedure—right or wrong—and encourage the student to defend the argument. At times I do the analysis myself.

At the point that most students are working problems as shown here (or even in their heads) some students are still confused when the series of steps is strung together as shown. They work one step and calculate an answer. They use that value to do the next step, and so on. I do not discourage this, but I do encourage students to follow the discussion of the more efficient process with the idea of adopting it when it appears sensible.

So far I have described initial work on development of proportional reasoning and how the factor-label algorithm is used to support that development. Much more must be done before one can reasonably expect students to generalize the logic of proportions to the wide range of problems that they will encounter in science.

SEQUENCING OF ACTIVITIES

Problems such as those already described are encountered from the first week in the course until the end of the semester. As indicated in the previous section, the first problems are kept simple. The problems deal with situations that students already understand (cost of apples), and the information in the problem statement is limited to what is required for the problem. Later, problems are made more complex by including information that is not needed. Occasionally, problems are presented with insufficient information, and students are expected to recognize this deficiency and ask for the additional data. (Such problems are also included on weekly quizzes and hour exams. On exams, missing information is supplied on an attached data sheet.) Still later, problems involving unfamiliar content are encountered.

When the unfamiliar content is first encountered, problems are again reduced to necessary information only. If students must deal with unfamiliar content at the same time they are learning to apply newly learned logical operations, working memory is overloaded, they are unsuccessful, and the desired learning does not occur.

OTHER EXPRESSIONS OF PROPORTIONS

GRAPHS. Several times during the course, students prepare graphs from data collected in the laboratory. I collect sample data, prepare an overhead projection of the graph, and use the graph as a focus of discussion during lecture. Students are invited to search for relationships revealed by the graph. If the graph is a straight line, the slope of the line and possible interpretations of the slope are given particular attention.

One of the first instances of such activity occurs when density is introduced. Students collect ordered pairs corresponding to the mass and volume of glass as 16 glass rods are measured. They then plot total mass on the *y*-axis and total volume on the *x*-axis as shown in Figure 16.1. During class discussion, the slope of the resulting line is calculated and attention is called to the fact that this slope is constant. In other words, the ratio of mass to volume of glass was constant in the experiment. The slope is interpreted as "the mass of a piece of glass having a volume of one cubic centimeter", and the term *density* is introduced. Throughout the semester students study several proportional relationships, plot values for the variables involved, calculate the slope of the resulting straight line, and give a physical interpretation to the number obtained; for example, 2.09 g of zinc chloride are formed for each gram of zinc that reacts.

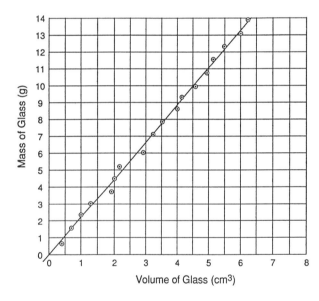

Figure 16.1. Graph of proportional relationship.

ALGEBRAIC FORMULAS. My final activity related to proportional reasoning takes place when the behavior of gases is discussed near the end of the course.

A discussion of gases is included in the remedial course in preference to several other topics that are equally important (and perhaps more interesting) because this discussion is a convenient way to examine mathematical equations that describe direct and inverse proportions.

The lecture before gases are introduced is devoted to the volume of a cylinder, which is a familiar idea. Students are first asked what is meant when we say, "*A* is directly proportional to *B*"?

The first response is usually something such as, "*A* is directly proportional to *B* if *A* increases when *B* increases". If the suggestion goes unchallenged, I make a table like the following:

```
A B
1 5
2 6
3 7
4 8
```

"Okay. *A* increases as *B* increases. Is the value of *A* directly proportional to *B*?" Some students say "yes", but others insist that "it has to increase proportionally".

"Well, okay. How do I know when it is doing that?"

Eventually the better students (those who understand the concept but have not developed a language for talking about it) insist that the first table of values does not represent a direct proportion, but the following one does.

```
A B
1 5
2 10
3 15
4 20
```

When pushed to express the difference mathematically, some student produces the statement, "$A/B = k$", where k is constant for a particular set of values. In this case, k is 0.2.

I then ask whether the volume of water in a cylinder is proportional to its height. Students say that it is, and I ask how we can verify that it is. Measurements are taken for volume and height as an unmarked cylinder is filled to different levels. The ratio is calculated for each data set, and students are asked to decide if the value is constant.

This question is not trivial. The measurements have error, and the calculated values are not numerically the same. For the first time, students begin to appreciate the fact that our earlier fuss about uncertainty in measurement and significant digits has a purpose. Until uncertainty is considered, the question about constancy of the ratio cannot be answered.

Next students are asked if the volume of a cylinder is proportional to its diameter. Again the answer is "yes", again measurements are taken (this time by using a variety of cylinders filled to the same height), and the ratio of volume to diameter is calculated. It is immediately obvious that the ratio is *not* constant.

Students are then asked if they recall a formula for calculating the volume of a cylinder. Someone does, and it is written and rearranged to the following form:

$$V = \pi/4 d^2 h$$

The rest of the lecture focuses on the formula and its relation to the preceding activity.

In the first case, the diameter was constant and the equation can be written as

$$V = kh \text{ or } V/h = k$$

where $k = \pi/4d^2$.

In the second case, the height was constant, so the equation can be written as

$$V = kd^2 \text{ or } V/d^2 = k$$

where $k = \pi/4h$.

Obviously, the volume is *not* proportional to the diameter, but it *is* proportional to the *square* of the diameter. By using the experimental data still on the board, this relationship is quickly verified. Similar discussions of the equations describing the behavior of ideal gases takes place in subsequent lectures.

DISEMBEDDING LOGIC FROM ITS CONTEXT

A major task in teaching logical operations is getting the logic disembedded from the context in which it is used. Many of the activities described in the preceding section are designed to facilitate this process. Two conditions seem to be critical for disembedding:

1. The more opportunities the learner has to see the logic or strategy applied to dissimilar problems or situations, the more likely it will be disembedded and available for application to still unanticipated problems.
2. The more the learner is encouraged to focus on the logic or strategy as opposed to the specific problem in which the logic or strategy is applied, the more likely the learner will disembed the logic or strategy from the application at hand.

The slope of a straight line can be used in many ways to give rate-of-change information. This procedure is a general one that can be used for $C = \pi D$, $F = ma$, $D = m/v$, $P = f/a$, or any other direct proportion. However, if this procedure is to be useful, the proportional relationship and the properties of a proportional relationship must somehow be disembedded from the formula or context in which it is seen.

By repeatedly plotting data for linear relationships, calculating the slope of the resulting straight line, and writing an equation to describe the relationship as illustrated in the discussion of Figure 16.1, students begin to see its generalizability. If the instructor focuses on the *form* of the relationship at each occurrence rather than focusing on the specific relationship being studied, generalization will be enhanced.

Often just the opposite is done. For example, many beginning chemistry textbooks emphasize very specific algorithms for working very specific problems: weight–weight, mole–mole, and mole–mass conversions in stoichiometry, for example. In addition, examinations generally focus on the concepts being developed rather than the logical relationships inherent in those concepts. Such an emphasis diverts the student's attention from the logical relationships.

My experience is that students seldom see the underlying logical relationship when it is first pointed out. For example, few students appreciate the point of calculating the slope of the line in Figure 16.1 if this is the first time that proportional relationships have been treated in this manner. When another proportional relationship ($C = \pi D$, for example) is later treated in the same manner (i.e., the circumference and diameter of several circles are measured, a graph of the ordered pairs is made, and the resulting slope is calculated), students often see no relationship between the two exercises. (The *only* relationship, of course, is the *underlying* proportional one.) However, after repeated exposure to similar exercises in a variety of contexts, students begin to see the feature that remains constant across the examples, and they may comment, "All of these exercises show variables that are related in the same way." At this point, the logical relationship appears to become disembedded from the particular context in which it is seen, and generalization to a variety of contexts is made possible.

EXAMPLE II: AN INTERVIEW WITH FAE

Fae was a premed major at the time of this interview. Fae had taken a freshman chemistry course the previous semester and failed. When I tried to determine why Fae was failing the course, Fae explained that Fae *did* understand the material, was not having difficulty with any other courses, and seemed to just freeze on my exams. To what extent Fae *believed* this, I do not know. What I do know is that Fae did not understand what I was trying to teach, and I wanted to know why. I began, as I usually do, by looking at exam results to see if there were any clues. Here are three questions that Fae missed, and Fae's answers are in bold type.

---

1. In the metric system the cubic centimeter is a unit of
   a) mass
   b) density
   c) **length**
   d) area
   e) volume
2. A box measures 27 cm high, 3 cm deep, and 4.6 cm long. How many liters will it hold?
   a) **372.6 L**
   b) 3.7×10^2 L
   c) 4.0×10^2 L
   d) 4×10^{-1} L
   e) 0.3726 L
3. The density of glop is 6.21×10^{-1} g/cm^3. What is the volume of 4.90×10^2 grams of glop?
   a) **304 cm^3**
   b) 78.9 cm^3
   c) 12.5 cm^3
   d) 12.7 cm^3
   e) none of these

As usual, Fae's difficulty is impossible to infer from these responses. However, these responses do provide a point of departure for an interview. My first question is usually, "How did you arrive at that answer?" Students often realize where they went wrong, but Fae did not. I handed Fae a chalk box and asked about it.

"If I asked you to tell me the volume of this box, what metric units would you use?"

"Grams?" Fae asked.

"How would you measure the grams?"

"On a balance." (No question implied.)

"And what would you be telling me about the box if you tell me the volume in grams?"

"How heavy it is." (Again, said confidently.)

So far, so good. Apparently, Fae had volume associated with amount of matter, and Fae knew that you measure amount on a balance, that the units are grams, and that the number of grams provides a measure of *heaviness*. But of course Fae had confused volume, one measure of "amount of stuff", with mass, a very different measure of "amount of stuff".

To say that Fae was totally confused is unfair. Fae had, after all, associated volume, mass, and the units of grams with "amount of matter". However, Fae's construction of knowledge apparently had not progressed to the point that Fae differentiated the various ways that we have of describing amount and the various units that we use in the process.

Where one goes from this point in an interview is a matter of choice. I decided to get back to the cubic centimeters of the original question, so I traded Fae a centimeter cube for the chalk box.

"What is this?" I asked.

"A square."

Once again, the evidence suggested that Fae had begun to construct knowledge about shapes, but either the concepts held or the language used lacked the precision needed for accurate communication. At this point, I decided to distinguish between a square (a plane figure) and a cube (the three-dimensional shape) Fae held.

I then returned the chalk box to Fae and asked what metric units could be used to tell me the distance between two corners of the box. Fae indicated centimeters, and I asked what that told me about the box—length, area, mass, or what? Fae indicated length, and I was satisfied that Fae had a reasonable concept of length and the units used to describe length. However, when I focused attention on a surface of the box and asked what units Fae would use to describe how large the surface was, Fae was unable to tell me. *Neither could Fae say what we would be describing (i.e., length, area, mass, etc.) when we talked about the size of that surface.*

By now it seemed clear that several fundamental concepts related to measurement were unclear to Fae, and I saw little point in going further. (Also, Fae's problems in the course were obviously not a simple matter of freezing on the exam!)

I will relate one additional episode that took place somewhat later. I handed Fae a lead sinker and a large, Styrofoam ball that weighed more than the sinker, even though it did not feel as though it should.

"Which is heavier?"

"The sinker." (The wrong, but expected, response.)

The two objects were then placed on opposite sides of a double-pan balance, and the side with the Styrofoam ball went down. Sides were reversed; again, the side with the ball went down.

"Why do you think you got fooled?"

"The sinker just felt heavier."

"Can you think of any way that you could compare Styrofoam and lead so that you would not get fooled?"

"Compare pieces of the same size."

"How could you get pieces of the same size?"

"Put little pieces of lead on the balance until you get a gram of it. Then do the same for the Styrofoam". (Note that what Fae meant by "same size" was apparently "same mass" rather than "same volume".)

Fae then pointed out that the piece of Styrofoam would be bigger than "the same size" piece of lead. We talked about this statement at length, and Fae seemed to be beginning to construct an acceptable idea of density, even though Fae's expression was the inverse of what scientists call density.

The reason for sharing this later episode is that it suggests that Fae was capable of constructing sensible ideas on the basis of experience, even though Fae had not constructed the accepted ideas for area and volume. Nothing seemed to be wrong with Fae's *ability* to learn, even though a great deal was wrong with *what* was learned.

USING INFORMATION FROM INTERVIEWS

The procedure that I followed with Fae is far too time-consuming to use with every student, and it would not be cost-effective if the student being interviewed were the only one to benefit. This is not the case, however. Information gained from such interviews is used in several ways:

1. Misconceptions revealed by individual interviews may be held by other students. To test this hypothesis, I develop activities that can be administered to the entire class. The activity may be a demonstration in lecture, followed by a show of hands from those students who interpret the demonstration in one way or another. The suggested interpretations, of course, are designed to reveal the misconception. Misconceptions concerning area that were revealed in individual interviews have led to a laboratory exercise in which students are asked to determine one millionth of a sheet of paper. Calculations required to complete the exercise reveal which students misunderstand area and the relationship between linear and square measures. (One cm^2 is **not** equal to 10 mm^2!) An individual interview revealing misconceptions about what is represented by chemical symbols led to a standard test question: "Explain in words and draw pictures to illustrate what is represented by Cl, $2Cl$, Cl_2, and Cl^-."

2. Conceptual difficulties revealed in interviews are shared with teaching assistants during staff meetings. They are asked to be alert for signs of similar misunderstanding, and we discuss possible causes and cures for the difficulty.

3. Interviews frequently reveal that things I have said (or not said) have led to reasonable, but incorrect inferences on the part of the student. The offending instructional procedure is subsequently modified.

PRACTICALITY

My diagnostic procedures (particularly the contact with individual students) are often dismissed by others as totally unrealistic. They are not. Although certain aspects of my procedures could not be implemented by everyone, alternatives can be found by anyone who believes that diagnosis and remediation of learning difficulties is important. Some modification of current teaching practice may be required, but it can be done, even in large courses. (I have had approximately 250 students in a course.) Although I had several assistants who helped with remediation, I did most of the individual diagnosis. Keep in mind that the sequence followed in diagnosis rapidly reduces the number of students who receive attention. Of the 250 students in the course, 75–125 will score below 70% on the first exam. Of these students, 50–75 know that they have put forth little effort, and I did nothing with these students. (My attitude was that their failure should not concern me unless it concerns them.) Of the 25–50 students left, 10–20 received no more than 5 minutes of attention concerning their study habits. I seldom made initial appointments with more than 20 students. Of these students, no more than 10 received additional attention. Even this amount of attention is not insignificant, but neither is it impossible.

EXAMPLE III: USING THE LABORATORY

During the time that I coordinated the field trial of the Intermediate Science Curriculum Study, I visited classrooms to find out what students were doing. When I asked students, I got two qualitatively different kinds of response. In some class-rooms, conversations went something like this:

"What are you doing today?"

"Activity 6-7."

"And why are you doing that activity?"

"Because it follows Activity 6-6."

"Why do you think the authors put those activities in the book?"

(Silence and a blank stare.)

In other classrooms the conversations were very different:

"What are you doing today?"

"We're trying to see how the nature of a surface affects the force needed to move a block of wood."

"And why are you doing that activity?"

"Well, we're studying forces, and one of the things we're trying to find out is what affects how much force it takes to pull this block of wood. Yesterday we found out that the angle of the pull makes a difference, and the day before we learned that adding weight to the block makes it harder to pull. This is just one more thing we're looking at."

When I tried to figure out what was different about the two classrooms, the major difference appeared to be the way that the teacher interacted with the students. In the classroom where the first conversation took place, the teacher either sat at his desk, checking and correcting the answers written in student record books, or walked around the room doing the same thing verbally. In the second classroom, the teacher did very little checking of answers. When she did, she showed little interest in whether it was right or wrong. Rather, she wanted to know how the student decided that it was right, how that answer was related to information the student had developed through previous activities, and what questions the student thought would be sensible to ask next. I came to realize that the *kind* of dialogue between teachers and students is far more important than the *amount*.

DESCRIBING OBSERVATIONS

The following dialogue approximates my conversation with a college student doing an experiment to separate the dyes in ink by using paper chromatography.

"How's the experiment going?"

"Fine."

(Looking at the chromatography paper in the beaker and calling attention to the dot representing the original ink spot at the bottom of column C in Figure 16.2.) "Why is this ink separating into the different colors?"

"Well, the solvent is causing the ink to decompose."

"Let me see if I understand what you mean by decompose. When chemists use the term *decompose*, they mean something like what happens when you do an electrolysis of water. (This process had been demonstrated and discussed in a recent lecture.) Recall that water actually breaks apart into hydrogen gas and oxygen gas, materials that are very different from water. Is that what you mean is taking place with the ink?"

"Well, uh, I'm not sure."

"What else might be taking place?"

"Well, I guess the solvent could be . . . well, like filtering out things."

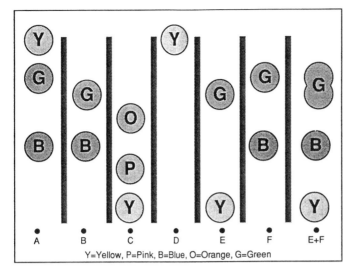

Y=Yellow, P=Pink, B=Blue, O=Orange, G=Green

Figure 16.2. Diagram of paper chromatogram.

"Can you tell me what you mean by filtering out?"

"Well, kind of decomposing into parts."

"Hmmm. Let me describe two things and see if you can tell me which possibility the ink is like.

"I can take peanuts, cereal, and pretzels and make a party mix. I can then take the party mix and separate it by simply picking out the peanuts and putting those in one pile, picking out the cereal and putting them in another pile, and so on. What chemists say to describe that process is that the party mix represents a *mixture* and that we are *separating* the mixture into components.

"Chemists contrast this process with the change in water that I described a minute ago, *decomposition*. Chemists think the two processes are fundamentally different. In decomposition, they think you have one kind of substance that breaks apart into components that have very different properties—so different that you would not dare call them the same as the original substance. For example, water certainly looks nothing like oxygen gas in the air and neither does it behave in the same way. If you do not believe it, dive into a swimming pool and try to breathe water. I think you won't last long! You can, however, breathe the oxygen, one of the things that is produced when the water breaks apart. That's different from the separation of the party mix.

"If you put a handful of party mix in your mouth, you will get the taste of the peanuts, the taste of the cereal, and so on. In the party mix situation, the properties seem to be there but all mixed up and you separate them; in the water situation, you literally produce something with new properties. Now, after that long explanation, which of these is taking place on the chromatography paper?"

"Well, I really think it's more like the party mix situation."

"That's what I think is happening, too. Now, let me focus on something else. I see here that there is a yellow dot right at the bottom of column C (Figure 16.2)—where you put this ink and it started to separate. Over here in column E is another yellow dot at the bottom. How do you interpret that information?"

"I do not know. That they break down into the same things?"

"Now do you mean 'break down' or separate? Remember, 'break down' is the term that chemists use to describe what happens to water during electrolysis. Chemists would not say 'break down' to describe what happens to party mix."

"Well, they separate."

"All right, what does that say about the components of these various inks? Do they have anything in common?"

"Well, I guess you'd have to conclude that these inks all have some yellow dye in them that's the same."

"That's exactly what we think is going on. And why do we make that conclusion?"

"Well, they are both yellow."

"That's right, and what that says is that there is a property of these substances that is the same, which leads us to believe that the substances are the same. Is there any other property of that dye, other than being yellow, that appears to be common?"

"Well, I do not see anything."

"All right. Now let me focus your attention on something else. Besides this yellow color that's down here near the bottom, notice that the ink in column A produced a yellow color way up at the top (Figure 16.2). The colors look pretty much the same. Are these same dyes? In other words, is the yellow up here near the top of the paper the same as that yellow down at the bottom?"

"Well, I do not think so."

"Why not?"

"One of them went up to the top and the other one stayed down here at the bottom."

"I agree. The two yellow dyes have a different property. They are not moved along at the same rate by the solvent. According to the rules of the game that chemists play, we say that if the properties (i.e., all of the *characteristic* properties) are the same, we assume it's the same substance. If the characteristic properties are *not* the same, we assume it must be a different substance.

"You have to worry about what's characteristic and what is not, but I'll tell you that the rate at which things move along this sheet *is* a characteristic property. We could prove that by taking dyes that we know are different, and doing this experiment. That's been done, and each substance *does* move at a slightly different rate as the solvent moves across the paper. Therefore, that property can be used to identify the substance; how fast the ink moves is a *characteristic* property."

"I see. That makes sense."

"Well, give this some thought as you're thinking about the experiment and interpreting your results. Looks like you're getting good results. I've talked with you long enough! Let me go and pester some other poor soul!"

SHORTCOMINGS OF THE EXCHANGE

I do not present this exchange as an ideal. I talked too much, and the student talked too little. I am not at all confident that the party mix and electrolysis examples were helpful. A computer-animated simulation of the processes might have been more instructive. A well-planned instructional sequence would look much more like the learning cycle or generative learning models outlined in Chapter 6 than the impromptu exchange shown here.

The point of including this exchange is to show how a laboratory setting can provide opportunities to direct students' thinking to critical issues in the lesson to see how they are thinking about the phenomena they observe—perhaps more correctly, to *encourage* them to think about what they see. Far too often, students go through laboratory exercises without even *trying* to make sense of their observations. Intellectual development cannot take place without some kind of sense-making activity.

SUMMARY

This chapter is a cursory description of events that take place over a period of 16 weeks. The description is not a prescription for developing proportional reasoning, problem solving, or stoichiometry. Exactly how that development should take place in other classrooms depends on many factors that only the instructor can know—student attitudes, level of cognitive functioning, the content covered, and the way the course is organized. Still, these general characteristics illustrated by the examples taken from my remedial college chemistry course seem to be supported by current research:

1. Successful development of generalized intellectual skills depends on many variables: attitudes, subject-related facts and skills, and previously developed intellectual skills. Real learning activities involve *all* of these variables. Consequently, instructional episodes aimed at developing generalized intellectual skills are likely to deal with concept learning, principle learning, language, and other domain-specific knowledge as well as the generalized skill.
2. Development proceeds in steps under the influence of existing logical operations. Students construct increasingly abstract and formalized procedures while maintaining confidence in the validity of new procedures by comparing results with those derived from established operations.
3. When the focus of a lesson is logical development, the context should be familiar so that attention can be devoted to logical operation rather than domain-specific knowledge.
4. As students gain confidence in new operations, new content and new contexts should be introduced so that the new operation can be generalized.
5. Algorithms should be taught and practiced so that they can be applied automatically, reduce the load on working memory, and allow the student to attend to logical operations.
6. Executive strategies for checking the validity of the logic applied in a problem should be developed along with the operational knowledge. Algorithms that facilitate such checking (e.g., factor-label) should be used whenever appropriate.
7. If we are serious about intellectual development, we must reduce the content presented in courses to allow time for students to apply chemical ideas in many contexts and to develop the strategic knowledge needed to apply information to novel problems.
8. Symbolic representation (mathematical and chemical equations) is useful only to the extent that we are able to make valid and reliable inferences about its meaning. We assign meaning to symbols such as English sentences or mathematical equations by making inferences based on past

experience, syntactic knowledge of the language, and contextual cues (*see* Chapter 13 on language). Instruction designed to help students connect logical operations with the symbols that call for their application is a necessary part of developing generalized intellectual skills.

9. Encountering a logical operation (e.g., proportional reasoning) in one context (e.g., calculating the best buy at the supermarket) is not sufficient for general application of that operational knowledge. Focusing on the same logical relationship in a variety of contexts (e.g., in defining concepts such as molarity, in interpreting the slope of a graph, and in interpreting mathematical equations such as the ideal gas law) will enhance generalization.

10. Intellectual skills such as logical operations and executive strategies used in problem solving can be developed, but they are most likely to develop when students encounter meaningful problems in many contexts over an extended period of time.

Things in the Affective Domain

17

Values in Science Teaching

The controversial nature of values has led us to shy away from discussing them in schools, and that approach is wrong. Values and beliefs are potent factors in all learning. If we are to deal with problems of learning, we must deal with values and attitudes that derive from them.

ATTITUDES AND VALUES

Attitudes and values are sometimes confused. *Attitudes* reflect many interrelated feelings and beliefs focused on some entity—a person, an object, an event, or a practice. *Values* represent the standards and principles that we use to judge the worth of those entities. We have attitudes about science, grades, school, and family. Those attitudes are influenced by many things—experiences in school, comments by parents and friends, relationships with people associated with the object of the attitude, and our values. We are sometimes conscious of the values behind our actions, for example, when we refuse to lie because we value honesty. But values can function unconsciously. For example, we may befriend an unpopular associate without any thought to the value we place on human dignity.

Like attitudes, values embody feeling, but values are more than emotions. "Values have intellectual meaning that can be defined and clarified. And values can be compared, related to one another, and consciously applied as criteria" (Shaver and Strong, 1982, p 18). Values are fundamental to our frames of reference, and relatively few values influence numerous attitudes.

ORIGINS OF TIMIDITY ABOUT TEACHING VALUES

Historically, the inculcation of values was an accepted part of American education (Amundson, 1991, pp 17–18). However, events during the past quarter century have left teachers confused and reticent. No single event explains current apprehension about teaching values in schools, but all of the following contributed.

1. The civil rights movement, which moved from the rural South to the urban North, forced all Americans to contemplate the gap between the ideals expressed in the American Creed[1] and the realities of everyday life. As the 1954 Supreme Court decision overturning the separate but equal doctrine was implemented, teachers became keenly aware of discrepancies between the attitudes expressed by parents and the values these parents professed.

2. The apparent success of African Americans in gaining civil rights encouraged women, homosexuals, Native Americans, and others to seek redress for real or perceived discrimination. Attitudes that had prevailed for centuries were called into question along with the values that allowed those attitudes to prevail.

3. America's faith in its institutions was shaken. World War II left Americans confident about the material and moral superiority of their country. America had conquered her enemies, and through the Marshall Plan she was compassionately helping them to rebuild. But the Cold War produced doubts. When Francis Gary Powers' U-2 plane was shot down over Russia in 1960, our president's repeated denials that we had ever violated Soviet air space were revealed as lies. The Bay of Pigs debacle in 1961, the Cuban missile crisis in 1962, the assassination of John F. Kennedy in 1963, evidence of deepening involvement in Vietnam despite official denials throughout the 1960s, the shooting of student protesters by the National Guard at Kent State, the assassination of Martin Luther King, and the assassination of Robert Kennedy all raised questions about the integrity and capability of our hallowed social institutions. Schools were not exempt.

4. Political asylum granted first to Cuban refugees, later to Vietnamese, Cambodians, and Laotians, and still later to political refugees from Central and South America increased our cultural diversity. Increased civil rights won by African and Native Americans made the Caucasian majority more aware of cultural differences that were previously hidden through segregation. As these differences became more evident, attitudes and policies unconsciously shaped by values rooted in Protestant Christianity seemed embarrassingly parochial in the face of new diversity in religious belief and cultural practice. The melting pot seemed to be freezing into isolated clans and cultural enclaves that looked past one another without understanding.

The social upheaval brought about by these and other events affected schools directly. Parents who had always known that schools served the needs of their children and that teachers protected parental interests were less sure. Many parents were able to believe children's stories of mistreatment and discrimination—some true, but most lies or self-serving distortions—that parents living in less chaotic times would have immediately discounted. Claims by reactionary groups that schools were undermining traditional family and religious values became plausible to people who, in an earlier age, would have dismissed them out of hand.

Under what seemed to be constant attack from the left or the right, many teachers questioned whether they were, in fact, inadvertently indoctrinating children

---

[1]This term, coined by Myrdal in his 1944 book, *An American Dilemma*, refers to ideals commonly accepted in the United States. "These ideals of the essential dignity of the individual human being, of the fundamental equality of all men, and of certain inalienable rights to freedom, justice, and a fair opportunity represent to the American people the essential meaning of the nation's early struggle for independence" (Myrdal, 1964, p 4).

with values that were a matter of personal conviction. As Grant described it, "[Teachers] no longer trusted their own moral compasses" (cited in Amundson, 1991, p 19). Teachers, without pangs of conscience, side-stepped values issues just to play it safe.

DEMOCRATIC VALUES

In view of the social upheaval that characterized the 1960s, current timidity about teaching values is understandable, but it is not defensible. Schools are instruments of society, and a common understanding of the democratic values upon which our society is built is critical to its continued existence. As Shaver and Strong (1982) put it, "For a teacher not to deal with values explicitly is, in fact, a dereliction of duty" (p 74).

WHAT ARE OUR DEMOCRATIC VALUES?

Most teachers can readily accept the teaching of basic societal values, but during the last quarter century doubts have increased about just what those values are. Shaver and Strong argue that, although personal values differ widely, an American Creed exists (Myrdal, 1964) that holds our pluralistic country together. The American Creed is based on a belief that every person has worth. Because every person has worth, there must be sanctity to life, and society must commit itself to such ideals as equal opportunity, equal protection under the law, religious freedom, and freedom of expression—ideals that make human dignity possible.[2]

Amundson (1991), Lickona (1991), and other recent writers on values education agree with Shaver and Strong that values *must* be taught in schools and that the appropriate values for schools to address are those that support human dignity and democratic ideals. Lickona (1991) identified respect and responsibility as key values for schools to teach and "honesty, fairness, tolerance, prudence, self-discipline, helpfulness, compassion, cooperation, courage, and a host of democratic values" as specific values that are "forms of respect and/or responsibility or aids to acting respectfully and responsibly" (p 45).

Amundson (1991) cited the work of R. Freeman Butts:

Some values seem primarily desirable, cohesive, and unifying elements in a democratic political community. The "unum" values, which Butts describes as "the obligations of citizenship", include: justice, equality, truth, authority, participation, [and] patriotism.

A second group of values seem primarily to promote desirable pluralistic and individualistic elements. These "pluribus" values, which Butts calls the "rights of citizenship", include diversity, privacy, freedom, due process, human rights, [and] property (pp 24–25).

DETERMINING LOCALLY ACCEPTED VALUES

Although considerable agreement exists concerning the values that should be addressed in schools, all of the authors cited here point out the importance of teach-

---

[2]I would add to the values of the American Creed those values of intellectual freedom that characterize science. These values are addressed on pp 232–236.

ers listing the values that they want to teach, and the authors recommend involving teachers, administrators, other school staff, parents, students, and community representatives in the process. Amundson reported the "common core of values" resulting from such a process in Baltimore County, Maryland. The list is as follows: compassion, courtesy, critical inquiry, due process, equality of opportunity, freedom of thought and action, honesty, human worth and dignity, integrity, justice, knowledge, loyalty, objectivity, order, patriotism, rational consent, reasoned argument, respect for others' rights, responsibility, responsible citizenship, rule of law, self-respect, tolerance, and truth (p 27).

In dealing with these societal values we must be careful not to confuse basic values with attitudes that derive from them. Both advocates and opponents of abortion value life, but their attitudes toward abortion differ because of differences in their understanding of when life begins and their respective beliefs about limits on individual rights imposed by a higher moral authority. Those who value religious freedom disagree about public prayer, and those who value equal opportunity argue for and against quotas for admitting members of minority groups to jobs and higher education. Teachers must explicitly support democratic values, but they must respect the rights of students to reach individual positions about the implications of those values for daily living. Grant said, "[Teachers] need not agree on capital punishment, for example, but they ought to be clear about honesty" (cited in Amundson, 1991, p 26).

> Basic societal values are so fundamental to the reason for the existence of public schools that they are not only to be supported but even inculcated, *with increasing attention to their rational bases as students gain in intellectual ability.* . . . [I]n the context of value conflict, pluralism, and respect for human dignity, teachers must accept and respect the differing decisions that students come to—as long as they are based on careful consideration of the basic values of the society (Shaver and Strong, 1982, p 143; my italics).

"Respecting differing decisions that students come to" should not mean that one decision is as good as another. Our society is not only based on democratic values, it is based on reason as well. Science relies so heavily on the intellectual processes that we associate with reason that science teachers have a special obligation to teach those "values of science" that characterize rational thought.

VALUES OF SCIENCE

EDUCATION AND THE SPIRIT OF SCIENCE

In 1966 the Educational Policies Commission of the National Education Association published a small monograph in which it argued that the values of science should be taught as the foundation of education in a democratic society. These values, as listed in the NEA monograph, are

1. longing to know and to understand,
2. questioning of all things,
3. searching for data and their meaning,
4. demand for verification,
5. respect for logic,
6. consideration of premises, and
7. consideration of consequences.

Let us consider what these values mean and why they are fundamental to democracy as well as to science.

LONGING TO KNOW AND TO UNDERSTAND

Curiosity about the natural world is fundamental to all science. Science views knowledge as valuable in and of itself, but this value goes far beyond science. At no time in history has a longing to know and to understand been more important to our individual and collective prosperity. Increasingly, information is the object of commerce, as it has always been the main line of defense against anarchy or dictatorship. Those with limited knowledge and understanding have limited prospects for a comfortable standard of living, and they are in no position to govern themselves.

A longing to know appears to be innate. Young children seem to have an insatiable curiosity, and parents tire of their incessant questions. So do teachers, and many of the things done in schools probably undermine natural curiosity that could lead to increased understanding. Contrast the implicit message in these classroom vignettes:

> "Mr. Lee, I saw a documentary on the first moon landing last night. Every time they took a step, you could see particles flying away from their feet in a funny path. How come?"
>
> "I do not know, Kip. Class, get out your exercise books and do Activity 10 on falling bodies."
>
> "Mr. Lee, the book said that litmus paper should turn blue in this solution, but mine remained red."
>
> "Really? What do you think could explain that?"
>
> "I do not know. Perhaps I made the solution wrong."
>
> "That's one possibility. How could you check to see if that is a reasonable explanation?"
>
> "Well, I could check to see what others in the class got . . . and I could make up a new solution and do the test again."
>
> "That sounds good. Why don't you do both and let me know the results?"
> (Herron, 1977b, p 31)

In an effort to "cover the text", prepare students for standardized examinations, or follow a prescribed syllabus, students can easily be discouraged from exercising their curiosity and learning to value knowing and understanding.

Suggesting that we encourage students' longing to know and to understand is not an argument to explore every question students ask. Classroom decisions—indeed, all decisions—frequently involve conflict between values. For example, lessons on sex, drugs, and even personal hygiene can elicit questions whose public exploration would constitute an invasion of privacy or create unnecessary conflict with some students' personal values.[3] Teachers can usually help students learn more about sensitive issues without classroom discussions that might be judged inappropriate because of privacy issues, because students lack interest, or because the issue is unrelated to the subject being taught.

---

[3]Personal values, according to Shaver and Strong (1982), "are those used in making daily life decisions. People seldom believe that they should impose these values on other people or use them generally to judge others' behavior" (p 22). The value placed on solitude and cleanliness are examples of personal values.

The fact that value conflicts spontaneously arise in the science classroom is good reason to examine values explicitly. The more we think about what we believe and why, the better prepared we are to resolve value conflicts and to defend our actions when decisions are made.

QUESTIONING OF ALL THINGS

The first part of this book makes clear that "absolute truth" is a figment of imagination. All knowledge is constructed under the influence of personal schemas. Conclusions of normal science are couched in the framework of current theories and beliefs, but revolutions in science only occur when those conclusions are challenged (Kuhn, 1970).

"Questioning all things" implies two things: that there is no human authority whose pronouncements must be accepted without question, and that every statement of scientific fact carries with it uncertainty. Democracy places on its leaders responsibility for making decisions on behalf of all citizens, but it never presumes that those decisions are infallible. The constant questioning of our leaders and the dissension caused by that questioning is a necessary protection against abuses of power.

As with most value decisions, encouraging students to question can produce conflict with other values—respect for authority and religious values, for example. Deliberately juxtaposing the Genesis account of creation with the scientific theory of evolution or encouraging students to challenge the school's dress code or discipline policies is not the best way to encourage development of this value. But neither is a refusal to discuss such issues when students raise them likely to be helpful. The key to successful questioning involves the next value in the list.

SEARCHING FOR DATA AND THEIR MEANING

Nothing is wrong with questioning a law or rule that you dislike, provided you are willing to examine the reasons for and against it. Neither is it wrong to claim that the Avogadro number is 6.023×10^{23} or that the charge on the electron is 1.602×10^{-19} coulombs if you have some understanding of how these values were derived, the assumptions inherent in the derivations, and the uncertainty associated with the measurements made. Only when you accept such statements as bald facts do you get into trouble. By encouraging students to search for data and their meaning, we help them avoid such trouble. Even in the case of religious doubt, the search for data and their meaning is far more likely to strengthen faith and religious conviction than to weaken it.[4]

DEMAND FOR VERIFICATION

As students gain understanding of the uncertainty associated with all knowledge, the value of verification becomes more salient. Problem solving, as described in Chapter

---

[4]In the late 1950s, discussion of evolution in my high school class seriously distressed a few students (and, I am confident, their parents) who had received very conservative religious training. I suggested that we contact scholars at a religious college to see what they had to say about the subject. Wheaton College, which had a reputation for sound scholarship but conservative religious views, responded with very helpful literature that focused on the religious truth in the Genesis story while accepting the scientific evidence for evolution. I am confident that resourceful teachers can find similar ways to encourage students to search for data and their meaning without unacceptably threatening their religious faith.

7, provides many opportunities to reinforce the value of verification because class discussions of student solutions usually reveal differences in results that require resolution.

Demand for verification is hollow without knowledge of procedures and an appreciation that, because knowledge can never be absolute, verification actually means increasing confidence in results by arriving at the same result by different routes.

Social truth is just as elusive as truth in science, and the meaning of verification is essentially the same. When a leader is vilified or a person is charged with a crime, the best we can do is seek evidence from as many independent sources as possible in the hope that our final decision is made with a confidence beyond reasonable doubt.

RESPECT FOR LOGIC

When students draw inferences that seem unreasonable to me, I admonish them for jumping to confusion. What they have told me—for example, that we remain on the surface of Earth because it is spinning—is illogical. But probing questions often reveal that inferences that seem illogical in the beginning are the result of tight chains of inference that conform to the "if . . . then . . . therefore" of hypothetico-deductive reasoning. For example, the student who claimed that we stick to the surface of Earth because it is spinning thought we live on the *inside* of a huge, hollow sphere. "It's like slinging a bucket of water round and round", the student explained. "The water does not fall out because the force from the spinning holds it in."

As in this case of "spin-induced gravity", we often encounter conflict between the values of truth and respect for logic. If we are not careful, we discourage respect for logic by placing too much emphasis on *right* answers. Consider this example:

> A teacher grading lab reports on an experiment based the grade on correct ordering of the metals from most to least active: Na > Mg > Zn > Fe > H > Cu > Ag. Most students who received full credit had reported that "Mg replaces Na from a solution of NaCl" and "Fe does not replace hydrogen from HCl(aq)." These observations were not discussed (Herron, 1977b).

Is a teacher encouraging respect for logic when full credit is given for correct answers that are contradicted by available data? What does it mean to respect logic? I like these words of the Educational Policies Commission: "Logical systems constitute agreed bases by which the validity of inferences may be judged. . . . Logic is used in connecting a thinker's concepts *in a manner open to evaluation by other persons*" (1966, p 18; my italics). If the chain of reasoning leading to inferences is not made evident, the logic used is impossible to judge.

CONSIDERATION OF PREMISES

As the Educational Policies Commission pointed out: "No amount of logical consistency will make valid any inferences or deductions which proceed from inadequate or faulty premises. Mere logical consistency does not constitute an adequate appraisal of a concept, proposition, or idea" (1966, p 18). Earth's spinning does not hold us to its surface, although we might reasonably infer that relation if the premise that we live inside a hollow sphere were true.

Rational discussion calls for conscious consideration of premises. Many debates during the Vietnam War era would have been less bitter if those engaged in verbal combat could have acknowledged that one started from the premise that a subjugated people were struggling for independence, whereas the other started from the premise that Vietnam was but the first in a series of dominoes falling under the influence of communist totalitarianism. Current debates over abortion, homosexuality, sex education, and prayer in schools suffer from similar ignorance of the premises on which an argument is based. What is often viewed as stark differences in values is, in fact, a difference in premises.

CONSIDERATION OF CONSEQUENCES

Science is sometimes argued to involve an uninhibited search for truth without consideration of the consequences. This assumption is not so. Scientists are part of humanity, and they know that what any human does has an effect on others. They know that they will be judged by others, and that the standard by which they will be judged is that they act responsibly.

So-called *STS issues*—issues related to the impact of science and technology on society as a whole—provide excellent opportunities to consider the consequences of various actions. However, most such issues are complex and involve economic, social, religious, and political issues as well as science. Science teachers are not usually informed sufficiently to guide students in considering consequences in all of these areas. We should not hesitate to admit the limits of our knowledge and seek assistance from experts in other fields.

In learning to consider the consequences of decisions, students must also learn that every consequence of an action cannot be anticipated. Alex Johnstone and Norman Reid (1981) developed several classroom simulations that focus attention on how difficult it is to anticipate consequences and to make historically defensible decisions. In one simulation, students work in small groups to decide which of several drugs they, as directors of a pharmaceutical company, will market. To make their decisions, students are given the best available information about the drugs. However, some groups are provided information that was available before thalidomide was marketed, whereas others are provided information that is available today. Invariably, students with the older information select thalidomide as the most promising drug to market; those with current information are reluctant to market any of the drugs, including aspirin. The potential risks seem far too high.

TEACHING VALUES

In 1977 I wrote a short article arguing that the values of science should be taught, and that the way they must be taught is through the implicit, hidden curriculum (Herron, 1977b). One of my students, Tom Havill (personal communication, 1979), took issue with me, arguing that

> Values are based on premises about the nature of man, his activity, and his reality. These premises are more often than not left unarticulated, unexamined, implicit. If values were made explicit, then the *consideration of their premises* (a value of the spirit of science) would be facilitated.
>
> I believe the goal of the teacher, therefore, is to
>
> (a) Examine his behavior for implicit values, and verbalize them explicitly.

(b) Test his values scientifically [by] using the values of science outlined above.

(c) Strive to make his behavior flow logically from a set of explicit values and premises.

(d) Explicitly (as well as implicitly) reveal these values and premises to students so that they can consciously test them in the light of their own experience.

(e) Help the students to behave as in (a)–(c) themselves.

Similar arguments are made by Amundson (1991) and Lickona (1991). Shaver and Strong (1982) stated the following:

> If your behavior as a teacher is to be as rational as possible, you need to bring your assumptions into the open, state them as clearly as possible, examine them for accuracy and consistency, and use them consciously as the basis for decisions about your instructional and other behavior toward students . . . (p 9)

I believe this advise is sound, and I have tried to follow it. The values outlined in the section of this chapter called Education and the Spirit of Science are ones that I hold implicitly. I have examined them as best I can,[5] and I have elaborated on them so that you, my *students*, can judge them in the light of your own experience.

VALUES CLARIFICATION

Other approaches to teaching values exist. Raths, Harmin, and Simon (1966) proposed a "values clarification approach". Values clarification focuses on the *process of arriving at value positions* rather than the product of that process.

In the process of values clarification, teachers set up situations that encourage students to consider such questions as how they might behave and what they prefer. Through the use of "clarifying responses" such as, "How did you feel when that happened?" or, "Are you glad about that?" teachers encourage students to think about their value positions. But teachers are urged to avoid "moralizing, criticizing, giving values, or evaluating", and they should avoid hinting whether a response is good, correct, or acceptable (Raths et al., 1966, pp 53–54; 1978, 55–56; cited in Shaver and Strong, 1982, p 136).

After reviewing research on values clarification, Shaver and Strong (1982) expressed several reservations about the approach. They call attention to its somewhat questionable definition of *value* and its relativism. Their most serious criticism is that

> all values are treated as equal. . . . Indeed, the question of which values a teacher should feel free to probe appears not to be of concern. . . . Equally important, suggested guidelines for selecting topics or issues do not address the matter of distinguishing between personal decisions not within the prerogative of the school and decisions about matters of societal concern. It is not that teachers are told never to deal with societal issues, but that the impression given is that such issues are not more important or no more within the school's proper domain than personal matters (p 141).

Other authors express similar reservations about values clarification:

---

[5]As Tom Havill pointed out in his letter to me, using the values of science to examine the values of science involves a certain circularity. However, this circularity seems to be unavoidable. If one is to examine one's values, are they not necessarily examined in terms of themselves?

Values clarification discussions made no distinction between what you might *want* to do (such as shoplift) and what you *ought* to do (respect the property rights of others). There was no requirement to evaluate one's values against a standard, no suggestion that some values might be better or worse than others. As one critic observed, "There's a big problem with any approach that does not distinguish between Mother Teresa and the Happy Hooker" (Lickona, 1991, p 11).

Many of the issues discussed in values clarification are more accurately described as attitudes, and Raths et al. are undoubtedly correct to recommend that teachers refrain from indoctrination when dealing with such issues. As pointed out earlier, many different attitudes are held by those who wholeheartedly endorse the values embodied in the American Creed. But teachers should encourage students to test their attitudes against basic democratic values, and they should help them use the values of science to judge whether the attitudes held are consistent with those values. Some of the strategies used in values clarification can facilitate that process.

Lickona (1991) cited another strength of values clarification: "Values clarification encouraged people to close the gap between espoused values and personal action. As long as the values are good ones—such as acting responsibly toward oneself and others—value–action consistency is certainly a worthwhile goal" (p 238).

DEVELOPING YOUR OWN RATIONALE

The values embodied in the American Creed and the Values of Science are ones that I feel obligated to teach. I can defend teaching them on the basis that they are essential to the democratic society in which I live and to intellectual freedom.

I hold other values that guide my life and, undoubtedly, influence my teaching; but these are personal values that I cannot defend imposing on others. This relationship does not mean that I hide them from students. To the contrary, I do not hesitate to let students know the values and beliefs that I hold, but I do so without suggesting that these are values and beliefs that everybody must hold. Such sharing is important. As Shaver and Strong (1982) pointed out:

> Students too often feel, legitimately, that the school and its staff are plastic, insulated, and isolated from the real world as the students see and feel it. To be authentic as a teacher, you will need to express beliefs, especially when students seek your opinion. But this expression should be an educational act in the context of the values of a democratic society, not an act of political indoctrination (p 80).

Perhaps you cannot defend teaching the values that I have described. Perhaps you can defend teaching many more. Whatever the case, you should develop your own list of values, and you should consciously teach them. Values are far too important to ignore. Again quoting Shaver and Strong:

> The teacher must be responsible to a conception of democracy—as attorneys are to be responsible to an ideal of justice and doctors are to be to the preservation of life and alleviation of suffering—and . . . this conception of democracy must supersede strident local interests and prejudices. The teacher is not the voters' servant or the servant of the pupils' parents. He or she is an agent of the society. But beyond that, each teacher is a professional with an obligation to promote education in the broad democratic context, not just to reflect the parents' or the voters' wishes (p 82).

SUMMARY

During the early history of education in the United States, values education was an important, conscious part of the curriculum. However, social changes that have taken place since World War II have discouraged teachers from openly teaching values. That outcome is wrong. Values represent schemas that have a powerful influence on how and what we learn. They cannot be ignored.

Two sets of values, democratic values and the values of science, should be consciously taught by chemistry teachers. However, each of us holds many personal values that may be shared with students, but these values should be shared without any attempt to impose those values on students.

Values clarification, a process of arriving at value positions without judging the merits of the end product, has been advocated as a substitute for teaching values in schools. Unfortunately, values clarification leaves the impression that one value is just as good as any other, and that impression is incompatible with our position as agents of a free, democratic society. In this country, democratic values are to be encouraged over totalitarian or anarchistic ones.

The techniques advanced by values clarification can still be used to teach values. Those techniques are consistent with democratic values, and they encourage students to close the gap between what they say they value and how they act, a worthwhile goal.

Readers should outline their own list of values that they are committed to teach, and they should enlist the aid of colleagues, parents, students, and community leaders in doing so. The bibliography appended to this chapter provides some excellent recommendations to guide the process.

BIBLIOGRAPHY

Amundson, K. J. *Teaching Values and Ethics*; American Association of School Administrators: Arlington, VA, 1991.

This publication is short, but it provides a great deal of useful information. The first chapter deals with the need for values education, followed by chapters that define values education, describe what others are doing, and suggest how to institute a values education program.

Lickona, T. *Educating for Character: How Our Schools Can Teach Respect and Responsibility*; Bantam Books: New York, 1991.

This is an excellent book for teachers. The first 50 pages present the case for values education, describe values that schools can legitimately teach, and discuss good character. Part II (200 pp) presents classroom strategies for teaching respect and responsibility. It includes an excellent chapter on cooperative learning. Part III (125 pp) is on school-wide strategies. It contains specific chapters on sex education and drugs and alcohol that would be of interest to chemistry teachers in secondary schools.

Kirschenbaum, H.; Simon, S. B. *Readings in Values Clarification*; Winston Press: Minneapolis, MN, 1973.

A variety of articles by several authors, some of which contain practical suggestions that could be used to develop values without the relativism normally associated with values clarification.

18

Motivation and Chemistry Teaching

Most teachers feel that they have little difficulty teaching students who want to learn. Still, the importance of what we teach seems so self-evident that we spend little time establishing a need to know. This dereliction is a mistake.

A few years after I went to Purdue I gave a lecture for a colleague who was out of town. About a third of the way through my lecture a student in the middle of the lecture hall raised his hand and, after being acknowledged, implored: "I'm here to become an engineer. Can you *please* tell me why I need to learn this chemistry crap?" The question was a fair one and could have been asked by an agriculture or biology major, a future doctor or nurse, or others who are required to take general chemistry without knowing why.

I had recently reviewed a textbook in chemical engineering that began with a series of mistakes committed by engineers who were ignorant of chemistry: expensive copper guttering falling off of a building because it was attached with iron nails, ceiling tiles falling into vats containing a solvent for the mastic used to attach the tiles, and so forth. After pointing out that I was not an engineer and did not know all of the ways that chemistry might be useful, I related what I had read to illustrate possible applications. The student thanked me with a note of sincerity, and I proceeded with the lecture.

We are inclined to project onto our students problems that are our own. When lessons are not understood, we claim that students are dumb; when students are not motivated, we claim that they are lazy, as though motivation were characteristic of learners rather than a state of being.

CHARACTERISTICS OF MOTIVATION

ACCOUNTING FOR DISINTEREST

Everyone is motivated. The problem is not that students lack motivation; it is that they are not motivated to do what *we* want them to do. As Thomas (1980) pointed out, "What drives individuals to seek out or avoid learning activities is the learners' perceptions of themselves, their perceptions of the value associated with the successful completion of the task, and their perceptions of the extent to which effort will result in achieving success" (p 223). Before we can improve motivation, we must determine which of these possibilities account for our students' disinterest.

MASLOW'S HIERARCHY OF NEEDS

In his 1954 book, *Motivation and Personality*, Maslow argued that we have basic needs that form a hierarchy. Until our most basic needs are met, they receive virtually all of our attention. However, as those needs are met, other needs come to the fore. For example, if we are starving, we do not worry much about anything other than food; but once satiated, we worry about other things and do not think about food at all.

According to Maslow, the most basic needs are *physiological needs*. These needs can be any number of things—coolness if we are too hot, food if we are hungry, sleep, sex, and so forth. If these needs are not met, nothing else seems to matter much. Next in the hierarchy are needs for *safety and security*. We do not want to feel threatened. A stable environment with only a moderate amount of uncertainty is needed. Next come the *belongingness and love needs*. We need to know that somebody cares for us. If we do not, we exhibit all sorts of pathologic behavior. *Esteem needs* follow. These include a sense of self-worth, self-confidence, strength, capability, and the like. Finally come *needs for self-actualization*. We need to do things that we feel we are meant to do, and we need to enjoy beauty and other aesthetic qualities of life.

The point of bringing up Maslow's work is to call attention to the fact that what we typically deal with in classrooms is related to self-actualization needs at the top of the hierarchy. Students probably will not attend to what we have in mind if they are hungry or afraid or feel unloved or worthless or incompetent. Although we may rightfully claim that we are teachers rather than social workers or counselors or police personnel, we will be ineffective teachers if we are insensitive to our students' more basic needs as well as those satisfied through education.

CONTROLLING BEHAVIOR

Ample evidence suggests that behavior can be controlled through basic needs. Starving people have been enticed to do things that they would not otherwise do, and spies giving up their most cherished secrets for sexual favors is commonplace in fiction if not in real life. To my knowledge, neither strategy has been used to entice students to learn chemistry, but punishment that threatens one's security or sense of belonging is common enough. Even more common are external rewards, ranging from gold stars and special privileges to money and grades.

EXTERNAL INFLUENCES

The use of rewards and punishment to control—I prefer *manipulate*—human behavior has been elevated, literally, to a science. There is no doubt that it works. Much as this book argues for a psychology of learning in opposition to behaviorism, principles of learning derived from behaviorism are valid.

Carl Rogers, in his 1969 book, *Freedom to Learn*, addressed the question of whether people are free to make their own choices or are controlled by others through external forces that shape behavior. Rogers acknowledged the validity of a wide range of behavioral research showing that human beings, as well as pigeons, respond to operant conditioning. Rogers summarized his review by saying

> I am impressed by the scientific advances illustrated in the examples I have given. I regard them as a great tribute to the ingenuity, insight, and persistence of the individuals making the investigations. They have added enormously to our knowledge. Yet for me they leave something very important unsaid (p 265).

INTERNAL INFLUENCES

Rogers then described a number of individuals who made conscious decisions to act in direct opposition to external conditioning:

> What I am trying to suggest in all of this is that I would be at a loss to explain the positive change which can occur in psychotherapy if I had to omit the importance of the sense of free and responsible choice on the part of my clients. I believe that this experience of freedom to choose is one of the deepest elements underlying change (p 268).

Evidence that operant conditioning works and evidence that people are free to make choices in opposition to external regulation may seem confusing and contradictory, but they are not. Both are true. Richard Crutchfield's classic experiment (cited in Rogers, 1969) illustrates the point.

Stripped to its bare essentials, Crutchfield's experiment was this: In groups of five, individuals were asked to give independent judgments about items flashed on a screen. However, the experiment was rigged so that each person was shown what were presumably the responses of the four others before he or she responded. In this way the experimenter could bring pressure on a person to conform to apparent consensus at variance with his or her own judgment.

> Crutchfield [found that] given the right conditions almost everyone will desert the evidence of his senses or his own honest opinion and conform to the seeming consensus of the group. For example, some high-level mathematicians yielded to the false group consensus on some fairly easy arithmetic problems, giving wrong answers that they would never have given under normal circumstances (cited in Rogers, 1969, p 263).

Crutchfield's experiment clearly shows that behavior is influenced by external forces such as other people's opinions, but this conclusion does not mean that we do not construct our own knowledge or that we are not free to act on our own. Indeed, a strong argument can be made that we *allow* opinions to be shaped by others *because it makes sense to do so.*

We all experience things that turn out quite differently than they first appear, and even though we know that the majority *can* be wrong, experience insists that, in an honest environment, a four-to-one majority is correct more often than not. Yielding can result from willful judgment just as easily as from external control.

The decision we make in the face of conflict between our own judgment and that of others depends on how often we have been right in the past and how those experiences have affected our self-confidence, and this situation is exactly what Crutchfield found. "The person who did not yield to pressure had a sense of competence and personal adequacy. . . . He was also a better judge of the attitudes of other people. . . . Those who made their own choices were . . . much more open, free, and spontaneous. They were expressive and natural, free from pretense, and unaffected. Where the conformist tended to lack insight into his own motives and behavior, the independent person had a good understanding of himself" (cited in Rogers, 1969, p 270). Because we wish to develop such self-confident, independent thinkers, we argue against external control of student behavior and for self-determination and intrinsic motivation.

SELF-DETERMINATION IN SCHOOLS

Assuming that all of our activity can be intrinsically motivated is unrealistic. I will never grade papers, reprimand students, or write books for the sheer joy of doing so. Our lives are filled with choices, many of which result from actions by others. An aging parent becomes ill and needs assistance. I do not want to leave my work and immediate family to provide that assistance, but I would feel guilty if I did not. I do not want to work through the weekend to complete a grant proposal, but I need funding to continue research that I consider to be important and enjoyable. All of these activities involve unpleasantness that I would rather avoid. But all of them provide satisfactions. Furthermore, many activities that I now do for reasons that are purely intrinsic were first done because I knew they were good for me or because I valued them as a means to an end: research, woodworking, and writing, to name a few.

It is just as unrealistic to assume that all activity by students in our classes will be intrinsically motivated. They, too, are faced with choices that prevent them from always acting from personal interest. An element of compulsion exists in the very nature of schools. Through much of schooling, students are required to attend, and at all levels they must pass prescribed courses to graduate. Teachers are expected to follow specified syllabi and maintain order in their classrooms. Still, within this environment, room for self-determination on the part of students exists.

SELF-DETERMINATION THEORY

The extent to which students' actions are self-determined rather than compelled by others is the subject of Deci, Vallerand, Pelletier, and Ryan's (1991) self-determination theory. The theory describes a continuum from external control to self-determination.

External regulation refers to threat of punishment or temptation of reward. Skinner's operant conditioning is a clear example of external regulation. Teachers should use external regulation as a last resort because research evidence suggests that, even when effective in producing desired behavior, external regulation leads to dependence rather than enhancing the intrinsic motivation that is associated with efficient learning.

Introjected regulation is based on guilt or shame. Although introjected regulation is internal, it is externally imposed. It is what urges me to visit my aging parents or spend hours polishing a letter so that others will think better of me. It may be a powerful motivator for students who have an ego rather than a task orientation (see discussion beginning on p 245), but it is far from the intrinsic motivation that is desired.

Identified regulation involves a sense of "I do not particularly like it, but I know it is good for me." Unlike external regulation, which tends to depreciate learning by displacing its value to some unrelated reward or the avoidance of punishment, identified regulation focuses on the inherent value of the learning itself. It is a kind of regulation that teachers and parents attempt to induce without pangs of guilt, and it is a kind of regulation that will lead to task rather than ego involvement.

Integrated regulation implies that the activity has become part of one's value system. It is very close to intrinsic motivation, but it falls short of learning for the sheer joy of it. *Intrinsic motivation* is characterized by interest in the activity itself, whereas integrated regulation is characterized by the activity's being personally important for a valued outcome (Deci et al., p 329). "I hate chemistry, but I know I need to understand it to be a doctor" or, "Grading papers is no fun, but it is important to give students feedback on their work." Like identified regulation, integrated regulation favors task rather than ego involvement.

Teachers may face situations in which external regulation appears necessary to avoid chaos. They do what they must, but teachers who understand that intellectual development is enhanced by independent activity based on intrinsic motivation will seek to move students toward higher levels of self-determination.

VALUE OF INTRINSIC MOTIVATION

In general, intrinsic motivation facilitates learning, whereas external control only leads to compliance. "Students who are intrinsically motivated for doing school work and who have developed more autonomous regulatory styles are more likely to stay in school, to achieve, to evidence conceptual understanding, and to be well adjusted than are students with less self-determined types of motivation" (Deci et al., 1991, p 332). Furthermore, studies have shown that

> when students received rewards such as monetary payments, good-player awards, or prizes for participating in an interesting activity, they tended to lose interest in and willingness to work on the activity after the rewards were terminated, relative to students who had worked on the activity in the absence of rewards. Similar results were found when people performed an interesting activity in order to avoid a negative consequence. . . . Studies have increasingly indicated that when evaluations are emphasized or made salient they will undermine intrinsic motivation, conceptual learning, and creativity. The same has been found for surveillance.
>
> Other external events designed to motivate or control people—including deadlines, imposed goals, and competition—have similarly been found to decrease intrinsic motivation (Deci et al., 1991, p 335).

EGO INVOLVEMENT VERSUS TASK INVOLVEMENT

Deci and his associates described the continuum from external regulation through integrated regulation as evolution from external control to self-determination. However, the same continuum may be viewed as one ranging from motivation tied to other people to motivation tied to the activities themselves. This distinction is basically the one that Nicholls (1989) addressed in his work.

Nicholls described two kinds of motivation: ego involvement and task involvement. "*Ego involvement* connotes the desire to enhance the self by establishing one's superiority relative to others. . . .*[T]ask involvement* implies a state where performing, understanding, or completing tasks is important in its own right" (pp 87–88; italics added).

EGO INVOLVEMENT

In ego involvement, learning is a means to an end—approval by the teacher, rewards, or enhanced status. In task involvement, the increased competence and understanding associated with learning is an end in itself. The rewards for ego involvement are extrinsic, and without those rewards, involvement will not continue. The rewards for task involvement are intrinsic, and involvement will continue so long as a sense of increasing competence and a perceived need for that competence exist.

Everyone is likely to compete from time to time and to take pride in being "the best" at something. In like manner, everyone is likely to engage in some activities because of intrinsic interest, without any consideration of possible rewards or how one's performance compares to the performance of others. In many situations, elements of both ego and task involvement will be evident, and the extent of each will depend on the situation at hand. However, people respond differently to the same situation. Nicholls used the terms *task orientation* and *ego orientation* to describe individuals who are prone to either task or ego involvement (1989, p 95).

TASK INVOLVEMENT IS BETTER

Clearly, task involvement is more likely to lead to intellectual development as outlined in the first part of this book than is ego involvement:

> The case for task involvement is consistent with the constructivist view of intellectual development. Piaget's concept of equilibration implies that intellectual development occurs when children sense inadequacies in their own knowledge. . . . A disequilibrium state is resolved by the construction of more adequate knowledge or new states of equilibrium. . . . In the constructivist view . . . sound intellectual development requires that children reorganize their own physical and social worlds to make them more meaningful. If learning is a means to an end (i.e., driven by ego rather than task involvement), the development of understanding will suffer (Nicholls, 1989, p 165).

Task involvement is also more consistent with the values outlined in Chapter 17. The value placed on equality in the American Creed and the longing to know and to understand at the heart of the values of science tie self-worth to competence and understanding rather than superiority relative to others.

Nicholls cited a 1972 study showing that "creative achievement in science and the arts is fostered by task orientation rather than ego orientation or other extrinsic factors" (1989, p 130). A more recent study by Nolen and Haladyna (1990) indicated that students with a task orientation are more likely to believe in the usefulness of learning strategies that require deep processing of information (monitoring of comprehension and memory, relating new information to prior knowledge, organization of information through outlining or note-taking, etc.) than are students with an ego orientation. Other empirical studies favor task involvement, but as Nicholls argued:

> [Even if] task involvement produced slower learning than ego or extrinsic involvement, there could still be grounds for favoring task involvement. Would

not something be wrong if children believed that the world is round or that two plus two equals four simply because "the teacher says it is" or because they get higher grades by learning such facts more quickly than others? . . . Even if learning could be advanced more rapidly by encouraging dependence on external authority or the desire to beat others, it should not be. . . . *If learning is to be meaningful, it must answer questions that are significant to the student* (p 166; my italics).

Although Deci et al. and Nicholls approach motivation from somewhat different perspectives, their work is complementary. Both theories stress the importance of motivation that emanates from within and focuses on learning for its own sake rather than motivation aimed at pleasing others, gaining status relative to others, or for the sake of external rewards. Both theories support the values outlined in the previous chapter and are consistent with the constructivist view of learning. But the question remains, "What can teachers do to foster task involvement and intrinsic motivation?" This question is the subject of the following chapter.

SUMMARY

Students who want to learn usually do; those who do not want to learn usually do not—at least they do not learn the things that adults want them to learn. Consequently, motivation is a major concern of every conscientious teacher.

All of us have certain physiological and emotional needs. Until those needs are adequately addressed, we have difficulty being concerned about anything else. Even though teachers are not social workers or counselors, we may occasionally need to see that students' basic needs are met before we can interest them in what we are trying to teach.

Behavior *can* be controlled—at least up to a point—by external influences, but that control is not the normal ideal. To the extent possible, behavior should be self-determined, and adults can encourage the development of self-determination through a series of steps beginning with external control and moving through introjected regulation, identified regulation, and integrated regulation to finally arrive at intrinsic motivation.

Intrinsic motivation facilitates intellectual development; external control leads to compliance. Furthermore, relying on external rewards for motivation leads to dependence. After the rewards are removed, the rewarded behavior ceases.

Many students' behaviors are motivated by a desire to please others or to be better than others (ego involvement), whereas other students' behaviors are motivated by interest in the activity itself (task involvement). As might be expected, task involvement facilitates the kind of intellectual development described in this book.

Theory is useful to the extent that it helps us rationalize what we know to be true, and the readers of this book already know a great deal about motivation. The exercises at the end of this chapter are designed to help you decide whether the theory presented in this chapter is useful. The exercises can be done alone, but they are likely to be more instructive if done in small groups that allow discussion of both the theory and individual perceptions of how motivation operates.

SUGGESTED ACTIVITIES

1. Recall from your experience learning situations in which motivation was high. (These situations may be ones in which you were very motivated or

ones in which your students were very motivated.) What, in your judgment, were the reasons? Is the theory presented in this chapter consistent with your experience? Are factors that seem to be important not addressed by the theory presented?

2. Can you recall situations in which you or your students originally engaged in a learning activity primarily because of external regulation but, over the course of the activity, the motivation became more internalized and activity continued because of its perceived value or inherent interest? If so, what factors promoted this internalization? Could you employ similar factors in your instruction to move students toward intrinsic motivation?

3. At the beginning of this chapter I mentioned an engineering student whose disdain for chemistry was somewhat mollified by examples of engineering mistakes that might have been avoided by understanding chemistry. Why might these stories have intensified his interest? List activities, illustrations, anecdotes, and the like that you might use to establish a need to know (*see* Herron, 1978b).

19

Promoting Task Involvement and Intrinsic Motivation

Inherent conflict exists in the suggestion that teachers should promote intrinsic motivation. *Intrinsic* motivation emanates from within; it implies students doing what *students* want to do. But *promoting* something implies *external* control; the teacher is trying to get students to behave as the *teacher* wants. Perhaps it is this conflict that has made it so difficult for progressive education, discovery learning, and other incarnations of the ideas developed in the preceding chapters to take hold in education. Unless teachers internalize the theory and use their professional judgment to work out applications that are appropriate for their classrooms, the ideas will not take hold now.

Task involvement and intrinsic motivation can be defended as desirable aspects of instruction. Both are consistent with a constructivist view of learning and values that we have an obligation to promote. But teachers should judge the following suggestions for promoting task involvement and intrinsic motivation in the context of their own classrooms. The unique dynamics in each classroom will dictate the best course of action to follow.

TEACHER CONTROL VERSUS STUDENT FREEDOM

THE NEED FOR ORDER

The problem in schools is not that students lack intrinsic motivation but that they lack intrinsic motivation to do those things prescribed by the school curriculum. Student freedom is constrained, and the teacher's challenge is to avoid chaos while engaging students in educational activity. It is no wonder that most classrooms are characterized by external control (on teachers as well as students). Teachers feel compelled to exert control to maintain a suitable learning environment.

If teachers are to foster intrinsic motivation, they must find ways to maintain order without external control. In other words, the motivation to behave must be intrinsic as well.

INVITING COOPERATION

Teachers who foster intrinsic motivation *and* maintain order typically try to convince students that unpalatable activities are either good for them (identified regulation) or will eventually lead to goals that students value (integrated regulation). The first step is to gain cooperation so that productive activity can take place. The opening speech may go something like this:[1]

"Okay, class, we are going to be together for the next several weeks. How enjoyable that will be depends on how well we get along. I have just one rule: That we respect one another and treat each other fairly. What constitutes respect and fair treatment in the eyes of one person differs from respect and fair treatment in the eyes of another. We need to decide what it means to us.

"John and Sally, go to the board. John, make a column labeled 'respectful' and another labeled 'disrespectful'. Sally, you make columns labeled 'fair' and 'unfair'. Now let's all think of things that go under these headings".

After getting the students' input, consensus is sought, and the data are summarized and kept for future reference. It is understood that anyone (student or teacher) violating the agreed-upon principles may be called to account by other parties to the agreement. It is also understood that when there are objections to behavior that is not covered by the agreement, discussion can be reopened to see if the agreement needs to be broadened.

Although the teacher in this scenario is directing the activity, there is ample opportunity for students to express their views and to argue for positions that are important to them. When violations of the agreement result in sanctions aimed at control, the offending student is more likely to reason, "I do not like it, but I know it is fair" (identified regulation), or "I do not want to do it, but I want the rest of the class to like me, so I will" (integrated regulation) than would be the case if the same rules were presented by the teacher without student input.

VALUE OF STUDENT PARTICIPATION IN DECISION MAKING

Students who participate in making decisions know that their ideas are valued, and they are more likely to value the contributions of others. Studies have shown that providing opportunities for students to participate in decisions encourages self-determined regulation and enhances learning and adjustment (Deci, Vallerand, Pelletier, and Ryan, 1991, p 336).

Involving students in decisions about instructional tasks has a similar effect. College students who were allowed to select tasks and to decide how much time to allot to each task were more intrinsically motivated than students who were told what to do and how long to spend (reported in Deci et al., 1991, p 336).

Regardless of how much teachers involve students in the selection of activities and other classroom decisions, there will be times when students are asked to do things that do not interest them or to do things in a manner they do not like. When rational justifications are given for the restrictions, students do not necessarily see them as restricting. Koestner et al. (1984) found that children's creativity and interest in painting were not diminished when they were told that rules governing the

---

[1]The exact procedure used to involve students in decisions and rule-making will depend on a number of factors, and the scenario outlined here is merely suggestive. Chapters 7 and 8 of Lickona (1991), Chapters 1–3 of Rogers (1969), and many other references provide specific examples of strategies that have been used successfully by teachers from elementary grades through graduate school.

use of paints were necessary to keep the colors clear and materials clean for other classes, but interest and creativity were diminished when the same rules were imposed without the justification (cited in Nicholls, 1989, p 170).

Acknowledging students' feelings of not liking an activity or the way it is being done also helps them to feel self-determined and encourages intrinsic motivation. Laboratory studies confirm what common sense would suggest: Showing interest in students and valuing them as persons with minds of their own—in short, treating them with dignity and respect—when asking them to participate in some event increases participation and learning (Deci et al., 1991, p 336).

INCREASING STUDENT FREEDOM

Adults have always disagreed about how much freedom students can handle. Many adults worry about involving students in classroom decisions as suggested above. Others argue for far less structure in schools and far greater freedom for students to manage their own learning. Thomas (1980) summarized the research evidence in two broad generalizations:

> 1) Provided that systematic procedures are followed for its implementation and a structured curriculum is provided for its maintenance, student-managed instruction has some important advantages over teacher-imposed control of instruction. These advantages include a more effective and individualized control of achievement-related and achievement-disrupting behaviors, a heightened sense of personal agency, and the possibility of a continued motivation to engage in learning activities.

> 2) . . . Environments that allow students to set their own standards, stress intra-student rather than interstudent competitiveness, emphasize the relationship between effort and achievement, and promote the use of student-generated incentives seem not only to produce the greatest short- and long-term achievement gains, but also are associated with a heightened sense of personal effectiveness among students (p 234).

THE WILLIAMS EXPERIMENT. There have been many successful schools (and programs within traditional schools) built around the idea that students learn best when they are permitted to direct their own learning. Rogers (1969) described a 1930 experiment in self-directed learning. H. D. Williams was given a class of the "worst boys" in a city of 300,000 people. The boys differed widely in ability, and Williams had little to work with other than a few art supplies, some text materials, and a chalkboard. According to Rogers:

> [Williams had two rules:] A boy must keep busy doing something, and no boy was permitted to annoy or disturb others. Each child was told, without criticism, of his results on an achievement test. Encouragement and suggestions were given only after an activity had been self initiated. Thus, if a boy had worked along artistic lines he might be given assistance in getting into a special art class. If activities in mathematics or mechanics had engaged his interest, arrangements might be made for him to attend courses in these subjects. The group remained together for four months. During this period the measured educational achievement (on the Stanford Achievement Test) of those who had been in the group for the major part of this period increased fifteen months on the average, and this improvement was evident in reading, arithmetic, and other subjects (p 160).

THE SUDBURY VALLEY EXPERIMENT. Much of the literature on schools that encourage student self-direction consists of glowing descriptions by the founder with little objective, third-party scrutiny. One exception is Gray and Chanoff's (1986) description of the Sudbury Valley School.

The Sudbury Valley School is a democratically administered primary and secondary school that has no learning requirements but supports students' self-directed activities. Children and adults come to a big house where they do things that interest them. In spite of its unconventional structure, Sudbury is accredited by the New England Association of Colleges and Schools.

Sudbury is governed by the "School Meeting", a formal assembly with a published agenda that meets weekly. Few people go unless there is something on the agenda that concerns them. There is also a "Judicial Committee" that deals with problems as they arise. Membership on the committee changes monthly, and everybody must serve on it from time to time.

Students at Sudbury come from middle-class homes. About a third enter the school before grade six; the rest spend an average of 2.5 years there after they drop out of regular school because of various problems. Sudbury had its first graduate in 1970, and its enrollment was 90 in 1984–1985. There are 13 faculty members, but only three work full-time. A total of 78 students had graduated in 1981.

Graduation occurs after students present a thesis to the School Meeting; the thesis is a justification of their contention that they are ready to take their place in society. Four people have left without trying to present a thesis; all but one who have presented have graduated.

Gray and Chanoff surveyed Sudbury graduates to determine how they felt about their education, what they had done after graduation, and how well Sudbury had prepared them. There is insufficient space to report their findings in detail, but the basic conclusion is that graduates have done okay. Whether they went to college or into the workforce, graduates felt that they were well prepared, and they had few complaints about their education.[2]

This brief description of the Sudbury School is not intended as a model for education. Rather, it is intended to show that giving students additional freedom of choice is not, in itself, detrimental.[3] Many other accounts of educational experiments reach similar conclusions (Williams, 1930; Neill, 1960; Wigginton, 1985).[4]

INCREASING MOTIVATION

USING EXISTING INTEREST

A great deal can be done to increase student motivation within a traditional school environment. One strategy is to capitalize on existing student interest. Three examples from personal experience illustrate ways that this can be done. In the first exam-

---

[2]Few Sudbury graduates have gone into science or other technical fields. It is possible that the extreme freedom of choice afforded at Sudbury results in an inadequate background in mathematics and science. However, this issue was not addressed in the research, and the failure of Sudbury graduates to pursue technical careers could be due to other reasons.

[3]Although not addressed by the research, it seems likely that Sudbury's treatment of students as responsible members of a community has as much to do with its success as does freedom of choice. Students have freedom of choice but not license to disregard their responsibility to themselves and to other members of the community.

[4]Of the references cited, Elliot Wigginton's (1985) account of the Foxfire experience is one of the most readable and heartwarming.

ple, the teacher is still in control; in the other examples, students are given increased opportunity for self-direction.

KITCHEN CHEMISTRY AND OTHER WORTHWHILE SCIENCE. My first organized attempt to capitalize on existing student interest was a cooperative venture with other science teachers at Versailles (Kentucky) High School. Students in the 9th-grade class came from schools throughout the county, and many had difficulty adjusting to the new school. Students' backgrounds in science ranged from virtually nothing to a sound, general knowledge. Most 9th-grade classes, including science classes, were fraught with discipline problems. No science teacher wanted to teach 9th-grade General Science, so we decided to share the responsibility. Each of five faculty members developed a six-week unit around topics that we perceived to interest students.[5] The five units were Kitchen Chemistry; The Physical Geology of Woodford County, Kentucky; Atomic Energy and Civil Defense; Electricity in the Home; and the Physiology and Psychology of Adolescents.[6] A concerted effort was made to relate the discussion to everyday events and to use those events to develop the concepts and principles that scientists have invented to explain them.

Although this program represents a modest attempt to capitalize on existing interests, the effort was a definite improvement over the previous course. When this effort was initiated more than 30 years ago, it did not occur to us to conduct a formal evaluation of the project. However, the anecdotal information supported our belief that student interest was high. Among the reports was this one from the principal: "I know you're doing something right. The period that 9th-graders take science[7] is the only time that I can get a cup of coffee! Every other period I'm handling 9th-grade discipline problems!

HIGH SCHOOL PHYSICAL SCIENCE. Part of my assignment at Kaiserslautern American High School was to teach a physical science course to seniors who needed science credit for graduation. Most students in the course were disaffected adolescents who were "practicing to become juvenile delinquents". They were not bad kids, but they were certainly not interested in school. The textbook on which the course was based presented a typical, expository treatment of science concepts and principles. It was awful!

My first attempt to ensure survival until the end of the year was to approach the principal with a rational argument for changing the course. Although sympathetic, he pointed out that my proposal for a new course would take a minimum of three years to work its way through the bureaucracy of the Army Dependent Schools. Because normal troop movements during the school year produced a 50% turnover in the student body, there was a concerted effort to offer the same courses in each Army Dependent School and to present courses at a uniform rate throughout the system.

---

[5]These perceptions were based on experience in previous classes. Admittedly, this approach is gross and undoubtedly misjudged the interest of many students.

[6]A sixth unit on library skills, laboratory skills, and strategies used to make sense of observations was presented by all five teachers at the beginning of the course. Students spent the first 12 weeks with the same instructor, studying the introductory unit and one of the other five units. After that, students rotated every six weeks to study the remaining units under each unit's author.

[7]All 9th-graders were scheduled for science at the same hour to facilitate the class rotation that took place at the end of each six-week unit.

I began the course by telling students that they had a great deal of interest in science, although they might not be aware of it. In response to their skepticism, I asked them what they liked, and one of the first responses was, "Rock music." Four of the students had a band, so I invited them to bring their instruments and play for us.[8]

Before class started, I connected a microphone to an oscilloscope and turned it on. Before the end of the first song, students, including the musicians, were paying attention to the sights on the scope as well as the sounds of the music. After some spontaneous experimentation on the part of the musicians to see how their actions affected the trace on the scope, I asked one and then another to play the same note so we could compare the traces made. Then we tried different notes. We tried the same note an octave higher and an octave lower. I found tuning forks, and we compared the note as it was produced by the tuning fork and the same note produced on each instrument.

I asked the musicians how they talked about the similarities and differences we were observing, and they knew a great deal. We explored the difference between "music" and "noise", and the musicians talked about what makes music "good" or "bad". Questions about keeping instruments in tune led to textbook readings and discussion about how sound travels and what affects its speed. The effect of temperature and humidity on the tuning of metal and wooden instruments provided an opportunity to anticipate topics that we would study later on.

I am convinced that, with a little ingenuity, I could have built the entire course around the students' interest in music, but even things that interest get old after a while.

Music turned out to be the most successful attempt to capitalize on my students' interests, but there were others. By the time we got to electricity, I knew that most boys and some girls in the class were interested in cars, so we dismantled the electrical system on my Volkswagen Beetle and used it to explore voltage, current, power, and their interrelationships. Using the coil, we talked about transformers and why points are needed in a car but not in the "coil" connected to the doorbell in their house.

We traced circuits in houses and cars and discussed their differences. We identified "broken" circuits as a common ailment in electrical appliances, and we repaired some that students brought from home. We wired lamps, and we talked about safety so that we would not get hurt in any of our activities.

During the year I deviated some from the standard syllabus to tie course content more closely to the expressed interests of the students, but we "covered the course", and the principal and I agreed that the increased interest and commitment to learning justified the slight perturbations in timing.

FISHING. Students in the high school chemistry courses that I taught were not always task involved or intrinsically motivated, but their classroom behavior was usually characterized by identified or internalized regulation. They were convinced that chemistry was "good for them", or they knew that to be the doctor, engineer, or other professional that they wanted to be, they needed to pass chemistry. They were the proverbial "good little boys and girls", and they did what I asked them to do. They were, however, more ego involved than task involved.

---

[8]Our classroom was located at the end of a sparsely populated wing of the building. I convinced the students that amplification was unnecessary, and I told the principal and nearby teachers what was going on in advance. It worked out fine.

My most successful ploy for fueling intrinsic motivation was what I call "fishing".

Fishing generally takes place in the lab. I circulate throughout the room, looking over shoulders to see what is going on. From time to time I interject a comment such as, "Hey, that's an interesting result", or "Hmmm, I wonder why that happened." I am usually ignored, but occasionally some student takes the bait and responds with something like, "You know, I wondered about that myself. I had expected . . ." Once I have a student on the hook like this, I play the fish for all it is worth:

Me: Gee, do you have any ideas?

S: Well, I thought it might have been because . . .

Me: Yeah, that does sound like a possibility. Any way you could find out?

S: Well, I might try . . .

Me: That's not a bad idea. Something else you might try is . . .

S: Could I really do that?

Me: Sure. Why not?

S: Well, for one thing I would not have time to finish the assigned experiment.

Me: That's no big deal. The rest of the class is doing that, so you can find out what happens from them. If you investigate this, everybody can learn something they would not find out doing the regular experiment. Everybody wins!

S: Okay, but I'm not sure I know exactly how to do what I want.

Me: Well, think about it and see what you come up with. When you're ready, tell me what you have in mind, and I'll see if it makes sense to me.

Once my high school students understood that it was permissible to deviate from the prescribed curriculum, some came up with questions they wanted to investigate. When the investigation could lead to new understanding—and I can think of no instance when it could not—it was safe, and we could get the materials and equipment needed, students were encouraged to do it.

At times questions came up during the prelaboratory discussions that set the stage for each experiment.[9] Students might disagree about the best way to do the experiment or contest whether a particular procedure would, in fact, yield the desired information. Disputes were sometimes resolved by some students following one procedure while others followed another.

These illustrations of strategies that I have used to increase intrinsic motivation and task involvement are certainly not meant to be prescriptive. As teaching strategies, they have no intrinsic value. Any other practice that increases intrinsic motivation and task involvement is just as worthwhile.

UTILITY AS A FACTOR IN MOTIVATION

Like the future engineer in the anecdote that opened Chapter 18, students frequently wonder why you are asking them to do all of these weird things. Most schoolwork is so divorced from everyday experience that students have difficulty seeing any use for it. As might be expected, research shows that intrinsic motivation is

---

[9]The second year that I taught high school chemistry, I used a trial edition of the *Chemical Bond Approach* (CBA, 1964), and I became a devoted disciple of the CBA laboratory philosophy. I still consider the CBA laboratory program to be the best ever developed. Every CBA laboratory activity begins with whole-class discussion of some question that has arisen from a demonstration, a previous investigation, or statements in the text. The prelab discussion focuses on ways the question might be investigated, what data would be required to answer it, and what techniques are available to collect the data. It is this kind of prelab discussion that is referred to here.

increased by a perception of utility. Deci et al. (1991) described several studies to support this generalization. In one representative study, "[college] students who learned text material in order to put it to use reported more intrinsic motivation for learning and showed greater conceptual understanding than did students who learned the material in order to be tested" (p 331). Learning to be tested is not any more effective in lower grades. "Grolnick and Ryan (1987) found that asking elementary students to learn material in order to be tested on it led to lower interest and poorer conceptual learning than did asking students to learn the material with no mention of a test, even though the test condition led to short-term (less than one week) gains in rote recall that had dissipated one week later" (cited in Deci et al., 1991, p 332).

By focusing on tests, grades, and class standing rather than on the inherent worth of what is being taught, teachers encourage ego rather than task involvement. In doing so, they defeat their purpose of enhancing learning.

PRAISE AND PUNISHMENT

After reviewing research on the effectiveness of external rewards in promoting learning, Thomas (1980) concluded that "External reward systems may be effective in inducing the will to learn on immediate tasks, but they have not been shown to be effective in leading to the sort of intrinsic motivation necessary to endure out-of-class learning" (pp 231–232). Ryan, Connell, and Deci (1985) reached similar conclusions: "Rewards such as money, good-player awards, prizes, and food can all undermine intrinsic motivation. . . . The danger is that to the extent something becomes linked with the receipt of rewards, it is subsequently less likely to occur in their absence" (p 19).

In spite of these findings it would be a mistake for teachers to stop praising and encouraging students for a job well done. We know from personal experience that encouragement and support promote persistence and hard effort, even at tasks that we elect to do for personal reasons. Positive feedback *does* enhance motivation, but it does not enhance *intrinsic* motivation unless it is given in a manner that supports the autonomy of students. "Congratulating students for having done well at a self-initiated educational activity is likely to promote feelings of competence and intrinsic motivation, whereas praising them for doing what they 'should' have done or what you told them to do is likely to lead to their feeling controlled, which in turn would reduce intrinsic motivation and strengthen nonautonomous forms of extrinsic motivation" (Deci et al., 1991, p 333). As Thomas (1980) points out, the issue is basically one of locus of control.

> Where the locus of control over learning-related behaviors is entirely vested in the teacher, where maximum structure is provided for carrying out learning activities, and where the motivation to perform is provided for through external rewards, praise, and/or fear of reprisal, there is little latitude or opportunity for students to develop a sense of agency and, subsequently, to become proficient at using learning strategies (p 236).

> Students who come to . . . accept that the locus of responsibility for attention, reinforcement, and, ultimately, success and failure is internal rather than external will be more likely to approach reading and mathematics tasks with that same sense of agency. . . . On the other hand, students who have had no experience with managing their own behavior and who have not learned to take

responsibility for success and failure do not see any connection between effort (in this case, the generation of learning strategies) and success on a learning task (p 235).

Ryan, Connell, and Deci (1985) summarize research on the effect of rewards, external evaluation, and other manifestations of external regulation in three propositions:

> **Proposition I**: . . . Any event that facilitates the perception of an internal locus of causality . . . will tend to enhance intrinsic motivation . . .

> **Proposition II**: . . . Any event that enhances perceived competence will tend to enhance intrinsic motivation. . . . When feedback is independent of effort and performance, regardless of its valence, it fails to facilitate . . . feelings of competence. . . . If a task is optimally challenging, it should generally enhance intrinsic motivation. . . . In optimal challenge, feedback will tend toward a positive valence, but will also contain elements of risk, failure, and negative feedback which are effectance relevant. Without these, there is no challenge . . .

> **Proposition III**: . . . It is not events per se that affect intrinsic motivation, but rather their meaning or functional significance [informational or controlling] for the individual (p 16).

The complexity of the classroom environment makes it impossible for teachers to act on the basis of rigid rules. But these propositions provide useful guidelines for the professional judgment that teachers must exercise in providing feedback so that intrinsic motivation and task involvement are enhanced.

PERSONALITY OF THE TEACHER

All of what has been said about promoting intrinsic motivation and task involvement implies something about the personality of the teacher. Teachers are a lot like people. They have biases, weaknesses, needs, and wants, just like other human beings. Some personality traits make it easier to interact with students in the manner suggested above; some make it virtually impossible.

Some teachers are insecure. They feel threatened by students in their classroom and administrators and parents outside of the classroom. Some have financial obligations that make them fear losing their jobs. It is difficult for such teachers to relinquish control, to encourage students to be autonomous, or to involve students in decision making.

External forces exerted on teachers to improve their performance are likely to be self defeating. The more teachers feel that their behavior is being controlled by others, the more they exert control over their students (Deci et al., 1991, p 340). This effect was confirmed in an experimental study by Deci, Spiegel, Ryan, Koestner, and Kaufman (cited in Ryan, Connell, and Deci, 1985). In the study, two groups of teachers were given identical instructions for teaching a lesson on spatial relations problems, but one group was told that it was their responsibility to ensure that students performed up to standards. The group that was reminded about the standards exerted more control over students. This happened despite the fact that research indicates that students perform better when teachers are less controlling (p 46).

COOPERATIVE LEARNING

Several authors have argued that intrinsic motivation and task involvement may be enhanced through cooperative learning (Johnson and Johnson, 1985; Nicholls,

1989; Ryan, Connell, and Deci, 1985). However, as with any other strategy suggested in this chapter, *how* it is done is more important than *what*. As Johnson and Johnson (1985) point out, "motivation to achieve goals is primarily induced through interpersonal processes" (p 250).

Johnson and Johnson (1985) argue that lack of interaction among individuals working on achievement-oriented tasks produces these undesirable results:

a. extrinsic motivation based on achieving to benefit only oneself
b. expectations for success or failure based on a monopolistic view of one's academic ability, one's effort to achieve, . . . and one's achievement history
c. incentives for achievement based on working for self-benefit . . .
d. a lack of epistemic curiosity [i.e., task involvement] and continuing interest in learning . . .
e. negative attitudes toward the subject being studied . . . [and]
f. lack of persistence . . . (p 255).

This list probably overstates the case. You can probably recall from personal experience instances in which there was little or no interaction among individuals, but intrinsic motivation and task involvement were high. The Sudbury School, described earlier in this chapter, would appear to be an example in which learning is highly individualistic but task involvement and intrinsic motivation are fostered.

On another point Johnson and Johnson make considerable sense:

Within a cooperative situation, disagreement over information, conclusions, theories, and opinions tends to lead to uncertainty, epistemic curiosity [i.e., motivation to search for more information concerning the topic], and a reevaluation of one's conclusions. . . . Within an individualistic situation, there is no opportunity for disagreement and, therefore, initial conclusions are not challenged and fixation on initial impressions is common . . . (1985, p 271)

Individuals working alone on a project are likely to encounter differences and apparent contradictions among the sources they consult, so there *is* opportunity for disagreement in an individualistic situation, but Johnson and Johnson are correct to suggest that it is far more likely in a cooperative situation, *assuming that the cooperative learning situation is properly designed*.

There are many designs for cooperative learning, and the amount of autonomy exercised by students as well as the amount of interaction among students can vary appreciably. The cooperative learning activity can be narrowly focused on isolated facts looked up by individuals in the group, or it can aim at synthesizing information from many sources to arrive at broad generalizations that represent group consensus. One would expect very different outcomes from such disparate learning activities, and research supports that expectation. Nicholls cites two studies, one in a school setting and one in a laboratory setting. Neither study used intrinsic motivation or degree of task involvement as criterion measures, but they do show that the outcome of cooperative learning varies, depending on its specific organization and focus.

The studies described by Nicholls looked at achievement. They suggest that when the measure of achievement is recall of simple information, arrangements that focus on isolated facts are superior; when the measure of achievement is more complex understanding, methods that encourage group interaction and negotiation of meaning are superior (1989, p 177).

SUMMARY

This chapter has suggested several strategies that teachers can use to foster self-determination and intrinsic motivation. However, they are only suggestions, and teachers who invent strategies that fit the age and educational level of their students, are appropriate for the subject they are teaching, and are compatible with their own personality will be far more successful than those who copy strategies used by others. The advice presented in this chapter is captured in these recommendations taken from McCombs and Whisler (1989):

1. understand and demonstrate real interest, caring, and concern for each student and his or her needs, interests, and goals;
2. challenge students to invest effort and energy in taking personal responsibility and being actively involved in learning activities;
3. relate learning content and activities to personal needs, interests, and goals;
4. help students define personal goals and their relationships to general learning goals;
5. structure general learning goals and activities such that each student can accomplish his or her personal goals and experience success;
6. provide students with opportunities to exercise personal control and choice over carefully selected task variables; such as type of learning activity, level of mastery, amount of effort, or type of reward;
7. highlight the value of student accomplishment, the value of students skills and abilities, and the value of the learning process and learning task; and
8. [praise] students' accomplishments and encourage them to reward themselves and develop pride in their accomplishments (pp 299–300).

Appendixes

Appendix A

CONCEPT ANALYSIS FOR *MIB*

DEFINITION

A *mib* is a plane geometric figure made by attaching a segment to the short leg of a right triangle so that it is an external, perpendicular bisector of the short leg.

CRITICAL ATTRIBUTES

    C1. The figure contains a right triangle.
    C2. There is a segment attached to the short leg.
    C3. The segment is external.
    C4. The segment is perpendicular to the short leg.
    C5. The segment bisects the short leg.

VARIABLE ATTRIBUTES

    V1. Size of the figure.
    V2. Orientation of the figure.
    V3. Length of the segment.
    V4. Whether there are additional segments.

SUPRAORDINATE CONCEPTS: plane figure, geometric figure
COORDINATE CONCEPTS: square, rectangle, right triangle
SUBORDINATE CONCEPTS: none

<u>EXAMPLES</u>   <u>NON-EXAMPLES</u>

1.   (V1, V2, V3)   1.   (C2)

2.   (V1)   2.   (C4)

3.   (V4)   3.   (C5)

4.   (V2)   4.   (C1)

5.   (V3)   5.   (C3)

Appendix B

CONCEPT ANALYSIS FOR *CONCEPT*

DEFINITION

A *concept* is a set of specific objects, symbols, or events that are grouped together on the basis of shared characteristics and that can be referenced by a particular name or symbol.

CRITICAL ATTRIBUTES

C1. It is a class or set.
C2. Members share common characteristics.
C3. The class is referenced by a particular label.

VARIABLE ATTRIBUTES

V1. The set can be almost anything: objects, symbols, events, attributes, etc.
V2. The reference label may be a name, a symbol, or any other kind of label.
V3. The set can have two or more members.

SUPRAORDINATE CONCEPT: knowledge
COORDINATE CONCEPTS: principle, rule, skill, cognitive strategy
SUBORDINATE CONCEPTS: defined concepts, concepts with no perceptible examples, concepts with no perceptible attributes

Examples	*Nonexamples*
1. sentence (V1, V3)	1. $2X + 4 = 10$ (C1, C3)
2. man (V1, V3)	2. Dudley Herron (C1)
3. hydrocarbon (V1, V3)	3. Hydrocarbons burn. (C1, C2, C3)
4. furniture (V1, V3)	4. the contents of my office (C2, C3)
5. mass (V1)	5. 72 kg (C1)
6. love (V1)	6. I love my wife. (C1, C2, C3)
7. pH (V2)	7. * (C1, V2)

COMMENTS: The number in parentheses following each example and nonexample is the number of the attribute I am focusing on by using that example or nonexample. In some cases, it is the collection rather than a specific instance that provides the desired focus.

Most examples were selected to show the variety of classes that may be considered concepts. Example *7* was selected to illustrate a concept that is referenced by a symbol other than a word. Nonexample 7 was paired with it to show a symbol that does not reference a concept. (The reader can undoubtedly think of a concept that is sometimes referenced by *, but unlike pH, it will not be an unambiguous label.) Nonexample 1 is presented as an instance of a sentence that is not a concept. (Nonexamples 3 and 6 could be used as well.)

Nonexamples 1, 3, and 6 are *examples* of the concept *sentence*, but they are not examples of the concept *concept*. (In other words, if *sentence* were the concept under analysis, nonexamples 1, 3, and 6 would be listed as examples.) Listing anything that could not be construed as an example of *some* concept would be difficult, given the way we construct knowledge.

Examples 2, 4, and 5, with their corresponding nonexamples, illustrate the difference in the concept (man, furniture, and mass) and *examples* of the concept (Dudley Herron, the contents of my office, and 72 kg).

Examples 3 and 6, with their corresponding nonexamples, illustrate the difference in concepts (hydrocarbons and love) and a principle concerning the concept (Hydrocarbons burn.) or a statement incorporating the concept (I love my wife.).

Appendix C

CONCEPT ANALYSIS FOR *ATOM*

DEFINITION

An *atom* is the smallest unit of an element that can exist either alone or in combination with other atoms and still have the microscopic properties of the element.

CRITICAL ATTRIBUTES

C1. It is the smallest unit of an element.
C2. It retains the microscopic properties of the element.

VARIABLE ATTRIBUTES

V1. It may exist alone or in combination.
V2. It may be combined with like atoms or unlike atoms.
V3. Any kind of element may be represented.
V4. It may exist in substances in any physical state.

SUPRAORDINATE CONCEPT: elementary particle
COORDINATE CONCEPTS: molecule, ion, proton, electron, neutron
SUBORDINATE CONCEPTS: sodium atom, hydrogen atom, carbon-12 atom, ground-state atom, isotope

Pseudoexamples

1. Animated sequence showing iron nail magnified to show pulsating spheres held together; one sphere identified. (V1,V3,V4)
2. Similar animated sequence showing copper penny. (V3)

Pseudo Nonexamples

1. Same animated sequence with group of spheres identified. (C1)

2. Similar sequence extended to show sphere coming apart with a part identified. (C2)

Pseudoexamples

3. Similar sequence showing vaporization of metal producing gaseous atoms, one sphere identified. (V4)
4. One ball from ball-and-stick model set. (V1,V3)
5. Molecular model; one ball identified. (V1,V2,V3)
6. Crystal lattice model; one ball identified. (V1,V2,V3)
7. Chemical symbol such as H or O. (V1)
8. Chemical formula with one symbol identified. (V1,V2)

Pseudo Nonexamples

3. Similar sequence showing vaporization of water; molecule identified. (C1)
4. Two balls connected. (C1)
5. Molecular model; entire model identified. (C2)
6. Crystal lattice model; unit cell identified. (C1)
7. Symbol such as H_2. (C2)
8. Chemical formula; entire unit identified. (C2)

Appendix D

FLAWED ANALYSIS FOR *ELEMENT*

Attributes	Critical	Irrelevant
1. Composed of atoms that are alike.	X	
2. Cannot be decomposed by ordinary chemical means.	X	
3. Collection of like atoms in solid, liquid, or gaseous form that have a net neutral charge.	X	
4. May be a member of the periodic chart.		X
5. May be a) a metal b) a nonmetal c) a semimetal		X
6. Its abundance may be a) common b) not very common		X
7. Can, in natural state, occur as a) free b) compound		X

Teaching Examples

1. Element 107 (4, do not want to postulate on further properties)
2. Mercury (4, 5a, 6b, 7b)
3. Chlorine (4, 5b, 6a, 7b)
4. Sulfur (4, 5b, 6a, 7a)
5. Boron (4, 5a, 6b, 7b)

6. Neon (4, 5b, 6b, 7a)

Teaching Nonexamples

1. Water (lacks 1, 2)

2. AgCl (s) (lacks 1, 2)
3. CO_2 (lacks 1, 2)
4. NaCl (aq) (lacks 1, 2, 3)
5. Iron ions isolated by membrane in water solution. (lacks 3)

Source: This appendix is reproduced with permission from Markle and Tiemann (1970).

Teaching Examples
7. Iron (4, 5a, 6a, 7a, b)
8. Cesium (4, 5a, 6b, 7b)

Testing Examples
1. Titanium (4, 5a, 6a, 7b)
2. Gallium (4, 5a, 6b, 7b)
3. Oxygen (4, 5b, 6a, 7a, b)
4. Cobalt (4, 5a, 6b, 7b)
5. Element 104 (4, do not want to postulate on further properties)

Testing Nonexamples
1. Salt (lacks 1, 2)
2. LiOH (aq) (lacks 1, 2, 3)
3. Stones (lacks 1, 2)
4. Steel (lacks 1, 2)
5. Sodium ions isolated by membrane in water solution. (lacks 3)

Appendix E

CONCEPT ANALYSIS FOR *ELEMENT*

DEFINITION

MACROSCOPIC. An *element* is a pure substance that cannot be separated into simpler macroscopic substances with characteristic properties that differ from the original substance.

MICROSCOPIC. An *element* is a substance composed of a single kind of atom.

CRITICAL ATTRIBUTES

C1. It is a kind of matter. (macroscopic)
C2. It is pure. (macroscopic)
C3. Separation into parts of macroscopic size results in no change in characteristic properties. (macroscopic)
C4. It is composed of a single kind of atom. (microscopic)

VARIABLE ATTRIBUTES

V1. It may exist as solid, liquid, or gas. (macroscopic)
V2. Examples may have very different properties. (macroscopic)
V3. It may react easily or with reluctance. (macroscopic)
V4. Atoms may exist in various size aggregates. (microscopic)

SUPRAORDINATE CONCEPT: Matter
COORDINATE CONCEPT: Compound
SUBORDINATE CONCEPTS: hydrogen, oxygen, mercury, iron, copper, helium, naturally occurring element; artificial element
EXAMPLES: (None with perceptible attributes) hydrogen, oxygen, sulfur, iron, helium
NONEXAMPLES: (None with perceptible attributes) water, carbon dioxide, hydrogen sulfide, iron oxide

USEFUL TEACHING ACTIVITIES

1. Present instances including heat, light, or other entities that are not examples of matter and ask, "Could this be an element? Why or why not?" (Ask probing questions until students insist that only those instances that are forms of matter can be elements.) (C1)
2. Present instances that include pure substances and mixtures and ask, "Can this be an element?" (Ask probing questions until students insist that only pure substances can be elements.) (C2)
3. Present instances that include elements and compounds and ask, "Can this be an element?" (Ask probing questions until students insist that only substances that (a) form no substances with new characteristic properties or (b) only form substances with new characteristic properties when the new substances weigh more than the original can be elements.) (C3)
4. Present instances of models that include (a) piles of disconnected, identical spheres, (b) piles of connected, identical spheres, (c) piles of disconnected, unidentical spheres, and (d) piles of connected, unidentical spheres and ask, "Can this pile of models representing atoms represent an element?" [Ask probing questions until students insist that those models (and only those models) with one kind of sphere can represent an element.] (C4, V4)
5. Present instances including chlorine, bromine, and iodine (or other examples of elements in each of the three states of matter) and ask, "Can this be an element?" (Ask probing questions until students insist that a substance may be an element whether it is in a gaseous, liquid, or solid state.) (V1, V2)
6. Present instances including elements with divergent properties and ask, "Can this be an element?" (Ask probing questions until students insist that instances with very different properties can be elements.) (V2, V3)
7. Present instances including chlorine and nitrogen or other examples that differ in reactivity and demonstrate reactivity. Ask, "Can this be an element?" (Ask probing questions until students insist that both reactive and unreactive substances can be elements.) (V3)

Appendix F

CONCEPT ANALYSIS FOR *MOLE*

DEFINITION

A *mole* is the amount of substance of a system that contains as many elementary entities as there are carbon atoms in 0.012 kg of carbon-12. The elementary entity must be specified and may be an atom, a molecule, an ion, an electron, etc., or a specified group of such particles.

CRITICAL ATTRIBUTES

C1. Mole refers to an amount of substance.
C2. A mole contains 6.02×10^{23} elementary entities.
C3. The elementary entity must be specified.

VARIABLE ATTRIBUTES

V1. Kind of elementary entity considered.
V2. Units used to describe amount. (Each unit used must have an associated principle that relates amount in terms of that unit to amount in terms of number of elementary entities; e.g., "There are 500 sheets in a ream of paper.")

SUPRAORDINATE CONCEPT: measures of amount of substance

COORDINATE CONCEPTS: mass, number of particles, volume

SUBORDINATE CONCEPT: mole of atoms, mole of molecules, mole of electrons

Examples	*Nonexamples*

Group I

(In these examples, students must apply rules of arithmetic and knowledge of relations that should be familiar. They serve to focus on C2, C3, V1, and V2.)

6.02×10^{23} cars (car)	6.02×10^{22} cars (car)
3.01×10^{23} bikes (wheel)	6.02×10^{23} bikes (wheel)
1.20×10^{21} reams of paper (sheet)	1.20×10^{21} gross of pencils (pencil)

Group II

(In these examples, students must apply the rule, "The atomic mass of an element in grams contains one mole of atoms." It is assumed that the rule will be new to students. The examples focus on C3, V1, V2, and the new principle.)

16 g oxygen (O atom)	15 g oxygen (O atom)
14 g nitrogen (N atom)	28 g nitrogen (N atom)
35.5 g chlorine (Cl atom)	35.5 g chlorine (Cl_2)

Group III

(In these examples, students must apply additional rules that relate molecular mass to moles of molecules. It is assumed that the rules will be new. The examples focus on C3, V1, V2, and the new rules.)

32 g oxygen (O_2 molecule)	16 g oxygen (O_2 molecule)
18 g water (H_2O molecule)	18 g water (H atom)
9 g water (H atom)	9 g water (H_2O molecule)

Group IV

(In these examples, students must apply still other rules and principles that constitute part of the content of chemistry. They focus on C1, C2, C3, V1, V2, and several principles.)

2 g oxygen (electron)	2 g oxygen (O atom)
22.4 L gas at standard temperature and pressure (gas particle)	22.4 L gas at room temperature and pressure (gas particle)
1.0×10^{23} carbon-12 (proton)	1.0×10^{23} carbon-12 (nucleon)

Appendix G

CONCEPT ANALYSIS FOR *MIXTURE*

DEFINITION

MACROSCOPIC: A *mixture* is an aggregate of two or more substances or phases of a single substance, each of which can be identified by one or more characteristic property.

MICROSCOPIC: A *mixture* is an aggregate of two or more kinds of particles (atoms, ions, or molecules) that are not bonded together by strong forces.

CRITICAL ATTRIBUTES

    C1. Contains two or more substances or phases.
    C2. Each component can be identified on some basis.
    C3. The mixture lacks constant macroscopic properties.
    C4. Contains two or more kinds of separate particles. (microscopic)

VARIABLE ATTRIBUTES

    V1. Components of the mixture.
    V2. Number of components.
    V3. Kind and number of phases.
    V4. Size of identifiable particles.
    V5. Ease of detecting variable properties.
    V6. Microscopic particles may be atoms, ions, radicals, or molecules. (microscopic)

SUPRAORDINATE CONCEPTS: substance, matter
COORDINATE CONCEPT: pure substance
SUBORDINATE CONCEPTS: homogeneous mixture, heterogeneous mixture

Examples	*Nonexamples*

Group I	
(These examples are heterogeneous; that is, direct observation reveals that a mixture is present. However, the nonexamples could easily be confused with the examples of Group II. These examples focus on C1–C3 and V1–V4.)	

Dimes and quarters	dimes; quarters
Cement	copper
Brick	aluminum
Dirt	sugar
Granite	diamond
Smoke in air	chlorine gas
Oil and water	oil; water
Ice and water	ice; water

Group II	
(These examples are homogeneous; that is, direct observation will not reveal that a mixture is present. Classification will require application of various principles. The examples focus on the various facts and principles involved and V5.)	

Milk	water
Gasoline	benzene
Air	nitrogen gas
Ethanol and water	water; ethanol
Wax	naphthalene
Solder	lead

Group III	
(Teaching activities like number 4 in Appendix E and pseudoinstances like those numbered 1–5 in Appendix C can be used to clarify the microscopic attributes of mixtures.)	

Appendix H

CONCEPT ANALYSIS FOR *WEIGHT*

AREA: Expressing relationships
TARGET CONCEPT LABEL: Weight

DEFINITION: Gives the name of the supraordinate concept and its criterial attributes of the target concept. (If there is no supraordinate concept, then all attributes of the target concept should be given.) Weight is the measure of how light or heavy an object is.

SUPRAORDINATE CONCEPT: measure
COORDINATE CONCEPTS: area, length
SUBORDINATE CONCEPT: pounds

CRITICAL ATTRIBUTES: Identify the target concept within the selected supraordinate concept (or coordinate concepts if a supraordinate has not been identified). *Weight* is how light or heavy an object is.

OTHER ATTRIBUTES that are relevant but not criterial for the target concept include the following: (The attributes of the supraordinate need not be specified.)
Other attributes relevant to weight are those of its supraordinate *measure;* e.g., has a unit of measure, is a measurement.

IRRELEVANT ATTRIBUTES: Irrelevant attributes of the target concept (attributes which vary among instances of the target concept) include the following:
1. Size of the objects.
2. Specific unit of weight; e.g., tons, pounds.
3. How much an object weighs.

CONCEPT EXAMPLES: Concept examples include the following: 10 pounds, 10 tons, 10 ounces.

Source: This appendix is reproduced with permission from Romberg, Steitz, and Frayer (1971).

CONCEPT NONEXAMPLES: Concept nonexamples include the following: 10 feet, 10 dozen, 10 square yards.

RELATIONSHIP WITH AT LEAST ONE OTHER CONCEPT. (This relationship should preferably be a principle. It should definitely *not* be a direct supraordinate–subordinate relationship, a relationship involving a critical attribute, or a relationship involving an example.)

A standard unit for weight is an ounce.

Appendix I

CONCEPT ANALYSIS FOR *MASS*

DEFINITION

Mass is the quantity of matter that a particle, body, or object contains.

CRITICAL ATTRIBUTES

C1. Mass is a property of matter.
C2. Anything with mass occupies space.
C3. There is an attractive force between masses.
C4. Mass imparts inertia to matter.

VARIABLE ATTRIBUTES

V1. The physical state of the matter
V2. The physical properties of the matter (e.g., density, conductivity, melting point).
V3. The chemical properties of the matter (e.g., pH, chemical activity).

SUPRAORDINATE CONCEPT: physical property
COORDINATE CONCEPTS: weight, volume, length, temperature
SUBORDINATE CONCEPTS: gravitational mass, inertial mass
CONCEPTS WITH WHICH THE ATTRIBUTE IS CONFUSED: weight, density, solid

HOW IS THE ATTRIBUTE DETECTED?

1. Physical perception (direct or indirect) that space is occupied.
2. Evidence (direct or indirect) that the body is attracted to another mass.
3. Evidence that the body has inertia.

HOW IS THE ATTRIBUTE MEASURED?

By physical comparison of the gravitational force exerted on the object being measured and the gravitational force exerted on an object taken by agreement as a standard mass. Common units of measurement are kilogram, gram, and slug.

POSSIBLE TEACHING ACTIVITIES

1. Present instances such as the following and ask students to identify those that have mass.

Have Mass	Lack Mass
a. horn (V1,V2)	a. sound (C1,C2,C3)
b. water (V1,V2)	b. heat (C1,C2,C3)
c. air (V1,V2,V3)	c. idea (C1,C2,C3)
d. orbiting spaceship (V1,V2)	d. vacuum (C3,C4)
e. moon (V1,V2,V3)	e. love (C1,C2,C3,C4)

2. For each of the instances listed above, ask for evidence that the entity does or does not have mass. (Student responses should indicate that (a) space is occupied and (b) the entity is attracted to another mass or (c) the entity has inertia.)

3. Pass around equal size pieces of lead and aluminum and ask, "Which has more mass? How do you know?"

4. Pass around pieces of polystyrene such as Styrofoam and of lead selected so that the polystyrene is about 20% heavier than the lead. Ask which has more mass, and when students respond "the lead", place the samples on each side of a platform balance and ask them to reconsider. Ask students why they were fooled. Ask students to distinguish between mass and density.

   (These are suggestions only. Other activities would be required to clarify the concept.)

POSSIBLE TEST QUESTIONS

1. Which of the following has mass?
   (a) an astronaut in space
   (b) the Goodyear blimp
   (c) a rock
   (d) all of the above
   (e) none of the above
2. Air has mass, but you cannot detect its mass by filling a bag with air and placing it on a balance. Which of the following provides evidence that air has mass? (Write a short justification for your choice in the space provided.)
   (a) If I exhale under water, the air pushes back the water to form a bubble.
   (b) Strong wind can destroy buildings and uproot trees.
   (c) Gravity holds air near the earth's surface.
   (d) all of the above
   (e) none of the above

3. Which of the following has mass?
   (a) a magnetic field
   (b) an electric field
   (c) an electron
   (d) all of the above
   (e) none of the above
4. Both space travelers and earthlings who eat too much must "watch their weight". As a physicist defines mass and weight, the proper expression would be "watch their *mass*". Using the space traveler as an example, explain why it is *mass* that concerns us rather than weight.

Appendix J

CONCEPT ANALYSIS FOR *CHEMICAL SYMBOL*
FIRST ATTEMPT

DEFINITION
A *chemical symbol* is an abbreviated notation for any element. The abbreviated notation is the only one of its kind for a given element.

CRITICAL ATTRIBUTES
C1. The first letter of the name of the element is capitalized.
C2. Lower-case letters following the capital letter are used to distinguish between two or more elements when their names begin with the same letter.
C3. No symbol contains more than two letters. (No longer true.)

VARIABLE ATTRIBUTES
V1. Size of letters.
V2. Physical state of the element represented.
V3. The amount of the element represented.

SUPRAORDINATE CONCEPT: symbolic representation
COORDINATE CONCEPTS: formula, equation
SUBORDINATE CONCEPTS: iron, oxygen, chlorine, lead

Examples	*Nonexamples*
1. He	1. he
2. H	2. Hyd
3. C	3. cu
4. Cu	4. SO
5. Na	5. So

SECOND ATTEMPT

DEFINITION

A *chemical symbol* is a letter or pair of letters used as an abbreviated notation for an element.

CRITICAL ATTRIBUTES

C1. A single letter or pair of letters.
C2. Represents a single element.
C3. Does not represent a molecule of the element. (It must be pointed out that *molecule* is not a well-defined concept. Some chemists describe a single atom of argon or helium as a molecule of those elements. If that concept of molecule is accepted, this is not a critical attribute of "symbol". The concept of *molecule* held by other chemists, including this author, has as a critical attribute, "a microscopic aggregate of two or more atoms.")

VARIABLE ATTRIBUTES

V1. The element represented may or may not exist.
V2. The symbol may or may not be generally accepted.

SUPRAORDINATE CONCEPT: symbolic representation
COORDINATE CONCEPTS: formula, equation
SUBORDINATE CONCEPTS: iron, oxygen, chlorine, lead

Examples
1. H (V1, V2)
2. O (V1, V2)
3. Li (C1)
4. Wo (symbol of a nonexistent element; V2)
5. Br (V1, V2)

Nonexamples
1. hydrogen (C1)
2. O_2 (C3)
3. Li^{3+} (C1)
4. H_2O (C2)
5. :Br. (C1)

Appendix K

CONCEPT ANALYSIS FOR *MELTING*

DEFINITION

Melting is the change of state from solid to liquid.

CRITICAL ATTRIBUTES

C1. The solid form of the substance decreases and the liquid form increases.
C2. The chemical composition of the substance is not changed.

VARIABLE ATTRIBUTES

V1. Whether the substance is pure or impure.
V2. Whether the substance is an element or a compound.

SUPRAORDINATE CONCEPT: physical state
COORDINATE CONCEPTS: boiling, freezing, dissolving, condensing
SUBORDINATE CONCEPTS: none

Examples
1. ice left at room temperature (V1,V2)
2. wax placed over heat (V1,V2)
3. sulfur placed over heat (V2)
4. butter in hot water (V1,V2)

Nonexamples
1. water left at –10 °C (C1)
2. wax warmed in hand until pliable (C1)
3. dry ice at room temperature (C1)
4. sugar cube in hot water (C1,C2)

Appendix L

SELECTED PIAGETIAN TASKS

CONSERVATION OF NUMBER

Make a line of 7 or 8 blue poker chips (or similar objects) and ask the child to use red chips to form a line that has just as many in it. (The child will normally make a one-to-one correspondence matching the line you have made.) After the child is satisfied that there are the same number of red and blue chips, spread out one of the rows and ask, "Now are there more red chips, more blue chips, or the same amount?"

If the child responds that there are more of one color than another, ask, "If these were pieces of candy wrapped in red and blue paper, which would you rather have, the red candies or the blue ones?" (After response) "Why?"

Children under the age of 5 or 6 apparently perceive that spreading out the pieces changes the number present. Older children readily accept that number is conserved during the transformation.

CONSERVATION OF SUBSTANCE

Form several balls from modeling clay and place them before the child. Ask the child to select two that have the same amount of clay. After being sure that the child is convinced that the two balls contain the same amount, discard the others. Give the child one of the two selected while you take the other. Roll your ball into a long sausage or flatten it into a pancake and ask, "Now do you have more clay, do I have more, or do we have the same amount?"

If the child responds that they are different, ask, "If this were chocolate candy, which piece would you rather have?"

Children under the age of 6 or 7 apparently believe that the amount of substance changes when the shape is deformed. Older children readily accept that amount is conserved during the transformation. (This task is easily altered to focus on conservation of mass rather than substance. Similar results are obtained, but conservation responses appear at a later age.)

DISPLACEMENT VOLUME

Make identical cylinders of aluminum and iron (or any other metal of greater density) and tie strings to them so that they can be lowered into a graduated cylinder partially filled with water. Let the child examine the metal cylinders and compare their size and weight.

After the child is convinced that the metal cylinders are of equal size but unequal weight, have the child compare the water level in two graduated cylinders. When the child is convinced that the water level in the two cylinders is the same, lower the aluminum cylinder into one of the graduated cylinders and have the child observe the new water level. Now say, "I want you to lower the other metal cylinder into this other graduated cylinder, but before you do, predict how high the water will rise. Will it be higher than it is in this one (pointing to the graduated cylinder containing the aluminum cylinder), lower, or the same height?"

After the prediction is made, have the child lower the metal into the graduated cylinder and check his or her prediction. If the prediction is not confirmed, ask the child if he or she can explain why.

Children under 12–14 normally predict that the heavier metal will cause the water to rise higher. Occasionally they assert that it *did* rise higher, after making the observation. They are unable to explain the observation that the levels are the same.[1]

COMBINATORIAL REASONING

Prepare the following solutions and place them in bottles labeled A, B, C, D, and g:

- A, a dilute solution of hydrogen peroxide;
- B, pure water;
- C, a dilute solution of sulfuric acid;
- D, a solution of $Na_2S_2O_7$; and
- g, a solution of KI in a dropper bottle.

Before bringing the child into the room, place water in one beaker and a mixture of A and C in another. Now bring the child into the room and show him the bottles of solution as well as the two beakers of liquid. Say, "The liquid in each of these beakers came from those bottles labeled A, B, C, and D. Now I want you to watch what happens when I place a few drops of liquid from this bottle labeled g into each of these beakers." (The water remains colorless and the other liquid turns yellow.)

After the child is clear about what has happened, say, "Now I'm going to pour out the liquid in the beakers and I want you to see if you can reproduce the yellow color."

Answer any questions that the child has about the procedure, and observe how the child proceeds. The observation of interest is how the child proceeds to check the various possibilities rather than ultimate success or failure.

Children under the age of 8–10 approach the task in random fashion. There is no apparent order to the liquids tried, and the child is unable to explain his or her success if it occurs by chance.

---

[1]It is not uncommon for older children and adults to predict that the heavier object will push the water level higher, but they immediately correct themselves when they see the result and explain that only the volume matters. There is some disagreement about how to interpret this result. My predilection is to assume that the original response is a casual prediction made without careful thought and that the explanation after seeing the final result is evidence of reasoning that assumes conservation.

Children of 10–12 try each of the liquids (A, B, C, and D) alone, but when they are unsuccessful, they are unable to proceed. If prodded to try two at a time, the combinations are not made systematically, and the child may be unable to explain his or her eventual success. If asked, "Are there any other possibilities for getting the yellow color?" the child is unable to answer.

After age 12–16, the various possibilities are tried systematically. When the successful combination is found, the child can explain what is required to produce the color. If asked, "Are there any other possibilities for getting the yellow color?" the child will proceed to try those combinations (including three at a time and four at a time) that have not been tried. (These combinations may be tried without any prompting if it was clear from the beginning that *all* possibilities for obtaining the yellow color should be sought.) The child will report the other combination (A, B, and C) that will produce a yellow color when g is added and will be quite confident that no other possibilities exist.[2]

PENDULUM TASK

Provide a pendulum support, string, and various weights for bobs. Demonstrate the pendulum by hanging a weight on the string, adjusting the string to a convenient length, pulling the bob back and releasing it. After the demonstration, invite the child to manipulate the pendulum and identify all of the things that could be changed about it. (The child should identify the weight hung on the end, the length of the pendulum, and the angle from which the bob is released.) If the child fails to identify any of the variables, focus on that variable by asking, "What about _____? Could we change that as well?"

After you are confident that the child has identified all of the variables that might be manipulated and that he knows how to manipulate them, say, "Now I want you to experiment with these things and see if you can find out what affects how fast the pendulum swings back and forth."

Observe the child as he or she experiments. The observations of interest are whether the child keeps constant all of the variables except the one whose effect is being tested, and whether he or she tests the effect of all variables.

Children under the age of 12–14 appear not to appreciate the logical necessity of "all other things being equal" and do not control variables as they do the experiment. Older children normally do.

---

[2]I have interviewed able subjects who appear to make tests at random but have no difficulty telling me which combinations produce the yellow color and confidently state that they have tried all possible combinations. I conclude that they use a systematic procedure that differs from the one I expected.

Appendix M

RECENT RESEARCH ON PIAGET'S STAGE THEORY

Piaget spent more than 50 years observing children from birth to adolescence to see how they responded to a variety of tasks. On the basis of those observations, Piaget developed a theory about how the intellect develops. At the heart of Piaget's theory stand the complementary processes of assimilation and accommodation through which knowledge schemas are constructed. This theory, as it is conceived today, was described in Chapter 5. It remains, along with information processing theory, one of the major theories driving research in developmental psychology and cognitive science in general.

In addition to describing how children construct knowledge, Piaget described tasks that are successfully completed at various ages, and he developed a theory to explain these age-related differences. In doing so, he tried to separate the effects of declarative knowledge such as word meaning, conceptual knowledge, and propositional knowledge, and the effects of what he called *logical operations*. Piaget used operational knowledge as the basis for his stage theory.

Appendix L describes a few of the tasks that Piaget presented to children at various ages. On the basis of their response to those tasks, he inferred which logical operations children used to interpret events in their environments. Eventually, Piaget organized his description of development into four general stages: *sensorimotor* (birth to approximately 18 months), *preoperational* (approximately 18 months to 6 years), *concrete operational* (approximately 6–12 years), and *formal operational* (approximately 12 years through adulthood). Each stage was characterized by certain operational knowledge, and developing the operational knowledge of each stage depended on first developing the logical operations that characterized the previous stage. In other words, the stages form a hierarchy.

The stage that concerns us most is formal operations, and it is discussed in Chapter 14. If you want to know more about Piaget's work, short synopses can be found in most books on child development or educational psychology. The annotated bibliography in Appendix N includes some of the more lucid books summarizing his work.

FACTORS THAT AFFECT RESPONSES TO PIAGETIAN TASKS

Piaget's tasks have been administered to thousands of children and adults throughout the world. The results of this testing in different cultures generally have confirmed Piaget's observations concerning the relative difficulty of his tasks; that is, the order in which individuals accomplish the various tasks appears to be the same in all populations. However, the age at which various tasks are accomplished varies considerably. In some cultures, all adults fail the more difficult (formal operations) tasks used by Piaget, and in Western cultures a substantial portion of adults fail many of the tasks.

The 1971 study in which McKinnon and Renner administered several of the Piagetian tasks to college students and reported that only about half were "formal operational" was mentioned in Chapter 14. A number of others reported similar findings (Elkind, 1962; Tomlinson-Keasey, 1972; Towler and Wheatley, 1971). However, these results do not pertain to all groups of college age. Results from a written test of logical operations suggest that a high proportion of students in a chemistry course for science and engineering majors use formal operations reasoning (Bodner, personal communication, 1985; Ward, Nurrenbern, Lucas, and Herron, 1981).

Why is there so much variation in the performance of adults on tasks like those shown in Appendix L? It is difficult to see how lack of familiarity with materials used in the tasks could explain such differences. However, research reported by Cole (1977) and Cole and Scribner (1974) leaves little doubt about the important influence of unfamiliar materials on performance of such tasks in non-Western cultures. Several researchers have found that "primitive" people who are unsuccessful on Piagetian tasks presented with Western apparatus are successful when "native" materials are substituted. Inferences that are obvious to Western adults are simply not made by those who are strangers to the apparatus.

The problem is illustrated nicely by a study described by Donaldson (1978). In the original study by Kendler and Kendler (1967), young children were given the task of learning to operate a machine to get a toy. Children first had to press a button to get a marble and then insert the marble into a hole to release the toy. Children learned how to insert a marble into the hole and they learned how to press the button to get a marble, *but that was insufficient for them to solve the task.*

As Donaldson pointed out, adults are likely to infer that *any* marble (including the one obtained by pressing the button) will serve to release the toy, but it does not follow that young children who have little knowledge of such things will necessarily make the same kind of inference. Hewson (1977) modified the Kendler experiment to teach the functional equivalence of marbles. This modification produced significant improvement in the performance of 4- and 5-year-olds (but not 3-year-olds) on the task.

As part of research conducted shortly after the Russian Revolution but not published in English until 1976, Luria presented adults with objects of varying hues and asked them to place the objects in groups of the same color. The uneducated adults refused by explaining that it was impossible to do. All were different, they explained.

Luria stressed the importance of culture in the development of reasoning and discussed his results in terms of cultural influences:

> [L]anguage scientists all note that the absence of special names for groups of colors, or the presence of a large number of subcategories for other colors, is due not to the physiological peculiarities of color perception but to the influence of culture: the "interest" people have in certain colors and lack of interest

in others. . . . For example, many languages of people living near the Arctic contain dozens of terms for shades of white (expressions for referring to different types of snow—a fact of practical importance), whereas hues of red and green—of no special importance—are lacking in their vocabulary (Luria, 1976; p 23).

This cross-cultural research clearly has implications for anyone exporting Western science to developing countries, but it also has implications for every science teacher. Science is an unfamiliar culture for many students. Inferences that are perfectly obvious to those who have been working in the field for many years are not obvious to beginners, regardless of their intelligence (*see*, for example, Tobias, 1990). Responses that appear to reflect faulty reasoning and lack of intellectual development are often due to lack of familiarity with the world of the scientist. Many of these problems were discussed in Chapter 13 in connection with language.

Joe Novak's conclusion (Chapter 14, pp 184–185) "that children have . . . cognitive abilities . . . far beyond those suggested by narrow interpretations of Piaget's work" is based on studies showing that alternate forms of some Piagetian tasks are 'passed' by much younger children than the ones who passed the original task. One example is found in tests of class inclusion. In a typical Piagetian task, an experimenter first establishes that a bunch of 16 flowers consisting of 10 red ones and 6 blue ones are "all flowers". Then the child is asked, "Are there more red flowers or more flowers?" Primary schoolchildren usually compare the subclasses (red flowers with blue flowers) rather than compare the subclass with the superordinate class (flowers) and reply "More red flowers." However, as Flavell, Miller, and Miller (1993) pointed out:

> [D]ifferent forms of class inclusion vary greatly in difficulty, and the surprising failure of older children [and even college students—*see* Rabinowitz, Howe, and Lawrence, 1989] on some versions must be set against the precocious success of much younger children on other versions. A striking demonstration in the latter category has been reported by C. L. Smith (1979), who showed that 4-year-olds can sometimes make valid inferences based on class inclusion representations. . . . The message from this discussion is that class inclusion is not the sort of knowledge that a child either "has" or "does not have". Rather, there clearly are different senses and different degrees "having", depending on the task and the response measure in question (pp 94–95).

The inconsistency in children's responses to questions that adults view as logically consistent has puzzled cognitive scientists. As Flavell, Miller, and Miller put it:

> Exactly how do the minds of older children, adolescents, and adults differ from those of young children? Despite the existence of thousands of research studies comparing the cognitive performance of early-childhood and older subjects, we still lack a wholly satisfactory answer to this basic question. *It is not that these studies fail to show marked age differences in cognitive performance; they almost always do. Rather, the problem is to know how best to describe and explain the age differences found* (p 132; emphasis added).

POSSIBLE EXPLANATIONS FOR PERFORMANCE DIFFERENCES

One of the most thorough discussions of research pertaining to Piaget's stage theory was provided by Susan Carey (1985). She considered five hypotheses that might explain differences in performance and concluded that "on two interpretations of 'think differently' . . . considered judgment dictates that young children and sophisticated adults think alike" (p 514). However, this conclusion should not be misinterpreted. Carey, like Flavell and the Millers, left no doubt that there are important differences in the way children and adults respond to a wide variety of tasks. After discussing the research on class inclusion tasks, Carey stated, "These results leave no doubt that the young child differs from the adult in ability to impose inclusion hierarchies on new materials and in ability to make various deductive inferences that depend on inclusion" (p 493).

After reviewing experiments designed to reveal the underlying *cause* of these differences in performance, Carey concluded:

> [T]here is no compelling evidence that the child's *basic representational format* [emphasis added] differs from the adult's in type of concepts, capacity to represent class-inclusion hierarchies, or ability to recognize at least some quantitative and deductive consequences of inclusion. Nonetheless, in many different situations the child fails to deal with classes and class-inclusion hierarchies as would an adult (p 495).

Carey reaches similar conclusions in regard to the question of whether the difference between adult and child performance is due to differences in ability to entertain and test hypotheses:

> We must decide whether the failures . . . provide evidence that younger children cannot entertain and evaluate hypotheses. I believe that they do not, for the simple reasons that [Piaget's tasks] confound knowledge of particular scientific concepts with scientific reasoning more generally. It is well documented . . . that before ages 10 or 11 or so the child has not fully differentiated weight, size, and density and does not have a clear conception of what individuates different kinds of metals. . . . If these concepts are not completely clear in the child's mind, due to incomplete scientific knowledge, then the child will of course be unable to separate them from each other in hypothesis testing and evaluation (p 498).

In discussing metaconceptual development Carey adds:

> That there is metaconceptual development during childhood . . . is beyond doubt. The child's conception of thought, language, memory, and learning change with age. . . . It remains to be demonstrated, however, just *how* metaconceptual change affects learning in other domains. For example, how does having the concepts *hypothesis, experiment, [and] confirmation* consciously available actually affect inductive reasoning? (p 501).

Carey acknowledged that important reorganizations of knowledge take place over time. In discussing children's answers to questions about animals, she stated:

> I interpret these changes as reflecting reorganization of knowledge about animal properties; for 4- to 7-year-olds these properties are primarily organized in terms of the children's knowledge of human activities. By age 10 they are organized in terms of biological function. Presumably, the main impetus for this reorganization is the acquisition of biological knowledge in school (p 507).

It is clear throughout Carey's article that the hypothesis she favors as an explanation for differences in child and adult performance is differences in domain-specific knowledge. However, she is not so naive as to imply that this is a problem that is easily overcome by simply telling children "the truth" about things:

> Let me explicitly dispense with the implicit *mere* in the second proposition, that children differ from adults merely in the accumulation of knowledge. There are

hosts of unsolved problems concerning the acquisition of knowledge. . . . I know of two bodies of literature specifically concerned with the acquisition of domain-specific knowledge. . . . [B]oth agree that the acquisition of domain-specific knowledge cannot be thought of as the mere accumulation of facts. Both groups emphasize the reorganization that is a crucial part of the process. . . . Most important is the emergence of higher order concepts. . . . Also important is the enrichment of connections . . . among the concepts that articulate the domain, giving the domain stability and inferential power (pp 512–513).

Jerome Bruner (1985) reinforced Carey's view that it is not *mere* information that is at work:

There may not be Great Big Stages, each with its own unique structure, but growth *is* characterized by *structural* changes. There are *paradigm* shifts in children's developing theories of the world, and although these may not be reducible to massive stage changes in basic axioms, neither are they matters of the simple accretion of information (p 600).

Others recommended caution in accepting Carey's favored hypothesis that performance of younger and older children can be explained by increases in domain-specific knowledge:

Our own intuitions are that the acquisition-of-expertise model will *not* account for all of cognitive growth. As Markham (1979) points out, when adults are in a novel situation they know a great deal more than do children about how to move quickly from novice to expert status. They are experts at becoming experts. They quickly detect what it is they do not understand, have more potential solutions in their cognitive bank from which to draw, and more easily see similarities between the current situations and other previously encountered situations (Flavell, Miller, and Miller, p 146).

[Carey] suggests . . . that the development of metacognitive abilities makes it possible for the child to apply classification abilities to the task of organizing newly acquired knowledge. Whether this is correct needs to be studied. But it is clear that metacognitions come as a function of development (at least in this culture), and they do not seem to be available to very young children (Gelman, 1985, pp 538–539).

I agree with Gelman's account . . . of cognitive development: the older child tends to be more explicit, has a better grasp of how to access and manipulate structures and routines in order to get a job done or a problem solved (Bruner, 1985, p 602).

As you can see from this short review, many issues pertaining to Piaget's stage theory remain unresolved. Until we have a better understanding of what is different between children's thinking and that of adults, it will be difficult to learn when various science concepts should be introduced or how they should be presented to maximize learning.

Appendix N

BOOKS RELATED TO INTELLECTUAL DEVELOPMENT

Flavell, J. H. (1963). *The developmental psychology of Jean Piaget*. Princeton, NJ: Van Nostrand.

    This is one of the earlier books on Piaget's work published in English. It is still an excellent reference, but it requires effort to get through it. An obvious limitation of this book is that it does not discuss recent research related to Piaget's work.

Flavell, J. H., Miller, P., & Miller, S. (1993). *Cognitive development* (3rd ed.). Englewood Cliffs, NJ: Prentice-Hall.

    This book is heavily influenced by Piaget's work, but it is influenced by other research as well. It provides information from many sources, and it is easier reading than Flavell's 1963 book. It reflects findings from recent research.

Furth, H. G. (1981). *Piaget and knowledge: Theoretical foundations* (2nd ed.). Chicago: University of Chicago.

    This book deals exclusively with Piaget's work and is generally more readable than Flavell's 1963 book. Furth has also written books specifically for teachers.

Ginsburg, H. & Opper, S. (1988). *Piaget's theory of intellectual development: An introduction* (3rd ed.). Englewood Cliffs, NJ: Prentice-Hall.

    As the title suggests, this book is written for those who want to know about Piaget's work, but not too much. It remains a popular introduction to Piaget.

Inhelder, B. & Piaget, J. (1958). *The growth of logical thinking from childhood to adolescence*. New York: Basic Books.

    This book describes formal operations and is the best book to read if you want to know what Piaget means by the term. The theoretical material can be skipped, and the rest of the book is very readable.

Karplus, R., Lawson, A., Wollman, W., Appel, M., Bernoff, R., Howe, A., Rusch, J., & Sullivan, F. (1977). *Science teaching and the development of reasoning*. Berkeley, CA: University of California.

This material was developed to present Piaget's theory of intellectual development to teachers. It provides an overview of the theory plus examples from textbooks and laboratory manuals to clarify the theory. There are separate books for biology, chemistry, and physics.

Lawson, A. (Ed.). (1979b). *The psychology of teaching for thinking and creativity.* Columbus, OH: ERIC Clearinghouse for Science, Mathematics, and Environmental Education.

This book contains several useful articles describing Piagetian and neo-Piagetian theories of development. It is readable.

References

The following references were cited in the book. A list of additional references follows this list; these additional references were not cited in the book, but they are useful references.

Abraham, M., & Renner, J. (1985). The sequence of learning cycle activities in high school chemistry. *Journal of Research in Science Teaching, 22*, 121–143.

Adams, M. (1980). Failures to comprehend and levels of processing in reading. In R. Spiro, B. Bruce, & W. Brown (Eds.), *Theoretical issues in reading comprehension* (pp 11–32). Hillsdale, NJ: Erlbaum.

Al-Kunifed, A., Good, R., & Wandersee, J. (1993, April). *Investigation of high school chemistry students' concepts of chemical symbol, formula, and equation: Students' prescientific conceptions.* Paper presented at the 66th Annual Meeting of the National Association for Research in Science Teaching, Atlanta, GA.

Albert, E. (1978). Development of the concept of heat in children. *Science Education, 62*, 389–399.

Allport, G. W. (1954). *The nature of prejudice.* Cambridge, MA: Addison-Wesley.

Amundson, K. J. (1991). *Teaching values and ethics.* Arlington, VA: American Association of School Administrators.

Anamuah-Mensah, J. (1986). Cognitive strategies used by chemistry students to solve volumetric analysis problems. *Journal of Research in Science Teaching, 23*, 759–769.

Anderson, J. (1980). *Cognitive psychology and its implications.* San Francisco: Freeman.

Anderson, R. D. (1965). Children's ability to formulate mental models to explain natural phenomena. *Journal of Research in Science Teaching, 3*, 326–332.

Anderson, T. (1980). Study strategies and adjunct aids. In R. J. Spiro, B. C. Bruce, & W. F. Brewer (Eds.), *Theoretical issues in reading comprehension* (pp 483–502). Hillsdale, NJ: Erlbaum.

Andersson, B. (1980). Some aspects of children's understanding of boiling point. In U. A. Archenhold (Ed.), *Cognitive development research in science and mathematics* (pp 252–259). Proceedings of an International Seminar, The University of Leeds, Leeds, England.

Andersson, B. (1986). Pupils' explanations of some aspects of chemical reactions. *Science Education, 70*, 549–563.

Arons, A. (1984). Education through science. *Journal of College Science Teaching, 13*, 210–220.

Aronson, E. (1978). *The jigsaw classroom.* Beverly Hills, CA: Sage.

Atkin, J. M. (1968). Behavioral objectives in curriculum design: A cautionary note. *The Science Teacher, 35*, 27–30.

Atkin, J. M., & Karplus, R. (1962). Discovery or invention. *The Science Teacher, 29*(5), 45–51.

Ausubel, D. (1963). *The psychology of meaningful verbal learning.* New York: Grune and Stratton.

Baker, E. (1969). Effects on student achievement of behavioral and non-behavioral objectives. *Journal of Experimental Education, 37*(4), 5–8.

Barclay, J. (1973). The role of comprehension in remembering sentences. *Cognitive Psychology, 4*, 229–254.

Barke, H. (1982). Probleme bei der verwendung von symbolen im chemie unterricht: Eine empirische untersuchung an schulern der sekundarstufe 1 [Problems associated with the application of symbols in chemistry teaching: An empirical study of secondary schoolchildren]. *Naturwissenschaften im Unterricht Physik/ Chemie, 30*(4), 131–133.

Barnes, D. (1976). *From communication to curriculum.* Harmondsworth, Middlesex, England: Penguin Books.

Bartlett, F. (1932). *Remembering: A study in experimental and social psychology.* Cambridge, England: Cambridge University Press.

Ben-Zvi, R., Eylon, B., & Silberstein, J. (1986). Is an atom of copper malleable? *Journal of Chemical Education, 69*, 64–66.

Beyerbach, B., & Smith, J. (1990). Using a computerized concept mapping program to assess preservice teachers' thinking about effective teaching. *Journal of Research in Science Teaching, 27*, 961–971.

Bleichroth, W. (1965). Was wissen unsere volksschulkinder vom atom? [What do our elementary schoolchildren know about the atom?] *Zeitschrift für Naturlehre und Naturkunde, 4*, 89–94.

Bloom, B. (1971). Mastery learning. In J. W. Block (Ed.), *Mastery learning: Theory and practice* (pp 47–63). New York: Holt, Rinehart and Winston.

Blosser, P. (1980). *A critical review of the role of the laboratory in science teaching.* Columbus, OH: ERIC Clearinghouse for Science, Mathematics, and Environmental Education. (ERIC Document Reproduction Service No. ED 206 445)

Bodner, G. M. (1991). I have found you an argument: The conceptual knowledge of beginning graduate students. *Journal of Chemical Education, 68*, 385–388.

Boikess, R. S., & Edelson, E. (1978). *Chemical principles.* New York: Harper and Row.

Boyd, C. (1966). A study of unfounded beliefs. *Science Education, 50*, 396–398.

Bransford, J. (1979). *Human cognition: Learning, understanding and remembering.* Belmont, CA: Wadsworth Publishing Company.

Bransford, J., Arbitman-Smith, R., Stein, B., & Vye, N. (1985). Improving thinking and learning skills: An analysis of three approaches. In J. Segal, S. Chipman, & R. Glaser (Eds.), *Thinking and learning skills: Vol. 1. Relating instruction to research* (pp 133–206). Hillsdale, NJ: Erlbaum.

Bransford, J., & McCarrell, N. (1974). A sketch of a cognitive approach to comprehension: Some thoughts about understanding what it means to comprehend. In

W. B. Weimer & D. S. Palermo (Eds.), *Cognition and the symbolic processes* (pp 189–229). Hillsdale, NJ: Erlbaum.

Bransford, J., Sherwood, R., Kinzer, C., & Hasselbring, T. (1985). *Havens for learning: Toward a framework for developing effective uses of technology.* (Technical report no. 85.1.1). Nashville, TN: Vanderbilt University, Learning Technology Center.

Briscoe, C., & LaMaster, S. (1991). Meaningful learning in college biology through concept mapping. *American Biology Teacher, 53,* 214–219.

Brown, A. (1980). Metacognitive development and reading. In R. J. Spiro, B. C. Bruce, & W. F. Brewer (Eds.), *Theoretical issues in reading comprehension* (pp 453–481). Hillsdale, NJ: Erlbaum.

Brown, J. S., Collins, A., & Duguid, P. (1989). Situated cognition and the culture of learning. *Educational Researcher, 18*(1), 32–41.

Brown, J. S., Collins, A., & Harris, G. (1978). Artificial intelligence and learning strategies. In H. F. O'Neil (Ed.), *Learning strategies* (pp 107–139). New York: Academic Press.

Bruner, J. (1985). On teaching thinking: An afterthought. In S. Chipman, J. Segal, & R. Glaser (Eds.), *Thinking and learning skills: Vol. 2. Research and open questions* (pp 597–608). Hillsdale, NJ: Erlbaum.

Bruner, J. S., & Postman, L. (1949). On the perception of incongruity: A paradigm. *Journal of Personality, 18,* 206–223.

Buell, R., & Bradley, G. (1972). Piagetian studies in science: Chemical equilibrium understanding from study of solubility: A preliminary report from secondary school chemistry. *Science Education, 56,* 23–29.

Bunce, D. M., & Heikkinen, H. (1986). The effects of an explicit problem solving approach on mathematical chemistry achievement. *Journal of Research in Science Teaching, 23,* 11–20.

Cachapuz, A., & Martins, I. (1987). High school students' ideas about energy of chemical reactions. *Proceedings of the Second International Seminar on Misconceptions and Educational Strategies in Science and Mathematics: Vol. III* (pp 60–68). Ithaca, NY: Cornell University.

Cantu, L., & Herron, J. D. (1978). Concrete and formal Piagetian stages and science concept attainment. *Journal of Research in Science Teaching, 15,* 135–143.

Caramaza, A., McCloskey, M., & Green, B. (1981). Naive beliefs in "sophisticated" subjects: Misconceptions about trajectories of objects. *Cognition, 9,* 117–123.

Carey, S. (1985). Are children fundamentally different kinds of thinkers and learners than adults? In S. Chipman, J. Segal, & R. Glaser (Eds.), *Thinking and learning skills: Vol. 2. Research and open questions* (pp 485–517). Hillsdale, NJ: Erlbaum.

Carroll, J. (1963). A model of school learning. *Teachers College Record, 64,* 723–733.

Carter, C. (1988). The role of beliefs in general chemistry problem solving (Doctoral dissertation, Purdue University, 1987). *Dissertation Abstracts International, 49*(5), 1107-A. (University Microfilms No. DA8814459)

Case, R. (1972). Validation of a neo-Piagetian capacity construct. *Journal of Experimental Child Psychology, 14,* 287–302.

Case, R. (1975). Gearing the demands of instruction to the developmental capacities of the learner. *Review of Educational Research, 45,* 59–87.

Case, R. (1978a). Intellectual development from birth to adulthood: A neo-Piagetian interpretation. In R. S. Siegler (Ed.), *Children's thinking: What develops* (pp 37–71). Hillsdale, NJ: Erlbaum.

Case, R. (1978b). Piaget and beyond: Toward a developmentally based theory and technology of instruction. In R. Glaser (Ed.), *Advances in instructional psychology* (pp 167–228). Hillsdale, NJ: Erlbaum.

Case, R. (1979). Intellectual development and instruction: A neo-Piagetian view. In A. E. Lawson (Ed.), *1980 AETS Yearbook: The psychology of teaching for thinking and creativity* (pp 59–102). Columbus, OH: ERIC Clearinghouse for Science, Mathematics, and Environmental Education.

Case, R. (1985). A developmentally based approach to the problem of instructional design. In S. Chipman, J. Segal, & R. Glaser (Eds.), *Thinking and learning skills: Vol. 2. Research and open questions* (pp 545–562). Hillsdale, NJ: Erlbaum.

Cassels, J. R. T. (1976). *Language in chemistry: The effect of some aspects of language on 'O' grade chemistry candidates.* Unpublished master's thesis, University of Glasgow, Glasgow, Scotland.

Cassels, J. R. T. (1980). *Language and thinking in science: Some investigations with multiple-choice questions.* Unpublished doctoral dissertation, University of Glasgow, Glasgow, Scotland.

Cassels, J. R. T., & Johnstone, A. H. (1980). *Understanding of non-technical words in science.* London: Royal Society of Chemistry.

CBA, Chemical Bond Approach Project. (1964). *Investigating Chemical Systems.* St. Louis, MO: Webster Division, McGraw-Hill.

Chi, M. T. H., & Bassok, M. (1989). Learning from examples via self-explanations. In L. Resnick (Ed.), *Knowing, learning, and instruction: Essays in honor of Robert Glaser* (pp 251–282). Hillsdale, NJ: Erlbaum.

Chi, M. T. H., Feltovich, P. J., & Glaser, R. (1981). Categorization and representation of physics problems by experts and novices. *Cognitive Science, 5,* 121–152.

Chiappetta, E. (1976). A review of Piagetian studies relevant to science instruction at the secondary and college level. *Science Education, 60,* 253–261.

Clagett, M. (1967). *Giovanni Marliani and Late Medieval Physics* (Chapter 4). New York: Columbia University. (Original work published AMS Press 1941)

Cliburn, J. W. (1990). Concept maps to promote meaningful learning. *Journal of College Science Teaching, 19,* 212–217.

Clough, E.-E., & Driver, R. (1985). What do children think about pressure in fluids? *Research in Science and Technological Education, 3*(2), 133–144.

Clough, E.-E., & Driver, R. (1986). A study of consistency in the use of students' conceptual frameworks across different task context. *Science Education, 70*(4), 473–496.

Cognition and Technology Group at Vanderbilt. (1990). Anchored instruction and its relationship to situated cognition. *Educational Researcher, 19*(6), 2–10.

Cognition and Technology Group at Vanderbilt. (1992a). The Jasper experiment: An exploration of issues in learning and instructional design. *Educational Technology Research and Development, 40,* 65–80.

Cognition and Technology Group at Vanderbilt. (1992b). Designing learning environments that support thinking: The Jasper series as a case study. In M. Duffy, J. Wowyck, & D. Johassen (Eds.), *Designing environments for constructive learning* (pp 1–28). New York: Springer-Verlag.

Cognition and Technology Group at Vanderbilt. (1992c). The Jasper series as an example of anchored instruction: Theory, program description, and assessment data. *Educational Psychologist, 27*(3), 291–315.

Cognition and Technology Group at Vanderbilt. (1992d). Instruction in science and mathematics: Theoretical basis, developmental projects, and initial findings.

In R. Duschl & R. Hamilton (Eds.), *Philosophy of science, cognitive psychology, and educational theory and practice* (pp 244–273). New York: State University of New York Press.

Cognition and Technology Group at Vanderbilt. (1992e). Anchored instruction approach to cognitive skills acquisition and intelligent tutoring. In W. Regian & V. J. Shute (Eds.), *Cognitive approaches to automated instruction* (pp 135–170). Hillsdale, NJ: Erlbaum.

Cognition and Technology Group at Vanderbilt. (1993). The Jasper series: Theoretical foundations and data on problem solving and transfer. In L. A. Penner, G. M. Batsche, H. M. Knoff, & D. L. Nelson (Eds.), *The challenges of mathematics and science education: Psychology's response* (pp 113–152). Washington, DC: American Psychological Association.

Cole, M. (1977). An ethnographic psychology of cognition. In P. N. Johnson-Laird & P. C. Wason (Eds.), *Thinking: Readings in cognitive science* (pp 468–482). Cambridge, England: Cambridge University Press.

Cole, M., & Scribner, S. (1974). *Culture and thought: A psychological introduction.* New York: John Wiley & Sons.

Cole, S. (1992). *Making science: Between nature and society.* Cambridge, MA: Harvard University Press.

Collins, A., Brown, J. S., & Newman, S. (1989). Cognitive apprenticeship: Teaching the crafts of reading, writing, and mathematics. In L. B. Resnick (Ed.), *Knowing, learning, and instruction: Essays in honor of Robert Glaser* (pp 453–494). Hillsdale, NJ: Erlbaum.

Consalvo, R. (1969). Evaluation and behavioral objectives. *American Biology Teacher, 31,* 230–232.

Copes, J. (1981). *Hands-on chemistry: A laboratory manual.* New York: Random House.

Cosgrove, M., & Osborne, R. (1985). Lesson frameworks for changing children's ideas. In R. Osborne & P. Fryberg (Eds.), *Learning in science: The implications of children's science* (pp 101–111). Auckland, New Zealand: Heinemann.

Cowan, J. (1972). Student reaction to the use of detailed objectives. In K. Austwick & N. Harris (Eds.), *Aspects of educational technology VI* (pp 272–276). London: Sir Isaac Pitman & Sons.

Craik, F., & Lockhart, R. (1972). Levels of processing: A framework for memory research. *Journal of Verbal Learning and Verbal Behavior, 11,* 671–684.

Cros, D., & Maurin, M. (1986). Conceptions of first-year university students of the constituents of matter and the notion of acids and bases. *European Journal of Science Education, 8,* 305–313.

Dansereau, D. (1985). Learning strategy research. In J. Segal, S. Chipman, & R. Glaser (Eds.), *Thinking and learning skills: Vol. 1. Relating instruction to research* (pp 209–239). Hillsdale, NJ: Erlbaum.

de Leeuw, L. (1978). Teaching problem solving: The effect of algorithmic and heuristic problem-solving training in relation to task complexity and relevant aptitudes. In Lesgold, A., et al. (Eds.), *Cognitive psychology and instruction* (pp 269–275). New York: Plenum Press.

Deci, E. L., Vallerand, R. J., Pelletier, L. G., & Ryan, R. M. (1991). Motivation and education: The self-determination perspective. *Educational Psychologist, 26,* 325–346.

Dennis, W. (1957). Animistic thinking among college and high school students in the Near East. *Journal of Educational Psychology, 48,* 193–198.

Dewey, J. (1926). *Democracy and education: An introduction to the philosophy of education*. New York: Macmillan. (Original work published 1916)

Dimant, R. J., & Bearison, D. J. (1991). Development of formal reasoning during successive peer interactions. *Developmental Psychology, 27*, 277–284.

diSibio, M. (1982). Memory for connected discourse: A constructivist view. *Review of Educational Research, 52*, 149–174.

Doise, W., & Mugny, G. (1984). *The social development of the intellect*. (A. St. James-Emler, N. Emler, & D. Mackie, Trans.). New York: Pergamon. (Original work published 1981)

Donaldson, M. (1978). *Children's minds*. Glasgow, Scotland: Fontana/Collins.

Dooling, D., & Lachman, R. (1971). Effects of comprehension on retention of prose. *Journal of Experimental Psychology, 88*, 216–222.

Doran, R. (1972). Misconceptions of selected science concepts held by elementary school students. *Journal of Research in Science Teaching, 9*, 127–137.

Dornsife, C. (1992). *Beyond articulation: The development of Tech Prep programs*. Berkeley, CA: National Center for Research in Vocational Education, University of California at Berkeley. (Available from the National Center for Research in Vocational Education, Materials Distribution Service, Macomb, IL: Western Illinois University.)

Driscoll, D. (1978). More on the ionization quiz: Comments on ionization. *Journal of Chemical Education, 55*, 465.

Driver, R. (1981). Pupils' alternative frameworks in science. *European Journal of Science Education, 3*, 93–101.

Driver, R. (1983). *The pupil as scientist*. Milton Keynes, England: The Open University Press.

Driver, R., & Easley, J. (1978). Pupils and paradigms: A review of literature related to concept development in adolescent science students. *Studies in Science Education, 5*, 61–84.

Duchastel, P., & Merrill, P. (1973). The effects of behavioral objectives on learning: A review of empirical studies. *Review of Educational Research, 43*, 53–69.

Duncan, I., & Johnstone, A. (1973). The mole concept. *Education in Chemistry, 10*, 213–214.

Dunn, C. (1983). The influence of instructional methods on concept learning. *Science Education, 67*, 647–656.

Educational Policies Commission. (1966). *Education and the spirit of science*. Washington, DC: National Education Association.

Elkind, D. (1962). Quantity conceptions in college students. *Journal of Social Psychology, 57*, 459–465.

Ellerton, N., & Ellerton, H. (1987). Mathematics and chemistry problems created by students. *Proceedings of the Second International Seminar on Misconceptions and Educational Strategies in Science and Mathematics: Vol. III* (pp 131–136). Ithaca, NY: Cornell University.

Erickson, G. (1979). Children's conception of heat and temperature. *Science Education, 63*, 221–230.

Erickson, G. (1980). Children's viewpoints of heat: A second look. *Science Education, 64*, 323–336.

Eysenck, M. W. (1989). Human learning. In K. J. Gilhooly (Ed.), *Human and machine problem solving* (pp 289–315). New York: Plenum Press.

Fasching, J., & Erickson, B. (1985). Group discussions in the chemistry classroom and the problem-solving skills of students. *Journal of Chemical Education, 62,* 842–846.

Ferris, F. (1959). An achievement test report. *Science Teacher, 26,* 576–579.

Feuerstein, R., Miller, R., Hoffman, M., Rand, Y., Mintzker, Y., & Jensen, M. (1981). Cognitive modifiability in adolescence: Cognitive structure and the effects of intervention. *Journal of Special Education, 15,* 269–287.

Feuerstein, R., Rand, Y., & Hoffman, M. (1979). *The dynamic assessment of retarded performers.* Baltimore: University Park Press.

Feuerstein, R., Rand, Y., Hoffman, M., & Miller, R. (1980). *Instrumental enrichment.* Baltimore: University Park Press.

Finke, R. A., Ward, T. B., & Smith, S. M. (1992). *Creative cognition: Theory, research, and applications.* Cambridge, MA: The MIT Press.

Finley, F. N., & Stewart, J. (1982). Representing substantive structures. *Science Education, 66,* 593–611.

Fisher, K. M. (1990). Semantic networking: The new kid on the block. *Journal of Research in Science Teaching, 27,* 1001–1018.

Fisher, K. M., Faletti, J., Patterson, H., Thornton, R., Lipson, J., & Spring, C. (1990). Computer-based concept mapping: SemNet software: A tool for describing knowledge networks. *Journal of College Science Teaching, 19,* 347–352.

Flavell, J. H. (1963). *The developmental psychology of Jean Piaget.* Princeton, NJ: Van Nostrand.

Flavell, J. H. (1977). *Cognitive development.* Englewood Cliffs, NJ: Prentice-Hall.

Flavell, J. H. (1985). *Cognitive development* (2nd ed.). Englewood Cliffs, NJ: Prentice-Hall.

Flavell, J., Miller, P., & Miller, S. (1993). *Cognitive development* (3rd ed.). Englewood Cliffs, NJ: Prentice-Hall.

Frank, D. (1986). Implementing instruction to improve the problem-solving abilities of general chemistry students (Doctoral dissertation, Purdue University, 1985). *Dissertation Abstracts International, 47*(1), 141-A.

Frazer, M. (1982). Nyholm Lecture: Solving chemical problems. *Chemical Society Reviews, 11,* 171–190.

Frazer, M., & Sleet, R. (1984). A study of students' attempts to solve chemical problems. *European Journal of Science Education, 6,* 141–152.

Fuchs, H. (1987). Thermodynamics: A "misconceived" theory. *Proceedings of the Second International Seminar on Misconceptions and Educational Strategies in Science and Mathematics: Vol. III* (pp 160–167). Ithaca, NY: Cornell University.

Furth, H. G. (1981). *Piaget and knowledge: Theoretical foundations* (2nd ed). Chicago: University of Chicago Press.

Gabel, D. (1981). *Facilitating problem solving behavior in high school chemistry* (NSF Technical Report, RISE, SED, 79-20744). Bloomington, IN: Indiana University, School of Education. (ERIC Document Reproduction Service No. ED 210 192)

Gabel, D. (1993). Use of the particle nature of matter in developing conceptual understanding. *Journal of Chemical Education, 70,* 193–194.

Gabel, D., & Enochs, L. (1987). Different approaches for teaching volume and students' visualization ability. *Science Education, 71,* 591–597.

Gabel, D., & Samuel, K. (1986). High school students' ability to solve molarity problems and their analogue counterparts. *Journal of Research in Science Teaching, 23*, 165–176.

Gabel, D., & Sherwood, R. (1980). Effect of using analogies on chemistry achievement according to Piagetian level. *Science Education, 64*, 709–716.

Gabel, D., & Sherwood, R. (1984). Analyzing difficulties with mole-concept tasks by using familiar analogue tasks. *Journal of Research in Science Teaching, 21*, 843–851.

Gabel, D., Sherwood, R., & Enochs, L. (1984). Problem-solving skills of high school chemistry students. *Journal of Research in Science Teaching, 21*, 221–233.

Gagné, R. (1977). *The conditions of learning* (3rd ed.). New York: Holt, Rinehart and Winston.

Gardner, M., Greeno, J., Reif, F., Schoenfeld, A., diSessa, A., & Stage, E. (Eds.). (1990). *Toward a scientific practice of science education* (pp 31–54). Hillsdale, NJ: Erlbaum.

Gardner, P. L. (1972). *Words in science*. Melbourne, Australia: Australian Science Education Project.

Geddis, A., & Jaipal, K. (1993, April). *Chemical equilibrium: A case study in teaching for understanding*. Paper presented at the 66th Annual Meeting of the National Association for Research in Science Teaching, Atlanta, GA.

Gelman, R. (1985). The developmental perspective on the problem of knowledge acquisition: A discussion. In S. Chipman, J. Segal, & R. Glaser (Eds.), *Thinking and learning skills: Vol. 2. Research and open questions* (pp 537–544). Hillsdale, NJ: Erlbaum.

Gennaro, E. (1981). Assessing junior high students' understanding of density and solubility. *School Science and Mathematics, 81*, 399–404.

Gilbert, J., Watts, D., & Osborne, R. (1982). Students' conceptions of ideas in mechanics. *Physics Education, 17*, 62–66.

Ginsburg, H., & Opper, S. (1988). *Piaget's theory of intellectual development* (3rd ed.). Englewood Cliffs, NJ: Prentice-Hall.

Glassman, S. (1967). High school students' ideas with respect to certain concepts related to chemical formulas and equations. *Science Education, 51*, 84–103.

Glasson, G., & Lalik, R. (1993). Reinterpreting the learning cycle from a social constructivist perspective: A qualitative study of teachers' beliefs and practices. *Journal of Research in Science Teaching, 30*, 187–207.

Gleit, C., & Ellington, G. (1978). Performance objectives in college chemistry. *Journal of College Science Teaching, 7*, 175–178.

Goetz, E., & Armbruster, B. (1980). Psychological correlates of test structure. In R. Spiro, B. Bruce, & W. Brewer (Eds.), *Theoretical issues in reading comprehension* (pp 201–220). Hillsdale, NJ: Erlbaum.

Gorodetsky, M., & Gussarsky, E. (1986). Misconceptions of the chemistry equilibrium concept as revealed by different evaluation methods. *European Journal of Science Education, 8*, 427–441.

Gorodetsky, M., & Gussarsky, E. (1987). The roles of students and teachers in misconceptualization of aspects in "chemical equilibrium." *Proceedings of the Second International Seminar on Misconceptions and Educational Strategies in Science and Mathematics: Vol. III* (pp 187–193).

Gorodetsky, M., & Hoz, R. (1985). Changes in group cognitive structure of some chemical equilibrium concepts following a university course in general chemistry. *Science Education, 69,* 185–199.

Gray, P., & Chanoff, D. (1986). Democratic schooling: What happens to young people who have charge of their own education? *American Journal of Education, 94,* 182–213.

Greenbowe, T. (1984). An investigation of variables involved in chemistry problem solving (Doctoral dissertation, Purdue University, 1983). *Dissertation Abstracts International, 44,* 3651-A. (University Microfilms No. DA8407543)

Greenbowe, T., Herron, J. D., Lucas, C., Nurrenbern, S., Staver, J., & Ward, C. (1981). Teaching preadolescents to act as scientists: Replication and extension of an earlier study. *Journal of Educational Psychology, 73,* 705–711.

Grubb, W., Davis, G., Lum, J., Pihal, J., & Morgaine, C. (1991). *The cunning hand, the cultured mind: Models for integrating vocational and academic education.* Berkeley, CA: National Center for Research in Vocational Education.

Haberman, M. (1978). Behavioral objectives: Bandwagon or breakthrough. *The Journal of Teacher Education, 19*(1), 91–94.

Hackling, M., & Garnett, P. (1985). Misconceptions of chemical equilibrium. *European Journal of Science Education, 7,* 205–214.

Hakerem, G., Dobrynina, G., & Shore, L. (1993, April). *The effect of interactive, three-dimensional, high-speed simulations on high school science students' conceptions of the molecular structure of water.* Paper presented at the 66th Annual Meeting of the National Association for Research in Science Teaching, Atlanta, GA.

Hall, J. (1973). Conservation concepts in elementary chemistry. *Journal of Research in Science Teaching, 10,* 143–146.

Harlen, W. (1968). The development of scientific concepts in young children. *Educational Research, 11,* 4–13.

Haupt, G. (1952). Concepts of magnetism held by elementary schoolchildren. *Science Education, 36,* 162–168.

Hayes, J. (1981). *The complete problem solver.* Philadelphia: The Franklin Institute Press.

Haywood, H. C. (Ed.). (1992). Special issue: Interactive assessment. *Journal of Special Education, 26*(3), 233–335.

Haywood, H. C., & Tzuriel, D. (1992). *Interactive assessment.* New York: Springer-Verlag.

Heath, R., & Stickell, D. (1963). CHEM and CBA effects on achievement in chemistry. *Science Teacher, 30,* 45–46.

Helm, H. (1980). Misconceptions in physics amongst South African students. *Physics Education, 15,* 92–97 and 105.

Herron, J. D. (1971). The effect of behavioral objectives on student achievement in college chemistry. *Journal of Research in Science Teaching, 8,* 385–391.

Herron, J. D. (1975). Piaget for chemists. *Journal of Chemical Education, 52,* 146–150.

Herron, J. D. (1976). Commentary on 'Piagetian cognitive development and achievement in science.' *Journal of Research in Science Teaching, 13,* 355–359.

Herron, J. D. (1977a). Are chemical terms well defined? *Journal of Chemical Education, 54,* 758.

Herron, J. D. (1977b). Implicit curriculum—Where values are really taught. *The Science Teacher, 44*(3), 30–31.

Herron, J. D. (1978a). Response to "Are chemical terms well defined?" *Journal of Chemical Education, 55*, 393–394.

Herron, J. D. (1978b). Establishing a need to know. *Journal of Chemical Education, 55*, 190.

Herron, J. D. (1978c). Role of learning and development: Critique of Novak's comparison of Ausubel and Piaget. *Science Education, 62*, 593–605.

Herron, J. D. (1978d). Piaget in the classroom: Guidelines for application. *Journal of Chemical Education, 55*, 165–170.

Herron, J. D. (1981a). *Understanding chemistry: A preparatory course.* New York: Random House.

Herron, J. D. (1981b). *Instructor's manual for understanding chemistry: A preparatory course.* New York: Random House.

Herron, J. D. (1990). Research in chemical education: Results and directions. In Gardner, M., Greeno, J., Reif, F., Schoenfeld, A., diSessa, A., & Stage, E. (Eds.), *Toward a scientific practice of science education* (pp 31–54). Hillsdale, NJ: Erlbaum.

Herron, J. D., Agbebi, E., Cottrell, L., & Sills, T. (1976). Concept formation as a function of instructional procedures or what results from ineffective teaching. *Science Education, 60*, 375–388.

Herron, J. D., Cantu, L., Ward, R., & Srinivasan, V. (1977). Problems associated with concept analysis. *Science Education, 61*, 185–199.

Herron, J. D., & Greenbowe, T. (1986). What can we do about Sue: A case study of competence. *Journal of Chemical Education, 63*, 528–531.

Hertz-Lazarowitz, R., & Shachar, H. (1992). Teachers' verbal behavior in cooperative and whole-class instruction (pp 77–94). In S. Sharan (Ed.), *Cooperative learning: Theory and research.* New York: Praeger.

Hewson, M. (1984). The influence of intellectual environment on conceptions of heat. *European Journal of Science Education, 6*, 245–262.

Hewson, M. (1986). The acquisition of scientific knowledge: Analysis and representation of students' conceptions concerning density. *Science Education, 70*, 159–170.

Hewson, S. (1977). *Inferential problem solving in young children.* Unpublished doctoral dissertation, Oxford University, Oxford, England.

Hipsher, W. (1961). Study of high school physics achievement. *Science Teacher, 28*, 36–37.

Hofstein, A., & Lunetta, V. (1982). The role of laboratory in science teaching: Neglected aspects of research. *Review of Educational Research, 52*, 201–217.

Hull, D., & Parnell, D. (1991). *Tech Prep associate degree: A win/win experience.* Waco, TX: Center for Occupational Research and Development.

Hutchinson, R. (1985). Teaching problem solving to developmental adults: A pilot project. In J. Segal, S. Chipman, & R. Glaser (Eds.), *Thinking and learning skills: Vol. 1. Relating instruction to research* (pp 499–513). Hillsdale, NJ: Erlbaum.

Indiana Department of Education. (1992, August). *Tech Prep training manual.* Indianapolis, IN: Author.

Inhelder, B. (1976). Information processing tendencies in recent experiments in cognitive learning—Empirical studies. In B. Inhelder & H. A. Chipman (Eds.), *Piaget and his school: A reader in developmental psychology* (pp 121–133). New York: Springer-Verlag.

Inhelder, B., & Piaget, J. (1958). *The growth of logical thinking from childhood to adolescence.* New York: Basic Books.

Jenkins, J. J. (1979). Four points to remember: A tetrahedral model of memory experiments. In L. S. Cermak & F. I. M. Craik (Eds.), *Levels of processing in human memory* (pp 429–446). Hillsdale, NJ: Erlbaum.

Jensen, A. R. (1969). How much can we boost IQ and scholastic achievement? *Harvard Educational Review, 39,* 1–123.

Johnson, D. W. (1981). Student–student interaction: The neglected variable in education. *Educational Researcher, 10,* 5–10.

Johnson, D. W., & Johnson, R. (1992). Cooperative learning and achievement. In S. Sharan (Ed.), *Cooperative learning: Theory and research* (pp 23–37). New York: Praeger.

Johnson, D. W., & Johnson, R. T. (1985). Motivational processes in cooperative, competitive, and individualistic learning situations. In C. Ames & R. Ames (Eds.), *Research on motivation in education: Vol. 2. The classroom milieu* (pp 249–286). Orlando, FL: Academic Press.

Johnson, D. W., Johnson, R., & Maruyana, G. (1983). Interdependence and interpersonal attraction among heterogeneous and homogeneous individuals. *Review of Educational Research, 53,* 5–54.

Johnson, P., Ahlgren, A., Blount, J., & Petit, N. (1981). Scientific reasoning: Garden paths and blind alleys. In J. Robinson (Ed.), *Research in science education: New questions, new directions* (pp 87–114). Columbus, OH: ERIC Clearinghouse for Science, Mathematics, and Environmental Education.

Johnson-Laird, P., & Wason, P. (1977). A theoretical analysis of insight into a reasoning task. In P. Johnson-Laird & P. Wason (Eds.), *Thinking: Readings in cognitive science* (pp 143–157). Cambridge, England: Cambridge University Press.

Johnstone, A. (1980). Nyholm Lecture: Chemical education research: Facts, findings and consequences. *Chemical Society Reviews, 9,* 365–380. (An abbreviated version appeared in *Journal of Chemical Education*, 1983, *60,* 968–971.)

Johnstone, A., & El-Banna, H. (1986). Capacities, demands and processes: A predictive model for science education. *Education in Chemistry, 23,* 80–84.

Johnstone, A., MacDonald, J., & Webb, G. (1977). Misconceptions in school thermodynamics. *Physics Education, 12,* 248–251.

Johnstone, A., & Reid, N. (1981). Interactive teaching materials in science. *SASTA Journal, 812,* 4–15.

Kahle, J. B., Parker, L. H., Rennie, L. J., & Riley, D. (1993). Gender differences in science education: Building a model. *Educational Psychologist, 28,* 379–404. (This entire issue is devoted to gender equity issues.)

Kamii, C. (1979). Teaching for thinking and creativity: A Piagetian point of view. In A. E. Lawson (Ed.), *1980 AETS Yearbook: The psychology of teaching for thinking and creativity* (pp 29–58). Columbus, OH: ERIC Clearinghouse for Science, Mathematics, and Environmental Education.

Karmiloff-Smith, A., & Inhelder, B. (1977). If you want to get ahead, get a theory. In P. N. Johnson-Laird & P. C. Wason (Eds.), *Thinking: Readings in cognitive science* (pp 293–306). Cambridge, England: Cambridge University Press.

Karplus, R. (1977). Science teaching and the development of reasoning. *Journal of Research in Science Teaching, 14,* 169–175.

Karplus, R., Lawson, A., Wollman, W., Appel, M., Bernoff, R., Howe, A., Rusch, J., & Sullivan, F. (1977). *Science teaching and the development of reasoning: A workshop.* Berkeley, CA: Regents of the University of California.

Karplus, R., & Thier, H. (1967). *A new look at elementary school science*. Chicago: Rand McNally.

Katona, G. (1940). *Organizing and memorizing*. New York: Columbia University Press.

Kendler, T., & Kendler, H. (1967). Experimental analysis of inferential behavior in children. In L. Lipsitt & C. Spiker (Eds.), *Advances in child development and behavior: Vol. 3* (pp 157–190). New York: Academic Press.

Kibler, R., Barker, L. L., & Miles, D. T. (1970). *Behavioral objectives and instruction*. Boston: Allyn & Bacon.

Klausmeier, H., Ghatala, E., & Frayer, D. (1974). *Conceptual learning and development: A cognitive view*. New York: Academic Press.

Knight, G. P., & Bohlmeyer, E. (1992). Cooperative learning and achievement: Methods for assessing causal mechanisms. In S. Sharan (Ed.), *Cooperative learning: Theory and research* (pp 1–22). New York: Praeger.

Koestner, R., Ryan, R. M., Bernieri, F., and Holt, K. (1984). Setting limits on children's behavior: The differential effects of controlling vs. informational styles on intrinsic motivation and creativity. *Journal of Personality, 53*, 233–248.

Kramers-Pals, H., Lambrecht, J., & Wolff, P. (1983). The transformation of quantitative problems to standard problems in general chemistry. *European Journal of Science Education, 5*, 275–287.

Kubli, F. (1983). Piaget's clinical experiments: A critical analysis and study of their implications for science teaching. *European Journal of Science Education, 5*, 123–139.

Kuhn, D. (1981). The role of self-directed activity in cognitive development. In I. Sigel, D. Brodzinsky, & R. Golinkoff (Eds.), *New directions in Piagetian theory and practice* (pp 353–357). Hillsdale, NJ: Erlbaum.

Kuhn, T. (1970). *The structure of scientific revolutions* (2nd ed.). Chicago: University of Chicago Press.

Kulik, J., & Jaksa, P. (1977). PSI and other educational technologies in college teaching. *Educational Technology, 17*, 12–19.

Kumar, D., White, A., & Helgeson, S. (1993, April). *Effect of HyperCard and traditional performance assessment methods on expert–novice chemistry problem solving*. Paper presented at the 66th Annual Meeting of the National Association for Research in Science Teaching, Atlanta, GA.

Lambiotte, J., Dansereau, D. F., Cross, D., & Reynolds, S. (1989). Multirelational semantic maps. *Educational Psychology Review, 1*, 331–367.

Landa, L. (1975). Some problems in algorithmization and heuristics of instruction. *Instructional Science, 4*, 99–112.

Larkin, J. (1981). Understanding and problem solving in physics. In J. Robinson (Ed.), *Research in science education: New questions, new directions* (pp 115–130). Columbus, OH: ERIC Clearinghouse for Science, Mathematics, and Environmental Education.

Laughon, P. (1990). The dynamic assessment of intelligence: A review of three approaches. *School Psychology Review, 19*, 459–470.

Lawson, A. (1975). Developing formal thought through biology teaching. *American Biology Teacher, 37*, 411–419.

Lawson, A. (1979a). The developmental paradigm. *Journal of Research in Science Teaching, 16*, 510–519.

Lawson, A. (1979b). *1980 AETS yearbook: The psychology of teaching for thinking and creativity.* Columbus, OH: ERIC Clearinghouse for Science, Mathematics, and Environmental Education.

Lawson, A. (1980). Relationships among level of intellectual development, cognitive style, and grades in a college biology course. *Science Education, 64,* 95–102.

Lawson, A. (1982). The relative responsiveness of concrete operational seventh grade and college students to science instruction. *Journal of Research in Science Teaching, 19,* 63–77.

Lawson, A. (1985). A review of research on formal reasoning and science teaching. *Journal of Research in Science Teaching, 22,* 569–617.

Lawson, A., Abraham, M., & Renner, J. (1989). A theory of instruction: Using the learning cycle to teach science concepts and thinking skills. *National Association for Research in Science Teaching Monograph, 1.*

Lawson, A., & Wollman, W. (1976). Encouraging the transition from concrete to formal cognitive functioning: An experiment. *Journal of Research in Science Teaching, 13,* 413–430.

Lazonby, J., Morris, J., & Waddington, D. (1985). The mole: Questioning format can make a difference. *Journal of Chemical Education, 62,* 60–61.

Lee, K. (1982). Fourth graders' heuristic problem solving behavior. *Journal for Research in Mathematics Education, 13,* 110–123.

Lehman, J., Kahle, J., Nordland, F. (1981). Cognitive development and creativity: A study of two schools. *Science Education, 65,* 197–206.

Lewis, D. (1965). Objectives in the teaching of science. *Educational Research, 7*(3), 186–198.

Lickona, T. (1991). *Educating for character: How our schools can teach respect and responsibility.* New York: Bantam Books.

Lidz, C. S. (Ed.). (1987). *Dynamic assessment.* New York: Guilford.

Lloyd, C. V. (1990). The elaboration of concepts in three biology textbooks: Facilitating student learning. *Journal of Research in Science Teaching, 27,* 1019–1032.

Lowe, D. (1975). *A guide to international recommendations on names and symbols for quantities and on units of measurement.* Geneva, Switzerland: World Health Organization.

Lucas, J. (1974). The teaching of heuristic problem solving strategies in elementary calculus. *Journal for Research in Mathematics Education, 5,* 36–45.

Luria, A. (1976). *Cognitive development: Its cultural and social foundations.* Cambridge, MA: Harvard University Press.

MacDonald-Ross, M. (1973). Behavioral objectives: A critical review. *Instructional Science, 2*(1), 1–52.

Mahan, B. H. (1975). *University chemistry* (3rd ed.). Reading, MA: Addison-Wesley.

Mali, G., & Howe, A. (1979). Development of earth and gravity concepts among Nepali children. *Science Education, 63,* 685–691.

Markle, S., & Tiemann, P. (1970). *Really understanding concepts: Or in frumious pursuit of the jabberwock.* Champaign, IL: Stipes.

Maslow, A. H. (1954). *Motivation and personality.* New York: Harper and Brothers.

Mayer, R. E. (1989). Human nonadversary problem solving. In K. J. Gilhooly (Ed.), *Human and machine problem solving* (pp 39–56). New York: Plenum Press.

McClelland, G. (1975). Earthly mechanics: Two misapprehensions and a heresy. *Physics Education, 10*, 28–29.

McCombs, B. L., & Whisler, J. S. (1989). The role of affective variables in autonomous learning. *Educational Psychologist, 24*(3), 277–306.

McKinnon, J., & Renner, J. (1971). Are colleges concerned with intellectual development? *American Journal of Physics, 39*, 1047–1052.

Melton, R. (1978). Resolution of conflicting claims concerning the effect of behavioral objectives on student learning. *Review of Educational Research, 48*, 291–302.

Merrill, M., & Tennyson, R. (1977). *Teaching concepts: An instructional design guide*. Englewood Cliffs, NJ: Educational Technology Publications.

Mettes, C. T. C. W., Pilot, A., & Roosink, H. (1981). Linking factual and procedural knowledge in solving science problems: A case study in a thermodynamics course. *Instructional Science, 10*, 303–316.

Mettes, C. T. C. W., Pilot, A., Roosink, H., & Kramers-Pals, H. (1980). Teaching and learning problem solving in science. Part I: A general strategy. *Journal of Chemical Education, 57*, 882–885.

Mettes, C. T. C. W., Pilot, A., Roosink, H., & Kramers-Pals, H. (1981). Teaching and learning problem solving in science. Part II: Learning problem solving in a thermodynamics course. *Journal of Chemical Education, 58*, 51–55.

Milkent, M. (1977). It's time we started paying attention to what students don't know. *Science Education, 61*, 409–413.

Miller, G. (1956). The magical number seven, plus or minus two. *Psychological Review, 63*, 81–97.

Miller, N., & Harrington, H. J. (1992). A situational identity perspective on cultural diversity and teamwork in the classroom (pp 39–75). In S. Sharan (Ed.), *Cooperative learning: Theory and research*. New York: Praeger.

Mitchell, H., & Kellington, S. (1982). Learning difficulties associated with the particulate theory of matter in the Scottish Integrated Science Course. *European Journal of Science Education, 4*, 429–440.

Muth, K. D. (1991). Effects of cuing on middle-school students' performance on arithmetic word problems containing extraneous information. *Journal of Educational Psychology, 83*, 173–174.

Myrdal, G. (1964). *An American dilemma* (Rev. ed.). New York: McGraw-Hill. (Original work published 1944)

Nakhleh, M. (1993). Are our students conceptual thinkers or algorithmic problem solvers? *Journal of Chemical Education, 70*, 52–55.

Nakhleh, M., & Mitchell, R. (1993). Concept learning versus problem solving. *Journal of Chemical Education, 70*, 190–192.

National Council of Teachers of Mathematics. (1989). *Curriculum and evaluation standards for school mathematics*. Reston, VA: Author.

Neill, A. S. (1960). *Summerhill: A radical approach to child rearing*. New York: Hart.

Newell, A. (1977). On the analysis of human problem solving protocols. In P. N. Johnson-Laird & P. C. Wason (Eds.), *Thinking: Readings in cognitive science* (pp 46–61). Cambridge, England: Cambridge University Press.

Newell, A., & Simon, H. (1972). *Human problem solving*. Englewood Cliffs, NJ: Prentice-Hall.

Niaz, M. (1987). Relation between M-space of students and M-demand of different items of general chemistry and its interpretation based upon neo-Piagetian theory of Pascual-Leone. *Journal of Chemical Education, 64,* 502–505.

Niaz, M., Herron, J. D., & Phelps, A. J. (1991). The effect of context on the translation of sentences into algebraic equations. *Journal of Chemical Education, 68,* 306–309.

Niaz, M., & Lawson, A. E. (1985). Balancing chemical equations: The role of developmental level and mental capacity. *Journal of Research in Science Teaching, 22,* 41–51.

Nicholls, J. G. (1989). *The competitive ethos and democratic education.* Cambridge, MA: Harvard University Press.

Nichols, R., & Stevens, L. (1957). *Are you listening?* New York: McGraw-Hill.

Nolen, S. B., & Haladyna, T. M. (1990). Motivation and studying in high school science. *Journal of Research on Science Teaching, 27,* 115–126.

Norman, D., Gentner, D., & Stevens, A. (1976). Comments on learning schemata and memory representation. In D. Klahr (Ed.), *Cognition and instruction* (pp 177–196). Hillsdale, NJ: Erlbaum.

Novak, J. D. (1990). Concept mapping: A useful tool for science education. *Journal of Research in Science Teaching, 27,* 937–949.

Novak, J. D., & Gowin, D. B. (1984). *Learning how to learn.* Cambridge, England: Cambridge University Press.

Novak, J. D., & Musonda, D. (1991). A twelve-year longitudinal study of science concept learning. *American Educational Research Journal, 28,* 117–153.

Novick, S., & Mannis, J. (1976). A study of student perception of the mole concept. *Journal of Chemical Education, 53,* 720–722.

Novick, S., & Nussbaum, J. (1978). Junior high school pupils' understanding of the particulate nature of matter: An interview study. *Science Education, 62,* 273–281.

Novick, S., & Nussbaum, J. (1981). Pupils' understanding of the particulate nature of matter: A cross-age study. *Science Education, 65,* 187–196.

Nurrenbern, S. (1980). Problem solving behaviors of concrete and formal operational high school students when solving chemistry problems requiring Piagetian formal reasoning skills (Doctoral dissertation, Purdue University, 1979). *Dissertation Abstracts International, 40,* 4986A.

Nurrenbern, S. (1982). *Gases, a physical state: Concept vs. numbers.* Paper presented at the 55th Annual Meeting of the National Association for Research in Science Teaching, Lake Geneva, WI.

Nurrenbern, S., & Pickering, M. (1987). Concept learning versus problem solving: Is there a difference? *Journal of Chemical Education, 64,* 508–510.

Nussbaum, J., & Novick, S. (1981). Brainstorming in the classroom to invent a model: A case study. *School Science Review, 62,* 771–778.

Nussbaum, J., & Novick, S. (1982). *A study of conceptual change in the classroom.* Paper presented at the 55th Annual Meeting of the National Association for Research in Science Teaching, Lake Geneva, WI.

Olugbemiro, J. J., Alaiyemola, F. F., & Okebukola, P. A. O. (1990). The effect of concept mapping on students' anxiety and achievement in biology. *Journal of Research in Science Teaching, 27,* 951–960.

Osborne, R., & Cosgrove, M. (1983). Children's conceptions of the changes of state of water. *Journal of Research in Science Teaching, 20,* 825–838.

Osborne, R., & Freyberg, P. (1985). *Learning in science: The implications of children's science.* Auckland, New Zealand: Heinemann.

Pascual-Leone, J. (1970). A mathematical model for the transition rule in Piaget's developmental stages. *Acta Psychologica, 63,* 301–345.

Pascual-Leone, J., & Smith, J. (1969). The encoding and decoding of symbols by children: New experimental paradigm and a neo-Piagetian model. *Journal of Experimental Child Psychology, 8,* 328–355.

Patterson, M., Dansereau, D. F., & Newbern, D. (1992). Effects of communication aids and strategies on cooperative teaching. *Journal of Educational Psychology, 84,* 453–461.

Pereira-Mendoza, L. (1980). The effect of teaching heuristics on the ability of grade ten students to solve novel mathematical problems. *Journal of Educational Research, 73,* 139–144.

Perret-Clermoat, A. (1980). *Social interaction and cognitive development in children.* London: Academic Press.

Pfundt, H. (1981). The atom: The final link in the division process or the first building block? Pre-instructional conceptions about the structure of substances. *Chimica Didactica, 7,* 75–94.

Pfundt, H. (1982). Pre-instructional conceptions about transformations of substances. *Chimica Didactica, 8,* 25 pp.

Pickering, M. (1990). Further studies on concept learning versus problem solving. *Journal of Chemical Education, 67,* 254–255.

Picturesque word origins, with forty-five illustrations. (1933). Springfield, MA: G & C Merriam.

Polya, G. (1957). *How to solve it: A new aspect of mathematical method* (2nd ed.). Princeton, NJ: Princeton University Press.

Qin, Y., & Simon, H. A. (1990). Laboratory replication of scientific discovery processes. *Cognitive Science, 14,* 281–312.

Rainey, R. (1964). A comparison of the CHEM study curriculum and a conventional approach in teaching high school chemistry. *School Science and Mathematics, 64,* 539–544.

Raths, L., Harmin, M., & Simon, S. (1966). *Values and teaching: Working with values in the classroom.* Columbus, OH: Merrill.

Reimann, P., & Chi, M. T. H. (1989). Human expertise. In K. J. Gilhooly (Ed.), *Human and machine problem solving* (pp 161–191). New York: Plenum Press.

Renner, J. (1982). The power of purpose. *Science Education, 66,* 709–716.

Renner, J., & Stafford, D. (1972). *Teaching science in the secondary schools.* New York: Harper and Row.

Resnick, L. (1989). Introduction. In L. B. Resnick (Ed.), *Knowing, learning, and instruction: Essays in honor of Robert Glaser* (pp 1–24). Hillsdale, NJ: Erlbaum.

Rewey, K., Dansereau, D. F., Skaggs, L., & Hall, R. (1989). Effects of scripted cooperation and knowledge maps on the processing of technical material. *Journal of Educational Psychology, 81,* 604–609.

Robertson, W. C. (1990). Detection of cognitive structure with protocol data: Predicting performance on physics transfer problems. *Cognitive Science, 14,* 253–280.

Robertson, W. C., & Richardson, E. (1975). The development of some physical science concepts in secondary school students. *Journal of Research in Science Teaching, 12,* 319–329.

Rogers, C. R. (1969). *Freedom to learn.* Columbus, OH: Merrill.

Rohwer, W., Ammon, P., & Cramer, P. (Eds.). (1974). *Understanding intellectual development: Three approaches to theory and practice*. Hinsdale, IL: Dryden Press.

Romberg,T., Steitz, J., & Frayer, D. (1971). *Working paper no. 55: Selection and analysis of mathematics concepts for inclusion in tests of concept attainment*. Madison, WI: Wisconsin Research and Development Center for Cognitive Learning.

Rosch, E. (1977). Classification of real-world objects: Origins and representations in cognition. In P. N. Johnson-Laird & P.C. Wason (Eds.), *Thinking: Readings in cognitive science* (pp 212–222). Cambridge, England: Cambridge University Press.

Rosch, E., Mervis, C., Gray, W., Johnson, D., & Boyes-Braem, P. (1976). Basic objects in natural categories. *Cognitive Psychology, 8,* 382–439.

Rosenthal, R., & Jacobson, L. (1968). *Pygmalion in the classroom: Teacher expectation and pupils' intellectual development*. New York: Holt, Rinehart and Winston.

Roth, W.-M. (1990). Short-term memory and problem solving in physical science. *School Science and Mathematics, 90,* 271–281.

Rowe, M. B. (1974a). Pausing phenomena: Influence on the quality of instruction. *Journal of Psycholinguistics Research, 3,* 203–223.

Rowe, M. B. (1974b). Wait-time and rewards as instructional variables, their influence on language, logic, and fate control: Part I. Wait-time. *Journal of Research in Science Teaching, 11,* 81–94.

Rowe, M. B. (1974c). Relation of wait-time and rewards to the development of language, logic, and fate control: Part II. Rewards. *Journal of Research in Science Teaching, 11,* 291–308.

Rowe, M. B. (1974d). Reflections on wait-time: Some methodological questions. *Journal of Research in Science Teaching, 11,* 263–279.

Rowell, J., & Dawson, C. (1977). Teaching about floating and sinking: An attempt to link cognitive psychology with classroom practice. *Science Education, 61,* 245–253.

Rowell, J., & Dawson, C. (1983). Laboratory counterexamples and the growth of understanding in science. *European Journal of Science Education, 5,* 203–215.

Rumelhart, D. (1980). Schemata: The building blocks of cognition. In R. J. Spiro, B. C. Bruce, & W. F. Brewer (Eds.), *Theoretical issues in reading comprehension* (pp 33–58). Hillsdale, NJ: Erlbaum.

Ryan, R. M., Connell, J. P., & Deci, E. L. (1985). A motivational analysis of self-determination and self-regulation in education. In C. Ames & R. Ames (Eds.), *Research on motivation in education: Vol. 2. The classroom milieu* (pp 13–51). Orlando, FL: Academic Press.

Saltiel, E. (1981). Kinematic concepts and natural reasoning: Study of comprehension of Galilean frames by science students. *European Journal of Science Education, 3,* 110.

Savell, J., Twohig, P., & Rachford, D. (1986). Empirical status of Feuerstein's "Instrumental Enrichment" techniques as a method of teaching thinking skills. *Review of Educational Research, 56,* 381–409.

Sawrey, B. (1990). Concept learning versus problem solving: Revisited. *Journal of Chemical Education, 67,* 253–254.

Scantlebury, K., & Kahle, J. B. (1993). The implementation of equitable teaching strategies by high school biology teachers. *Journal of Research in Science Teaching, 30,* 537–545. (This entire issue is devoted to gender equity issues.)

Scardamalia, M. (1977). Information processing capacities and the problem of horizontal decalage: A demonstration using combinatorial reasoning tasks. *Child Development, 48*, 28–37.

Schmidt, H. (1987). Secondary school students' learning difficulties in stoichiometry. *Proceedings of the Second International Seminar on Misconceptions and Educational Strategies in Science and Mathematics: Vol. I* (pp 396–404). Ithaca, NY: Cornell University.

Schmidt, H. (1991). A label as a hidden persuader: Chemists' neutralization concept. *International Journal of Science Education, 13*, 459–471.

Schmidt, H. (1992). Conceptual difficulties with isomerism. *Journal of Research in Science Teaching, 29*, 995–1003.

Schoenfeld, A. (1979). Explicit heuristic training as a variable in problem solving performance. *Journal of Research in Mathematics Education, 10*, 173–187.

Scott, P. (1987). The process of conceptual change in science: A case study of the development of a secondary pupil's ideas relating to matter. *Proceedings of the Second International Seminar on Misconceptions and Educational Strategies in Science and Mathematics: Vol. II* (pp 404–418). Ithaca, NY: Cornell University.

Selley, N. (1978). The confusion of molecular particles with substances. *Education in Chemistry, 15*, 144–145.

Sharan, S. (Ed.). (1992). *Cooperative learning: Theory and research*. New York: Praeger.

Sharan, S., Kussell, P., Hertz-Lazarowitz, R., Bejarano, Y., Raviv, S., Sharan, Y., Brosh, T., & Peleg, R. (1984). *Cooperative learning in the classroom: Research in desegregated schools*. Hillsdale, NJ: Erlbaum.

Shaver, J. P., & Strong, W. (1982). *Facing value decisions: rationale-building for teachers* (2nd ed.). New York: Teachers College Press.

Shayer, M., & Adey, P. (1981). *Towards a science of science teaching: Cognitive development and curriculum demand*. London: Heinemann Educational Books.

Shayer, M., & Wylam, H. (1981). The development of the concept of heat and temperature in 10–13 year olds. *Journal of Research in Science Teaching, 18*, 419–434.

Shulman, L. (1987). Knowledge and teaching. *Harvard Educational Review, 57*, 1–22.

Siegler, R. (1976). Three aspects of cognitive development. *Cognitive Psychology, 8*, 481–520.

Siegler, R., Liebert, D., & Liebert, R. (1973). Inhelder and Piaget's pendulum problem: Teaching preadolescents to act as scientists. *Developmental Psychology, 9*, 97–101.

Silberstein, J., Ben-Zvi, R., & Eylon, B. (1982). *A survey of misrepresentations of basic concepts in chemistry textbooks* (Technical Report C1/82). Rehovot, Israel: Weizmann Institute, Department of Science Teaching.

Sinnott, J. D. (Ed.). (1989). *Everyday problem solving: Theory and applications*. New York: Praeger.

Skemp, R. (1979). *Intelligence, learning, and action*. New York: John Wiley & Sons.

Slavin, R. (1990). *Cooperative learning: Theory, research, and practice*. Englewood Cliffs, NJ: Prentice-Hall.

Slavin, R., Sharan, S., Kagan, S., Lazarowitz, R. H., Webb, C., & Schmuck, R. (1985). *Learning to cooperate, cooperating to learn*. New York: Plenum Press.

Smoke, K. L. (1932). An objective study of concept formation. *Psychological Monographs, 42*(4), 46.

Solomon, J. (1982). How children learn about energy or Does the first law come first? *The School Science Review, 63,* 415–422.

Solomon, J. (1983). Learning about energy: How pupils think in two domains. *European Journal of Science Education, 5,* 49–59.

Starr, M. L., & Krajcik, J. S. (1990). Concept maps as a heuristic for science curriculum development: Toward improvement in process and product. *Journal of Research in Science Teaching, 27,* 987–1000.

Stauffer, R. (1975). *Directing the reading–thinking process.* New York: Harper and Row.

Stavy, R., & Berkovitz, B. (1980). Cognitive conflict as a basis for teaching quantitative aspects of the concept of temperature. *Science Education, 64,* 679–692.

Stein, J. (Ed.). (1975). *The Random House college dictionary* (Rev. ed.). New York: Random House.

Swan, M. (1980). Comparison of students' percepts of distance, weight, height, area, and temperature. *Science Education, 64,* 297–307.

Tessmer, M., Wilson, B., & Driscoll, M. (1990). A new model of concept teaching and learning. *Educational Technology Research and Development, 38,* 45–53.

Thomas, J. W. (1980). Agency and achievement: Self-management and self-regard. *Review of Educational Research, 50,* 213–240.

Thorndike, E., & Lorge, I. (1944). *The teacher's word book of 30,000 words.* New York: Columbia University, Bureau of Publications, Teachers College.

Tobias, S. (1990). *They're not dumb, they're different: Stalking the second tier.* Tucson, AZ: Research Corporation.

Tomlinson-Keasey, C. (1972). Formal operations in females ages 11 to 54 years of age. *Developmental Psychology, 6,* 364.

Towler, J., & Wheatley, G. (1971). Conservation concepts in college students: A replication and critique. *Journal of Genetic Psychology, 118,* 265–270.

Tyler, R. (1972). Some persistent questions on the defining of objectives. In E. Stones (Ed.), *Educational objectives and the teaching of educational psychology* (pp 179–187). New York: Methuen & Company.

U.S. National Commission on Excellence in Education. (1983). *A nation at risk: The imperative of educational reform: A report to the nation and the Secretary of Education.* Washington, DC: U.S. Department of Education.

Varagunam, T. (1971). Student awareness of behavioural objectives: The effect on learning. *British Journal of Medical Education, 5,* 213.

Viennot, L. (1979). Spontaneous reasoning in elementary dynamics. *European Journal of Science Education, 1,* 205–221.

Voelker, A. (1975). Elementary schoolchildren's attainment of the concepts of physical and chemical change: A replication. *Journal of Research in Science Teaching, 12,* 5–14.

Vygotsky, L. (1986). *Thought and language* (A. Kozulin, Trans.). Cambridge, MA: MIT Press. (Original work published 1934)

Wallace, J. D., & Mintzes, J. J. (1990). The concept map as a research tool: Exploring conceptual change in biology. *Journal of Research in Science Teaching, 27,* 1033–1052.

Ward, C., Nurrenbern, S., Lucas, C., & Herron, J. D. (1981). Evaluation of the Longeot test of cognitive development. *Journal of Research in Science Teaching, 18,* 123–130.

Wason, P. C. (1966). Reasoning. In B. Foss (Ed.), *New horizons in psychology, I* (pp 119, 120, 143, 145). Harmondsworth, Middlesex, England: Penguin.

Watts, M. (1982). Gravity: Don't take it for granted. *Physics Education, 17,* 116–121.

Wells, G., Chang, G. L. M., & Maher, A. (1992). Creating classroom communities of literate thinkers. In S. Sharan (Ed.), *Cooperative learning: Theory and research* (pp 95–121). New York: Praeger.

Wheatley, G. H. (1991). Constructivist perspectives on science and mathematics learning. *Science Education, 75,* 9–21.

Wheatley, G. H., & Wheatley, C. L. (1982). *Calculator use and problem solving strategies of grade six pupils* (Final Report). West Lafayette, IN: Purdue University. (ERIC Document Reproduction Service No. ED 219 250)

Wheeler, A., & Kass, H. (1978). Student misconceptions in chemical equilibrium. *Science Education, 62,* 223–232.

Whimbey, A., & Lochhead, J. (1986). *Problem solving and comprehension* (4th ed.). Hillsdale, NJ: Erlbaum.

Wickelgren, W. (1974). *How to solve problems: Elements of a theory of problems and problem solving.* San Francisco: W.H. Freeman.

Wigginton, E. (1985). *Sometimes a shining moment: The Foxfire experience.* Garden City, NY: Anchor Press/Doubleday.

Williams, H. D. (1930). Experiment in self-directed education. *School and Society, 30,* 715–718.

Williamson, V., & Abraham, M. (1995). The effects of computer animation on the particulate mental models of college chemistry students. *Journal of Research in Science Teaching, 32,* 521–534.

Winograd, T. (1976). A framework for understanding discourse. In P. Carpenter & M. Just (Eds.), *Cognitive processes in comprehension* (pp 63–88). Hillsdale, NJ: Erlbaum.

Winograd, T. (1977). Formalisms for knowledge. In P. N. Johnson-Laird & P. C. Wason (Eds.), *Thinking: Readings in cognitive science* (pp 62–71). Cambridge, England: Cambridge University Press.

Wolke, R. (1973). Unhand me, sir! Your objectives are naught but behavioral! *Journal of Chemical Education, 50,* 99–101.

Wollman, W., & Lawson, A. (1978). The influence of instruction on proportional reasoning in seventh-graders. *Journal of Research in Science Teaching, 15,* 227–232.

Woodward, A., Bjork, R., & Jongeward, R. (1973). Recall and recognition as a function of primary rehearsal. *Journal of Verbal Learning and Verbal Behavior, 12,* 608–617.

Yackel, E. (1984). Characteristics of problem representation indicative of understanding in mathematics problem solving (Doctoral dissertation, Purdue University, 1984). *Dissertation Abstracts International, 45,* 2021A.

Yackel, E., Cobb, P., Wood, T., Wheatley, G., & Merkel, R. (1990). The importance of social interactions in children's construction of mathematical knowledge. In T. Cooney (Ed.), *1990 Yearbook of the National Council of Teachers of Mathematics* (pp 12–21). Reston, VA: National Council of Teachers of Mathematics.

Yarroch, W. (1985). Student understanding of chemical equation balancing. *Journal of Research in Science Teaching, 22,* 449–459.

Young, J. (1972). Report of the curriculum committee on the proper use of performance objectives. *Journal of Chemical Education, 49,* 484.

Zipf, G. K. (1949). *Human behavior and the principle of least effort.* Cambridge, MA: Addison-Wesley.

Zoller, U. (1990). Students' misunderstandings and misconceptions in college freshman chemistry (general and organic). *Journal of Research in Science Teaching, 27,* 1053–1062.

Additional References

The following references were used as background reading in preparing the book, but they are not cited in the book. They are very useful references.

Adi, H., Karplus, R., Lawson, A., & Pulos, S. (1978). Intellectual development beyond elementary school VI: Correlational reasoning. *School Science and Mathematics, 78,* 675–683.

Allen, J., Barker, L., & Ransden, J. (1986). Guided inquiry laboratory. *Journal of Chemical Education, 63,* 533–534.

Ames, C., & Ames, R. (Eds.). (1985). *Research on motivation in education: Vol. 2. The classrooom milieu.* Orlando, FL: Academic Press.

Ames, R., & Ames, C. (Eds.). (1984). *Research on motivation in education: Vol. 1. Student motivation.* New York: Academic Press.

Anderson, J. (1976). *Language, memory and thought.* Hillsdale, NJ: Erlbaum.

Ausubel, D. (1964). The transition from concrete to abstract cognitive functioning: Theoretical issues and implications for education. *Journal of Research in Science Teaching, 2,* 261–266.

Bechtel, W., & Abrahamsen, A. (1991). *Connectionism and the mind: An introduction to parallel processing in networks.* Cambridge, MA: Basil Blackwell.

Bender, D., & Milakofsky, L. (1982). College chemistry and Piaget: The relationship of aptitude and achievement measures. *Journal of Research in Science Teaching, 19,* 205–216.

Bertini, M., Pizzamiglio, L., & Wapner, S. (Eds.). (1986). *Field dependence in psychological theory, research, and application: Two symposia in memory of Herman A. Witkin.* Hillsdale, NJ: Erlbaum.

Bhaskar, R., & Simon, H. (1977). Problem solving in semantically rich domains: An example from engineering thermodynamics. *Cognitive Science, 1,* 193–215.

Bloom, B. (Ed.). (1956). *Taxonomy of educational objectives: Cognitive domain.* New York: David McKay.

Bloom, B., & Broder, L. (1950). Problem solving processes of college students: An exploratory investigation. *Supplementary Educational Monographs.* Chicago: University of Chicago Press.

Blumenfeld, P. C., Soloway, E., Marx, R. W., Krajcik, J. S., Guxdial, M., & Palincsar, A. (1991). Motivating project-based learning: Sustaining the doing, supporting the learning. *Educational Psychologist, 26,* 369–398.

Bowen, C. W. (1990). Representational systems used by graduate students while problem solving in organic synthesis. *Journal of Research in Science Teaching, 27,* 351–370.

Brainerd, C. (1978). *Piaget's theory of intelligence.* Englewood Cliffs, NJ: Prentice-Hall.

Broadbent, D. (1975). The magic number seven after fifteen years. In A. Kennedy & A. Wilks (Eds.), *Studies in long term memory* (pp 3–18). New York: John Wiley & Sons.

Brown, A., & Ferrara, R. (1985). Diagnosing zones of proximal development. In J. Wertsch (Ed.), *Culture, communication, and cognition: Vygotskian perspectives* (pp 273–305). Cambridge, MA: Cambridge University Press.

Butler, R. (1987). Task-involving and ego-involving properties of evaluation: The effects of different feedback conditions on motivational perceptions, interest and performance. *Journal of Educational Psychology, 79,* 474–482.

Butler, R. (1988). Enhancing and undermining intrinsic motivation: The effects of task-involving and ego-involving evaluation on interest and performance. *British Journal of Educational Psychology, 58,* 1–14.

Case, R., & Fry, C. (1973). Evaluation of an attempt to teach scientific inquiry and criticism in a working class high school. *Journal of Research in Science Teaching, 10,* 135–142.

Chadran, S., Treagust, D., & Tobin, K. (1987). The role of cognitive factors in chemistry achievement. *Journal of Research in Science Teaching, 24,* 145–160.

Chiu, M.-H. (1993, April). *Developing problem-solving skills in chemical equilibrium—A constructive model.* Paper presented at the 66th Annual Meeting of the National Association for Research in Science Teaching, Atlanta, GA.

Choi, B., & Gennaro, E. (1987). The effectiveness of using computer-simulated experiments on junior high school students' understanding of the volume displacement concept. *Journal of Research in Science Teaching, 24,* 539–552.

Cobb, P., & Steffe, L. (1983). The constructivist researcher as teacher and model builder. *Journal of Research in Mathematics Education, 14,* 83–94.

Cohen, H. G. (1982). Relationship between locus of control and the development of spatial conceptual abilities. *Science Education, 66,* 635–642.

Collins, A. (1985). Teaching reasoning skills. In S. Chipman, J. Segal, & R. Glaser (Eds.), *Thinking and learning skills: Vol. 2. Research and open questions* (pp 579–586). Hillsdale, NJ: Erlbaum.

Cornwell, J. M., Manfredo, P. A., & Dunlap, W. P. (1991). Factor analysis of the 1985 revision of Kolb's learning style inventory. *Educational and Psychological Measurement, 51,* 455–462.

Dark, V. J., & Benbow, C. P. (1990). Enhanced problem translation and short-term memory: Components of mathematical talent. *Journal of Educational Psychology, 82,* 420–429.

Dasen, P. (1972). Cross-cultural Piagetian research: A summary. *Journal of Cross-Cultural Psychology, 3,* 23–29.

Davidson, G. V. (1990, April). Matching learning styles with teaching styles: Is it a useful concept for instruction? *Performance and Instruction, 29*(4), 36–38.

Deci, E. L., & Ryan, R. M. (1985). *Intrinsic motivation and self-determination in human behavior.* New York: Plenum Press.

Dewey, J. (1938). *Experience and education*. New York: Macmillan.

Drake, R. F. (1985). Working backwards is a forward step in the solution of problems by dimensional analysis. *Journal of Chemical Education, 62,* 414.

Elkind, D. (1969). Piagetian and psychometric conceptions of intelligence. *Harvard Educational Review, 39,* 319–335.

Epstein, H. (1978). Growth spurts during brain development: Implications for educational policy and practice. In J. Chall & A. Mirsky (Eds.), *Education and the brain: 1978 NSSE yearbook* (pp 343–370). Chicago: University of Chicago Press.

Epstein, H. (1979). Correlated brain and intelligence development in humans. In *Development and evolution of brain size* (pp 111–131). Orlando, FL: Academic Press.

Estes, W. (1975–1979). *Handbook of learning and cognitive processes* (Vols. 1–6). Hillsdale, NJ: Erlbaum.

Evans, J. (1974). Vocabulary problems in teaching science. *School Science Review, 55,* 585–590.

Feuerstein, R., Miller, R., Rand, Y., & Jensen, M. (1981). Can evolving techniques better measure cognitive change? *Journal of Special Education, 15,* 201–219.

Flavell, J., Botkin, P., Fry, C., Wright, J., & Jarvis, P. (1968). *The development of role-taking and communication skills in children*. New York: John Wiley.

Flexer, B., & Roberge, J. (1980). IQ, FD-I, and the development of formal operational thought. *Journal of Genetic Psychology, 103,* 191–201.

Fraisse, P., & Piaget, J. (Eds.). (1969). *Experimental psychology: Its scope and method: VII Intelligence*. London: Routledge and Kegan Paul.

Franklin, B., & Good, R. (1993, April). *Teacher use of knowledge of student-held physical science concepts prior to instruction: A case study of six science teachers*. Paper presented at the 66th Annual Meeting of the National Association of Research in Science Teaching, Atlanta, GA.

Friedel, A., Gabel, D., & Samuel, J. (1990). Using analogs for chemistry problem solving: Does it increase understanding? *School Science and Mathematics, 90,* 674–682.

Gabel, D., & Sherwood, R. (1980). The effect of student manipulation of molecular models on chemistry achievement according to Piagetian level. *Journal of Research in Science Teaching, 17,* 75–81.

Gagné, R. (1966). The learning of principles. In H. J. Klausmeier & C. W. Harris (Eds.), *Analysis of concept learning* (pp 81–95). New York: Academic Press.

Gelman, R. (1969). Conservation acquisition: A problem of learning to attend to relevant attributes. *Journal of Experimental Child Psychology, 7,* 167–187.

Gelman, R. (1977). How young children reason about small numbers. In N. Castellan, D. Pisoni, & G. Potts (Eds.), *Cognitive theory: Vol. 2* (pp 219–238). Hillsdale, NJ: Erlbaum.

Genyea, J. (1983). Improving students' problem-solving skills: A methodical approach for a preparatory chemistry course. *Journal of Chemical Education, 60,* 478–482.

Geschwind, N. (1981). Neurological knowledge and complex behaviors. In D. A. Norman (Ed.), *Perspectives on cognitive science* (pp 27–35). Hillsdale, NJ: Erlbaum.

Glick, J. (1978). Cognition and social cognition: An introduction. In J. Glick & K. A. Clarke-Stewart (Eds.), *The development of social understanding* (pp 1–9). New York: Gardner Press.

Goodenough, D. R. (1986). History of the field dependence construct. In M. Bertini, L. Pizzamiglio, & S. Wapner (Eds.), *Field dependence in psychological theory, research, and application: Two symposia in memory of Herman A. Witkin* (pp 5–13). Hillsdale, NJ: Erlbaum.

Goodstein, M., & Howe, A. (1978). Applications of Piagetian theory to introductory chemistry instruction. *Journal of Chemical Education, 55*, 171–173.

Greeno, J. (1976). Cognitive objectives of instruction: Theory of knowledge for solving problems and answering questions. In D. Klahr (Ed.), *Cognition and instruction* (pp 123–159). Hillsdale, NJ: Erlbaum.

Greeno, J. (1977). Process of understanding in problem solving. In N. Castellan, D. Pisoni, & G. Potts (Eds.), *Cognitive theory: Vol. 2* (pp 43–83). Hillsdale, NJ: Erlbaum.

Greeno, J. (1978a). A study of problem solving. In R. Glaser (Ed.), *Advances in instructional psychology: Vol. I* (pp 13–75). Hillsdale, NJ: Erlbaum.

Greeno, J. (1978b). Nature of problem-solving abilities. In W. K. Estes (Ed.), *Handbook of learning and cognitive processes: Vol. 5. Human information processing* (pp 19–90). Hillsdale, NJ: Erlbaum.

Greeno, J. (1979). Trends in the theory of knowledge for problem solving. In D. Tuma & F. Reif (Eds.), *Problem solving and education: Issues in teaching and research.* Hillsdale, NJ: Erlbaum.

Holding, D. H. (1989). Adversary problem solving by humans. In K. J. Gilhooly (Ed.), *Human and machine problem solving* (pp 83–122). New York: Plenum Press.

Holliday, W., & Barden, L. (1993, April). *Using science processes to teach problem solving and conceptual change.* Paper presented at the National Association for Research in Science Teaching, Atlanta, GA.

Holliday, W., & Harvey, D. (1976). Adjunct labeled drawings in teaching physics to junior high school students. *Journal of Research in Science Teaching, 13,* 37–43.

Howe, A., & Durr, B. (1982). Analysis of an instructional unit for level of cognitive demand. *Journal of Research in Science Teaching, 19,* 217–224.

Hoz, R., Tomer, Y., & Tamir, P. (1990). The relations between disciplinary and pedagogical knowledge and the length of teaching experience of biology and geography teachers. *Journal of Research in Science Teaching, 27,* 973–985.

Inhelder, B., & Chipman, H. A. (Eds.). (1976). *Piaget and his school: A reader in developmental psychology.* New York: Springer-Verlag.

Johnstone, A., & Kellett, N. (1980). Learning difficulties in school science: Towards a working hypothesis. *European Journal of Science Education, 2,* 175–181.

Johnstone, A., & Mughol, A. (1979). Testing for understanding. *School Science Review, 61,* 147–150.

Karplus, R., & Peterson, R. (1970). Intellectual development beyond elementary school II: Ratio, a survey. *School Science and Mathematics, 70,* 813–820.

Kempa, R., & Nicholls, C. (1983). Problem-solving ability and cognitive structure: An exploratory investigation. *European Journal of Science Education, 5,* 171–184.

Kolodiy, J. (1984). A Piaget-based integrated math and science program. *Journal of College Science Teaching, 13,* 297–299.

Krupa, M., Selman, R., & Jaquette, D. (1985). The development of science explanations in children and adolescents: A structural approach. In S. Chipman, J. Segal, & R. Glaser (Eds.), *Thinking and learning skills: Vol. 2. Research and open questions* (pp 427–455). Hillsdale, NJ: Erlbaum.

Larkin, J., & Reif, F. (1976). Analysis and teaching of a general skill for studying scientific text. *Journal of Educational Psychology, 68*, 431–440.

Larkin, J., & Reif, F. (1979). Understanding and teaching problem solving in physics. *European Journal of Science Education, 1*, 191–203.

Lawson, A., & Blake, A. (1976). Concrete and formal thinking abilities in high school biology students as measured by three separate instruments. *Journal of Research in Science Teaching, 13*, 222–235.

Lawson, A., & Karplus, R. (1977). Should theoretical concepts be taught before formal operations? *Science Education, 61*, 123–125.

Lawson, A., & Renner, J. (1975). Relationships of science subject matter and developmental level of learners. *Journal of Research in Science Teaching, 12*, 347–358.

Lida, C. S. (1991). *Practitioner's guide to dynamic assessment.* New York: Guilford.

Linn, M. (1982). Theoretical and practical significance of formal reasoning. *Journal of Research in Science Teaching, 19*, 727–742.

Lynch, P., Benjamin, P., Chapman, T., Holmes, R., McCammon, R., Smith, A., & Symmons, R. (1979). Scientific language and the high school pupil. *Journal of Research in Science Teaching, 16*, 351–357.

Maehr, M. L., & Midgley, C. (1991). Enhancing student motivation: A schoolwide approach. *Educational Psychologist, 26*, 399–427.

Mayer, R. E., & Gallini, J. K. (1990). When is an illustration worth ten thousand words? *Journal of Educational Psychology, 82*, 715–726.

Mayer, R. E., & Greeno, J. G. (1972). Structural differences between learning outcomes produced by different instructional methods. *Journal of Educational Psychology, 63*, 165–173.

Misiak, J. (1993, April). *College students' representations of proportional problems.* Paper presented at the 66th Annual Meeting of the National Association for Research in Science Teaching, Atlanta, GA.

Montessori, M. (1965). *The Montessori Method: Scientific pedagogy as applied to child education in the children's houses with additions and revisions by the author.* (A. George, Trans.). Cambridge, MA: Robert Bentley. (Original work published 1912)

Nagy, P., & Griffiths, A. (1982). Limitations of recent research relating Piaget's theory to adolescent thought. *Review of Educational Research, 52*, 513–556.

National Council of Teachers of Mathematics. (1991). *Professional standards for teaching mathematics.* Reston, VA: Author.

Nedelsky, L. (1965). *Science teaching and testing.* New York: Harcourt, Brace, & World.

Norman, D. A. (1981). Twelve issues for cognitive science. In D. A. Norman (Ed.), *Perspectives on cognitive science* (pp 265–295). Hillsdale, NJ: Erlbaum.

Novick, L. R., & Holyoak, K. J. (1991). Mathematical problem solving by analogy. *Journal of Experimental Psychology: Learning, Memory, and Cognition, 17*, 398–415.

Nucci, L. (1982). Conceptual development in the moral and conventional domains: Implications for values education. *Review of Educational Research, 52*, 93–122.

Nucci, L. (1985). Children's conceptions of morality, societal convention, and religious prescription. In C. Harding (Ed.), *Moral dilemmas: Philosophical and psychological issues in the development of moral reasoning.* Chicago: Precedent Press.

Nussbaum, J. (1979). Children's conceptions of the earth as a cosmic body: A cross age study. *Science Education, 63*, 83–93.

Pazzani, M. J. (1991). Influence of prior knowledge on concept acquisition: Experimental and computational results. *Journal of Experimental Psychology: Learning, Memory, and Cognition, 17,* 416–432.

Penfield, W. (1952). Memory mechanisms. *American Medical Association Archives of Neurology and Psychiatry, 67,* 178–198.

Perkins, D. N., & Salomon, G. (1989). Are cognitive skills context-bound? *Educational Researcher, 18*(1), 16–25.

Phillips, D. C. (1983). On describing a student's cognitive structure. *Educational Psychologist, 18,* 59–74.

Piaget, J. (1964). Development and learning. *Journal of Research in Science Teaching, 2,* 176–186.

Piaget, J. (1972). Intellectual evolution from adolescence to adulthood. *Human Development, 15,* 1–12.

Piattelli-Palmarini, M. (Ed.). (1980). *Language and learning: The debate between Jean Piaget and Noam Chomsky.* London: Routledge and Kegan Paul.

Pines, A. L., & West, L. (1986). Conceptual understanding and science learning: An interpretation of research within a sources-of-knowledge framework. *Science Education, 70,* 583–604.

Quinlan, P. T. (1991). *Connectionism and psychology: A psychological perspective on new connectionist research.* Chicago: University of Chicago Press.

Reif, F. (1981, May). Teaching problem solving: A scientific approach. *The Physics Teacher,* 310–316.

Reif, F. (1983). How can chemists teach problem solving? *Journal of Chemical Education, 60,* 948–953.

Reif, F., Larkin, J., & Brackett, G. (1976). Teaching general learning and problem-solving skills. *American Journal of Physics, 44,* 212–217.

Renner, J., Abraham, M., & Birnie, H. (1983). *Sequencing language and activities in teaching high school physics.* (NSF Final Report, Grant SED-8015814)

Resnick, L. (1987). *Education and learning to think.* Washington, DC: National Academy Press.

Reustrom, L. (1987). Pupils' conceptions of matter: A phenomenographic approach. *Proceedings of the Second International Seminar on Misconceptions and Educational Strategies in Science and Mathematics: Vol. III* (pp 400–419). Ithaca, NY: Cornell University.

Rose, S., & Blank, M. (1974). The potency of context in children's cognition: An illustration through conservation. *Child Development, 45,* 499–502.

Ross, J. (1974). Three batons for cognitive psychology. In W. B. Weimer & D. S. Palermo (Eds.), *Cognition and the symbolic processes* (pp 63–124). Hillsdale, NJ: Erlbaum.

Roth, W.-M., & Roychoudhury, A. (1993). The concept map as a tool for the collaborative construction of knowledge: A microanalysis of high school physics students. *Journal of Research in Science Teaching, 30,* 503–534.

Rowe, M. B. (1969). Science, soul, and sanctions. *Science and Children, 6,* 11–13.

Sayre, S., & Ball, D. (1975). Piagetian cognitive development and achievement in science. *Journal of Research in Science Teaching, 12,* 165–174.

Scardamalia, M., & Bereiter, C. (1991). Higher levels of agency for children in knowledge building: A challenge for the design of new knowledge media. *The Journal of the Learning Sciences, 1,* 37–68.

Schank, R., & Abelson, R. (1977). *Scripts, plans, goals, and understanding.* Hillsdale, NJ: Erlbaum.

Schoenfeld, A. (1981, April). *Episode and executive decisions in mathematical problem solving*. Paper presented at the 1981 American Education Research Association Annual Meeting, Los Angeles, CA.

Schoenfeld, A. (1982). Measures of problem solving performance and of problem solving instruction. *Journal of Research in Mathematics Education, 13*, 31–49.

Segal, J., Chipman, S., & Glaser, R. (Eds.). (1985). *Thinking and learning skills: Vol. 1. Relating instruction to research*. Hillsdale, NJ: Erlbaum.

Shuell, T. (1990). Phases of meaningful learning. *Review of Educational Research, 60*, 531–547.

Shymansky, J., Teagust, D., Thiele, R., Harrison, A., Waldrip, B., Stockmayer, S., & Venville, G. (1993, April). *A study of changes in student understanding of force, motion, work, and energy*. Paper presented at the 66th Annual Meeting of the National Association for Research in Science Teaching, Atlanta, GA.

Smedslund, J. (1961). The acquisition of conservation of substance and weight in children. *Scandinavian Journal of Psychology, 2*, 11–20, 71–87, 153–160, 203–210.

Spiro, R., Bruce, B., & Brewer, W. (Eds.). (1980). *Theoretical issues in reading comprehension*. Hillsdale, NJ: Erlbaum.

Stavy, R. (1987). Acquisition of conservation of matter. *Proceedings of the Second International Seminar on Misconceptions and Educational Strategies in Science and Mathematics: Vol. I* (pp 456–465). Ithaca, NY: Cornell University.

Strike, K., & Posner, G. (1982). Conceptual change and science teaching. *European Journal of Science Education, 4*, 231–240.

Sutton, C. (1975). *Language and communication in science lessons*. New York: McGraw-Hill.

Tobin, K. (1987). The role of wait time in higher cognitive level learning. *Review of Educational Research, 57*, 69–95.

Walker, R., Mertens, T., & Hendrix, J. (1979). Formal operational reasoning patterns and scholastic achievement in genetics. *Journal of College Science Teaching, 8*, 156.

Wandersee, J. H. (1990). Concept mapping and the cartography of cognition. *Journal of Research in Science Teaching, 27*, 923–936.

Ward, C., & Herron, J. D. (1980). Helping students understand formal chemistry concepts. *Journal of Research in Science Teaching, 17*, 387–400.

Williams, H., Turner, C., Debreuil, L., Fast, J., & Berestiansky, J. (1979). Formal operational reasoning by chemistry students. *Journal of Chemical Education, 55*, 599.

Witkin, H., & Goodenough, D. (1977, December). *Field dependence revisited*. (Research Bulletin RB-77-16) Princeton, NJ: Educational Testing Service.

Witkin, H., Goodenough, D., & Oltman, P. (1979). Psychological differentiation: Current status. *Journal of Personality and Social Psychology, 37*, 1127–1145.

Wollman, W. (1977). Controlling variables: Assessing levels of understanding. *Science Education, 61*, 371–383.

Zoller, U. (1993). Are lecture and learning compatible? *Journal of Chemical Education, 70*, 195–197.

Index

Index

Copy editing and indexing: Scott Hofmann-Reardon
Production: Paula M. Bérard
Acquisition: Barbara Pralle

Cover design by Marshall Henrichs, Lexington, MA
Typeset by Vincent Parker, Washington, DC
Printed and bound by Maple Press, York, PA